Map Projections:

Theory and Applications

Frederick Pearson, II, M.S., P.D.D.

Systems Engineer
Combat Systems
Naval Surface Warfare Center
Dahlgren, Virginia

Lecturer
Department of Civil Engineering
Virginia Polytechnic Institute
Falls Church, Virginia

 CRC Press
Taylor & Francis Group
Boca Raton London New York

CRC Press is an imprint of the
Taylor & Francis Group, an **informa** business

CRC Press
Taylor & Francis Group
6000 Broken Sound Parkway NW, Suite 300
Boca Raton, FL 33487-2742

© 1990 by Taylor & Francis Group, LLC
CRC Press is an imprint of Taylor & Francis Group, an Informa business

First issued in paperback 2019

No claim to original U.S. Government works

ISBN-13: 978-0-367-45085-4 (pbk)
ISBN-13: 978-0-8493-6888-2 (hbk)

Visit the Taylor & Francis Web site at
http://www.taylorandfrancis.com

and the CRC Press Web site at
http://www.crcpress.com

Library of Congress Card Number 89-23986

Library of Congress Cataloging-in-Publication Data

Pearson, Frederick, 1936-
    Map projections : theory and applications / Frederick Pearson II.
        p.   cm.
    Includes bibliographical references.
    ISBN 13: 978-0-8493-6888-2
    1. Map-projection.   I. Title.
GA110.P425   1990
526′.8--dc20   89-23986
                                                                                          CIP

This book is dedicated to my beloved Shirley.

PREFACE

This present volume is the logical step forward from two previous works by the author. The first was *Map Projection Equations,* a Naval Surface Warfare Center Technical Report. This was an attempt to collect the maximum number of existing map projections in a single source, using a standardized notation. Then came *Map Projection Methods,* in which an attempt was made at modernization of the method of presentation. After teaching from this text for three summers in the Geodetic Engineering program at Virginia Polytechnic Institute at Falls Church the author could see that further improvement was necessary. Based on questions and suggestions from the students in these classes, the present volume is a further attempt to move from the classical and theoretical to the modern and practical. In the present volume, projections of mainly historical value have been excluded. Projections of use in photogrammetry, remote sensing, and target tracking applications have been emphasized more. The presentation of the projections that are the core of most cartographic work has been clarified.

In Chapter 1, the problems involved in map projection theory and practice are introduced. The concept of an ideal map is explored, and the deviations from the ideal and their causes are discussed. The basic nomenclature of map projections is defined.

Chapter 2 begins the mathematical consideration of map projection techniques and applications. In this chapter, the mathematical groundwork is laid down. The mathematical level of the student needed to understand this text is calculus and trigonometry. More advanced topics will be summarized and developed in this chapter.

The chapter begins with a discussion of the differential geometry of an arbitrary curve in space. This is followed by consideration of an arbitrary spatial surface. Of more interest to map projections is the differential geometry of surfaces of revolution. The form of the basic transformation matrices are developed in detail. A number of mathematical entities of importance in mapping are discussed. These are the convention for azimuth, constant of the cone, convergence of the meridians, and the basic coordinate systems for the Earth and the map. The important spherical trigonometric formulae for rotations are introduced.

In Chapter 3, the surfaces of revolution of immediate interest to mapping are considered. These are the sphere and the spheroid taken as useful models of the figure of the Earth. Formulas are developed for linear and angular measurement on the spheroid and sphere.

Chapters 4, 5, and 6 are dedicated to the derivation of the various projections themselves. In these chapters are the equal area, conformal, and conventional projections of major use today. In every case, the result is a set of direct transformation equations in which the Cartesian plotting coordinates may be obtained from geographic coordinates. In selected cases, the inverse transformation from Cartesian to geographic coordinates is also included.

Distortion is evaluated mathematically in Chapter 7. Numerical estimates of distortion can be obtained from the equations developed in this chapter.

Chapter 8 includes extended examples of the use of map projection formulae in the solution of modern problems.

Since it should be obvious for the complexity of the equations in Chapters 4, 5, and 6, that no worker in the field of mapping would want to routinely evaluate them with a pocket calculator, the computerization of the equations is considered in Chapter 9. The equations for the direct and inverse transformations, as well as grid systems, are treated.

A general discussion on the uses and choices in the field of map projections rounds out the text in Chapter 10.

This volume is intended to be the text in a one semester course in the mathematical aspects of mapping and cartography. The level of mathematics is consistent with upper level undergraduate and graduate courses.

Special thanks go to John P. Snyder, formerly of the U.S. Geological Survey, who, over the years, has consistently challenged me to produce a better presentation of the material. I also thank him for his review of the present volume and his many helpful suggestions.

A word of thanks is due to many people. These include Dr. L. Meirovitch of Virginia Polytechnic Institute for his encouragement of my first attempt, Dr. J. Junkins of Texas A. M., for my second attempt, and Dr. W. G. Rich and Glenn Bolick of the Naval Surface Warfare Center, for the present volume. Thanks to B. McNamara of the U.S. Geological Survey for permitting the use of his mapping program to generate many of the tables, and to the author's wife, Shirley, for typing the manuscript.

THE AUTHOR

Mr. Frederick Pearson, M.S., M.S., P.D.D., received a B.S. in History from St. Louis University, St. Louis, Missouri in 1961, and M.S. in Astronomy from Yale University, New Haven, Connecticut in 1964, a Diploma in Business Administration from the Hamilton Institute in 1968, and M.S. in Engineering Mechanics from Virginia Polytechnic Institute in 1977, and a P.D.D. in Civil Engineering from the University of Wisconsin, Madison, Wisconsin, in 1981.

Mr. Pearson has experience in applied science and engineering. He developed star charts and satellite trajectory programs and designed a celestial navigation device for the Aeronautical Chart and Information Center. At Emerson Electric Company, he conducted studies in orbital analysis and satellite rendezvous, for which he won the AIAA Young Professional Scientist Award in 1968. Mr. Pearson has also done vibration analyses and design work on spin-stabilized missiles. For TRW systems, he conducted mission planning and devised emergency back-up procedures for the Apollo missions. At McDonnell-Douglas, he worked on the design of the control systems of the HARPOON Missile and the guidance for the Space Shuttle.

Mr. Pearson is employed at the Naval Surface Warfare Center where his work has ranged from the analysis of satellite orbits to the design of machine elements and structures. He is presently a Systems Engineer in the AEGIS Program, working on Infrared Sensors.

Mr. Pearson has taught Map Projections at Air Force Cartography School, and Astrodynamics and Orbit Determination in a TRW Employee Training Program. Also, he has taught Map Projections for the Civil Engineering Department of Virginia Polytechnic Institute.

Mr. Pearson's publications include: *Map Projection Equations*, NSWC, 1977; *Map Projection Methods,* Sigma, 1984; and *Map Projection Software,* Sigma, 1984.

TABLE OF CONTENTS

Chapter 1

INTRODUCTION

A map is a visual representation of a part, or all of the surface of the Earth. It is expected that the various terrestrial features shown on the map will be approximately in their true relationship. The success of the representation depends on the map projection chosen to produce the map.

Map projection is the orderly transfer of positions of places on the surface of the Earth to corresponding points on a flat sheet of paper, a map. Since the surface of a sphere cannot be laid flat on a plane without distortion, the process of transformation requires a degree of approximation and simplification.[3] This first chapter lays the groundwork for this subject by detailing, in a qualitative way, the basic problem and introducing the nomenclature of maps. Succeeding chapters consider the mathematical techniques and the simplifications required to obtain manageable solutions.

All projections introduce distortions in the map. The types of distortion are considered in terms of length, angle, shape, and area. This first chapter discusses the qualitative aspects of the problem, while Chapter 7 deals with it quantitatively.

An examination of the geometric shapes on a globe representing the Earth indicates basic figures. If these figures are maintained on a map, a great step is taken in limiting the adverse effects of distortion. Scale factor refers to the relative length of a line on the model for the Earth, and the same line as represented on the map. The concept of scale factor is instrumental in reducing Earth-sized lengths to map-sized lengths.

Map projections may be classified in a number of ways. The principal one is by the features preserved from distortion by the mapping technique. Other methods of classification depend on the plotting surface employed, the location of the points of contact of this surface with the Earth, the orientation of the plotting surface with respect to the direction of the polar axis of the Earth, and whether the plotting surface is tangent or secant to the Earth. Finally, maps can be classified according to whether or not a map can be drawn by purely graphical means.

This chapter also considers the plotting equations derived in Chapters 4, 5, and 6 in a general way. These equations are the basis of the orderly transformation from the model of the Earth to the map. The results of evaluating these equations are often given in plotting tables. The conventions incorporated in the plotting tables of this text are discussed.

I. INTRODUCTION TO THE PROBLEM[2]

Map projection requires the transformation of positions from a curved surface, the Earth, onto a plane surface, the map, in an orderly fashion. The problem occurs because of the difference in the surfaces involved.

The model of the Earth is taken as either a sphere or a spheroid (Chapter 3). These curved surfaces have two finite radii of curvature. The map is a plane surface, and a plane is characterized by two infinite radii of curvature. As is shown in Chapter 2, it is impossible to tranform from a surface of two finite radii of curvature to a surface of two infinite radii of curvature without introducing some distortion. The sphere and the spheroid are called nondevelopable surfaces. This refers to the inability of these surfaces to be developed (i.e., laid flat) onto a plane in a distortion-free manner.

Intermediate between the nondevelopable sphere and spheroid and the plane are surfaces with one finite and one infinite radius of curvature. The examples of this type of surface of interest to mapping are the cylinder and the cone. These surfaces are called developable. Both the cylinder and the cone can be cut and then developed (essentially unrolled along the finite radius of curvature) to form a plane. This development introduces no distortion, and, thus, these figures may be used as intermediate plotting surfaces between the sphere and spheroid and the plane. However, in any transformation from the sphere or spheroid to the developable surface some damage has already been done. The transformation from the nondevelopable to the developable surface invariably introduces some degree of distortion.

Distortion is inevitable, and a map with ideal properties is never attained. Consider the following properties of an ideal map:

1.   Areas on the map maintain correct proportion to areas on the Earth.
2.   Distances on the map must remain in true scale.
3.   Directions and angles on the map must remain true.
4.   Shapes on the map must be the same as on the Earth.

The best a cartographer can hope for is the realization of one or two of these ideal properties in a single map projection. The other ideal properties will then be subject to distortions, but hopefully, only to a controlled extent. There is always some compromise involved.

The projections of Chapters 4, 5, and 6 attempt to achieve one or more of these ideal properties at the expense of the others. The best projection is chosen for a specific application. Distortion in the other features is then considered to be tolerable.

II. BASIC GEOMETRIC SHAPES

A series of basic geometric shapes can be found on a gore cut from the model of the Earth. A gore is the tapered shape obtained by slicing the model of the Earth through its poles. This is represented in Figure 1. Consider the top half of the gore. The lower third of the half-gore can be represented on

Gore of
Sphere

Basic
Shapes

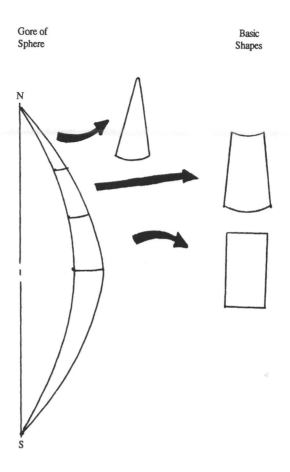

N

S

FIGURE 1.   Basic cartographic shapes.

a flat sheet of paper by a rectangular shape. The top third of the half-gore is easily represented on the map by a triangular shape. The middle third of the half-gore suggests a trapezoidal shape.

One of the easiest ways to minimize distortions is to choose a projection whose basic geometrical figures closely represent the shape of the corresponding figure on the model of the Earth. As is shown in Chapters 4, 5, and 6, this is easily realizable for projections designed to cover a limited portion of the globe. It is impossible when considering a single map of the whole Earth. Distortions set in and frustrate the attempt.

III. DISTORTION[5]

Distortion is the limiting factor in the process of map projection. Distortion is the twisting out of shape or size of a geometrical figure, thus incurring its misrepresentation. It is an untrue representation of area, linear dimensions,

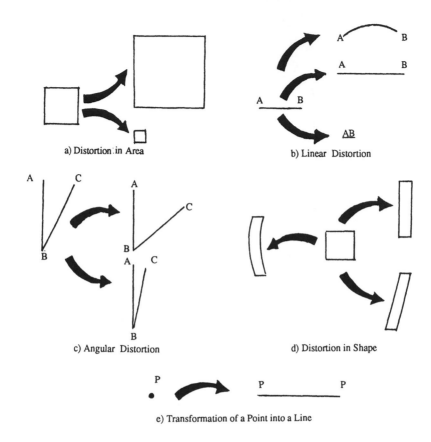

a) Distortion in Area

b) Linear Distortion

c) Angular Distortion

d) Distortion in Shape

e) Transformation of a Point into a Line

FIGURE 2.   Types of distortion.

angle, or shape. Distortion is caused by a specific method of map projection, and it is an inevitable part of the map projection process.

Distortion in area is shown in Figure 2a. In this case shape is maintained; but the area on the map is either enlarged or diminished from the true area on the model of the Earth.

Lines on the Earth are really in three-dimensional space, as fully explored in Chapter 2. However, on the map the line is constrained to two-dimensional space. Two processes of distortion occur with regard to lines. These are indicated in Figure 2b. First, a given line may either increase or decrease from true scale in the transformation from the model of the Earth to the map. The second potential distortion of a line is a transformation from the ideal straight line to a curved line. Often, while a cartographer is able to maintain true length in one direction on a map, he cannot do so in a second direction. It is more common for distortion to appear in both mutually perpendicular directions.

Angular distortion is also prevalent. This is illustrated in Figure 2c. Angles on the map will not necessarily be the same as their counterparts on the Earth.

As an example, azimuth on the model for the Earth may be represented as too large or too small on the map.

Distortion in shape can occur in a number of ways. This is shown in Figure 2d. One of these is a general distortion of shape, such as a square being transformed into a rectangle. A second type is a shearing effect in which a square is transformed into a rhombus. A third is a curving effect in which a rectangle appears as a truncated lune.

The most devastating distortion, which is experienced in some of the projections of Chapters 4 and 6, is the transformation of a point into a line. This is indicated in Figure 2e. When this occurs during a particular projection, the North and South Poles are transformed into a line of finite length. Then, severe distortions will follow in the qualities of shape, area, length, and angle.

Any actual map has a combination of these distortions. One or more of the ideal qualities may be preserved, but not all at the same time. As the individual projections are presented in the chapters that follow, the particular distortions inherent in each are indicated. The quantitative theory of distortion is presented in Chapter 7.

IV. SCALE[2]

Scale is the vehicle for transforming actual distances on the model for the Earth down to manageable sizes on the map. Two types of scale will be defined: principal scale and local scale.

Principal scale refers to the ratio of a true length distance on the map to the equivalent distance on the model for the Earth. It is given by

$$S = d_m/d_e \tag{1}$$

where d_m is the distance on the map, and d_e is the distance on the Earth. If the distance on both the map and the earth have the same units, then the scale is a dimensionless quantity. Scale is quoted as the reciprocal of some number, or as the ratio of one to some number; e.g., 1/10,000 or 1:10,000.

The terms large scale vs. small scale come from consideration of the fraction d_m/d_e. A scale of 1/10,000 is a large scale, and 1/1,000,000 is a small scale. A plan, or a map showing buildings, cultural features, or boundaries is usually 1/10,000 or larger. A topographic map, which gives roads, railroads, towns, contour lines, and other details, has a scale between 1/10,000 and 1/1,000,000. Maps of a scale smaller than 1/1,000,000 are atlas maps. These maps delineate countries, continents, and oceans. The scale factor, S, or principal scale, is used in the plotting equations of Chapters 4, 5, and 6.

The presence of distortion requires the definition of the second type of scale: local scale. While a line may be true scale at one area on a map, at other places on the map, where distortions are present, the scale will be different from true scale. This local scale will be larger or smaller than the principal scale, depending on the mechanism of the distortion. The local scale is given by

$$m = d_d/d_{TL} \tag{2}$$

where d_d is the actual length of the line element, and d_{TL} is the length it would have if it were true. The value in is generably given as a percentage. In Chapter 7, the mathematic relations for developing local scale are given for specific projections.

The local scale, as a function of distortion, and the principal scale may be quoted on the legend of a map. A more useful means is a graphical scale drawn in the map legend and specified for the latitudes and longitudes where it applies. As an example, for the Mercator projection (Chapter 5), a set of scales can be drawn as a function of latitude, which will ensure the correct basic distances for measuring.

A. Example 1

Given: $d_e = 200$ km, $S = 1:1,000,000$.
Find: d_m.

$$d_m = Sd_e$$

$$= \frac{(200)(1000)}{1,000,000} = 0.200 \text{ m}$$

B. Example 2

Given: $d_d = 0.0975$ m, $d_{TL} = 0.1000$ m.
Find: m.

$$m = \frac{d_d}{d_{TL}} = \frac{0.0975}{0.1000} = 0.975$$

$$\equiv 97.5\%$$

V. FEATURE PRESERVED IN PROJECTION[5]

Maps may be categorized by the feature preserved in the process of projection. This system of classification divides maps into three distinct types: equal area projections, conformal projections, and conventional projections.

The equal area projection preserves the ratio of areas on the Earth to corresponding areas on a map. Consider Figure 3a and Areas A and B on the Earth, and the corresponding areas a and b on the map. The equal area quality means that:

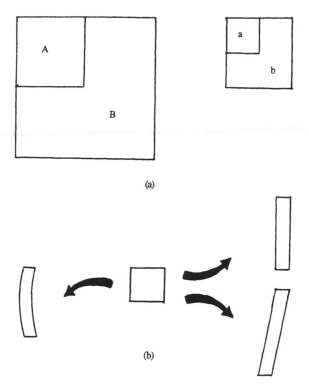

FIGURE 3. Representation of equal areas.

$$\frac{\text{Area a}}{\text{Area b}} = \frac{\text{Area A}}{\text{Area B}} \tag{3}$$

Practically, in achieving this, distortion is introduced in the projection in distance and angle. Angles usually suffer significant distortion. A contraction of distance in one direction will be offset by a lengthening of distance in another in order to maintain a constant area. This concept is illustrated in Figure 3b. A quadrilateral is distorted in three ways. Yet, in all cases, the area enclosed in the distorted figures is equal to that of the undistorted figure.

A conformal projection is one in which the shape of a figure on the Earth is preserved in its transfer to a map. This concept is valid at a point, but must not be extended over large areas. The condition for a conformal map may be stated as, at a point, the scale is the same in all directions. Another basic feature is that at a point, angles are the same on the map as they are on the model of the Earth. Thus, locally, angular distortion is zero. This feature is purchased at the price of distortion in size. Figure 4 illustrates the result of this conformal transformation. The resulting quadrilaterals are in the same shape after transformation. However, the size has either increased or diminished. For the figure, it is evident that this projection is orthomorphic (same form), another term used for this projection.

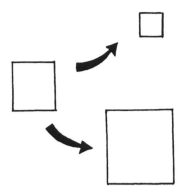

FIGURE 4.   Representation of conformal elements.

The term conventional projection refers to those projections which are neither equal area nor conformal. In these projections some other special feature is preserved at the expense of equal area of conformal properties. Or, as an alternative, a simple algorithm has been devised to obtain a useful projection which makes no pretense of holding any property distortion free. A large number of projections fit into the category of conventional projections. Examples are the gnomonic projection in which the great circle is portrayed as a straight line, and the azimuthal equidistant projection in which azimuth and range from the origin are true, and the polyconic projection in which all distances along a parallel circle are true.

VI. PROJECTION SURFACE[5]

The three plotting surfaces in use for practical map projections are the plane, the cone, and the cylinder. The plane is defined by two infinite radii of curvature, and the cone and cylinder by one finite and one infinite radius of curvature. It can be argued that all projection surfaces are conical, since the plane and the cylinder can be considered as the two limiting cases of the cone. This mathematical concept is used in some cases, but the three surfaces are considered as distinct in many others. Figure 5 shows each of these surfaces in relation to the sphere.

The planar projection surface can be used for a direct transformation from the Earth. The projections which result are called azimuthal (Figure 5a). Other names in use are zenithal or planar projections. Conical projections result when a cone is used as an intermediate plotting surface. The cone is then developed into a plane to obtain the map (Figure 5b). Cylindrical projections are obtained when a cylinder is used as the intermediate plotting surface (Figure 5c). As with the cone, the cylinder can then be developed into a plane.

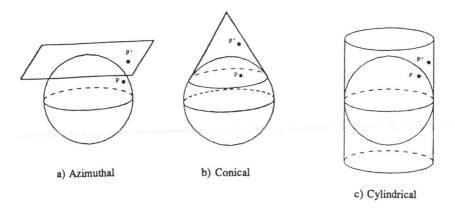

a) Azimuthal   b) Conical

c) Cylindrical

FIGURE 5. Projection surfaces.

In Figure 5, a representative position on the Earth, P, is shown transformed into a position on the projection surface, P' for each of the projection surfaces. Chapters 4, 5, and 6 explain the methods that affect such transformations and produce useful maps.

Two other types may be defined. The pseudocylindrical projection has straight and parallel latitude lines and meridians which are equally spaced. However, all meridians except the central meridian are curved. These are mathematically defined azimuthal projections. Pseudoconical projections have concentric circular arcs for parallels, but meridians are curved. These are still conical projections.

VII. ORIENTATION OF THE AZIMUTHAL PLANE[5]

Azimuthal projections may be classified by reference to the point of contact of the plotting surface with the Earth. Azimuthal projections may be classified as polar, equatorial, or oblique. When the plane is tangent to the Earth at either pole, we have a polar projection. When the plane is tangent to the Earth at any point on the equator, the projection is called equatorial. The oblique case occurs when the plane is tangent at any point on the Earth except the poles and the equator. Figure 6 indicates these three alternatives. In each case, T is the point of tangency. OT is the line from the center of the Earth to the point of tangency and is perpendicular to the mapping plane.

VIII. ORIENTATION OF A CONE OR CYLINDER[5]

Another classification can be defined for cones and cylinders. These plotting surfaces may be considered to be regular or equatorial, transverse, or oblique.

The regular projection occurs when the axis of the cone or cylinder coincides with the polar axis of the Earth. The transverse case has the axis

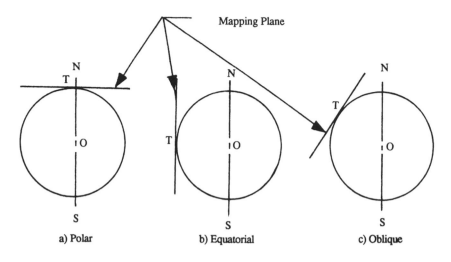

FIGURE 6.   Orientation of the azimuthal plane.

of the cone or cylinder perpendicular to and intersecting the axis of the Earth in the plane of the equator. The transverse Mercator is an example of this. If the axis of the cone or cylinder has any other position in space, besides being coincident with or perpendicular to the axis of the Earth, then an oblique projection is generated. In all cases, the axis of the cone or cylinder passes through the center of the Earth. Figure 7 demonstrates these three options for a cylindrical projection. The figure is an orthographic view of the sphere-cylinder pair.

IX. TANGENCY AND SECANCY

The plotting surface may be either tangent or secant to the model of the Earth. This is illustrated in Figure 8 for the azimuthal, cylindrical, and conical plotting surfaces. A plane is tangent to the model of the Earth at a single point. If a secant plane is used, one true length line is defined on the map at the line of secancy. For the conical and cylindrical projections, the line of tangency is a single true length line. In the secant case, two true length lines occur at the two lines of secancy. Note that secancy is not a geometric property, but only a conceptual one in many conic projections (e.g., Lambert conformal with two standard parallels).

X. PROJECTION TECHNIQUES[1]

Three types of projection techniques are available. These are the graphical, the semigraphical, and the mathematical methods.

In any graphical technique, some point O is chosen as a projection point, and the methods of projective geometry are used to transform a point P on

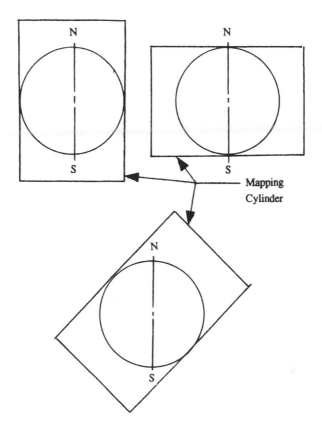

FIGURE 7.   Orientation of the mapping cylinder.

the Earth to a location P′ on the plotting surface. An example of this is indicated in Figure 9, where the point P on the Earth is transformed to the polar plane by the extension of line OP until it intersects the plane. In this example, O is arbitrarily chosen as the projection point. This figure represents the graphical technique for the stereographic projection. Since anything that can be done graphically can be described mathematically, the mathematical approach is followed in Chapters 4, 5, and 6.

In the semigraphical projections, a graphical construction is possible, but such complications as a varying projection point, as in the Mercator projection illustration in Figure 10, or a complex graphical scheme, as in the Mollweide, do not encourage the approach. A mathematical approach is the only reasonable one.

A third set of projections can be produced only from a mathematical definition. An example of this is the parabolic projection.

To obtain plotting coordinates with sufficient accuracy, only a mathematical approach can be used. This is done for the projections of Chapters 4, 5, and 6.

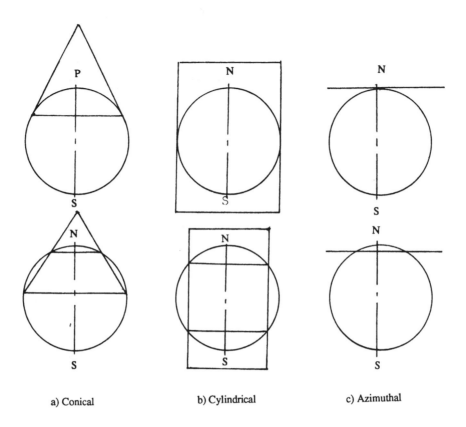

a) Conical   b) Cylindrical   c) Azimuthal

FIGURE 8.   Tangency and secancy of the mapping surfaces.

XI. PLOTTING EQUATIONS[4]

The subject of map projections certainly has intrinsic interest. Of more importance to workers in the field of map projections are equations which carry out the transformations numerically. The ultimate goal of this text is to derive Cartesian plotting equations which may be evaluated for a given latitude and longitude. These are the direct transformation equations. These equations may be written functionally as

$$\begin{Bmatrix} x \\ y \end{Bmatrix} = [T] \begin{Bmatrix} \phi \\ \lambda \end{Bmatrix} \tag{4}$$

where T is the transformation for a specific projection. After the basic concepts are developed in Chapters 2 and 3, the actual equations for mapping are derived in Chapters 4, 5, and 6.

Of nearly equal importance to workers in the field of mapping are the inverse transformations, that is, the transformation from Cartesian coordinates to geographic coordinates. In functional form, these may be written as

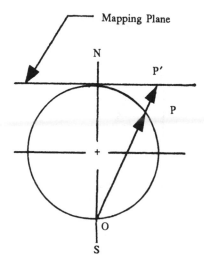

FIGURE 9.   Graphical projection onto a
plane, polar stereographic.

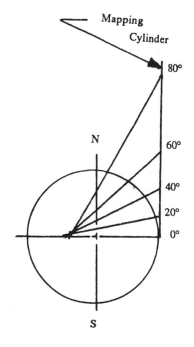

FIGURE 10.   A semigraphical projection:
the equatorial Mercator projection.

$$\left\{ \begin{matrix} \phi \\ \lambda \end{matrix} \right\} = [T]^{-1} \left\{ \begin{matrix} x \\ y \end{matrix} \right\} \tag{5}$$

These, too, are developed in Chapters 4, 5, and 6 for selected map projections.

It is evident from the derivations that the simple functional form is not attained. Instead, the plotting equations are rather complex transcendental relationships. The plotting equations for the direct transformation have certain essential characteristics, depending on the plotting surface chosen for the transformation. If a cylindrical plotting surface is used, the Cartesian coordinates result immediately.

If the plotting surfaces are polar azimuthal, or conical, the derivations are developed in polar coordinates. Then, these polar coordinates are transformed to Cartesian coordinates. For a polar azimuthal projection, assume the polar coordinates are ρ and θ. Then the transformation to Cartesian coordinates has the form

$$\left. \begin{matrix} x = \rho\sin\theta \\ \\ y = -\rho\cos\theta \end{matrix} \right\} \tag{6}$$

This convention places 0° longitude on the negative y axis. Then, the positive x axis corresponds to $\lambda = 90°$.

In a similar manner, if a conical projection is used, assume the polar coordinates are ρ and θ. Then the Cartesian coordinates are obtained from relationships of the form

$$\left. \begin{matrix} x = \rho\sin\theta \\ \\ y = \rho_o - \rho\cos\theta \end{matrix} \right\} \tag{7}$$

where ρ_o is the vector from the apex to the origin.

In Chapters 4, 5, and 6, the plotting equations are derived for the Northern Hemisphere. In Chapter 8, a simple technique is given for obtaining the Cartesian coordinates when it is desired to prepare a map in the Southern Hemisphere.

XII. PLOTTING TABLES[4]

Plotting tables are included for the majority of the projections treated in Chapters 4, 5, and 6. They are obtained by the systematic evaluating of the plotting equations for each projection. They are computed for a spherical model of the Earth. They are given for a product of the radius of the Earth times the scale factor equal to 1.

$$R \cdot S = 1 \tag{8}$$

Since R · S is a multiplicative factor in all of the direct transformation equations, a user of these tables need only to multiply the values in these tables by his or her own required product R · S to obtain values for his own applications.

While the spacing in the tables is too great to generate a usable grid, they can serve as a means to check a user-generated program for the direct transformation or to obtain a rough evaluation of results obtained from other means.

REFERENCES

1. **Adler, C. F.,** *Modern Geometry,* McGraw-Hill, New York, 1958.
2. **Deetz, C. H. and Adams, O. S.,** Elements of Map Projection, Spec. Publ. 68, U.S. Coast and Geodetic Survey, U.S. Gov't. Printing Office, Washington, D.C., 1944.
3. **Deetz, C. H.,** Cartography, A Review and Guide, Spec. Publ. 205, U.S. Coast and Geodetic Survey, U.S. Gov't. Printing Office, Washington, D.C., 1962.
4. **Snyder, J. P.,** Map Projections — A Working Manual, U.S. Geol. Surv. Prof. Pap., 1395, 1987.
5. **Steers, J. A.,** *An Introduction to the Study of Map Projections,* University of London, London, 1962.

Chapter 2

MATHEMATICAL FUNDAMENTALS

I. INTRODUCTION

The process of map projection requires the transformation from a set of two independent coordinates on the model of the Earth to a set of two independent coordinates on the map. The transformation involves the application of specific mathematical techniques. This chapter is devoted to the mathematical fundamentals of map projection.

The coordinate systems of use for defining positions on the model of the Earth and the map are given. Then, the convention for defining azimuth is treated. The next topic is grid systems.

The basic unifying concepts for map projections are explored in terms of differential geometry. The process is sequential, going from arbitrary curves in space, to arbitrary surfaces in space, to general surfaces of revolution.

The differential geometry of curves is discussed to introduce needed concepts, such as radius of curvature and torsion of a space curve. The differential geometry of surfaces discussion introduces the first and second fundamental forms, as well as parametric curves and the condition of orthogonality. The surfaces of interest in mapping are surfaces of revolution. The general surface is particularized to surfaces of revolution. The process of transformation from nondevelopable to developable surfaces is considered. Representations of arc length, angles, and area, as well as the definition of the normal to the surface, are also discussed.

The basic transformation matrices are derived in forms useful for both the equal area and the conformal projections. The conditions of equality of area and conformality are mathematically defined. Then, a rotation method is discussed in which transverse and oblique projections may be obtained from the regular case. The convergency of the meridians in angular and linear form is then considered. The concept of the constant of the cone is introduced for the tangent and the secant case.

II. COORDINATE SYSTEMS AND AZIMUTH[7]

Coordinate systems are necessary to specify the location of positions on both the Earth and the map. In both cases, we are dealing with two-dimensional systems. Discussion of the two-dimensional characteristic for the model of the Earth is deferred until Chapter 3. In this section, the nomenclature and conventions required for the understanding of the following sections are introduced.

The terrestrial coordinate system is demonstrated in Figure 1. The origin,

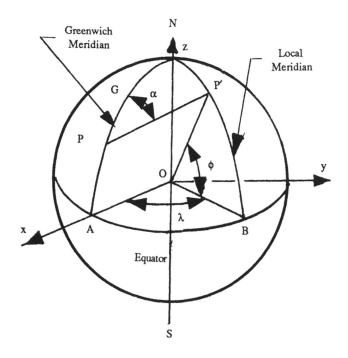

FIGURE 1.   Terrestrial coordinate system.

O, of the system is at the center of the Earth. The x and y axes form the
equatorial plane. The curve on the Earth formed by the intersection of this
plane with the surface of the Earth is the equator. The positive x axis intersects
the curve AGN. The curve AGN is a plane curve which is called the Greenwich
meridian. The positive z axis coincides with the nominal axis of rotation of
the Earth and points in the direction of the North Pole, N. The y axis completes
a right-handed coordinate system.

A point P on the surface of the Earth is uniquely located by giving the
latitude and longitude of the point. The latitude and longitude are independent
angular coordinates.

A meridian is a curve formed by the intersection of a ficticious cutting
plane containing the z axis and the surface of the Earth. The Greenwich
meridian has already been mentioned. There is an infinity of meridians,
depending on the orientation of the cutting plane. All meridians are perpendicular to the equatorial plane.

The use of latitude and longitude locates a point on a meridian and then
locates the meridian with respect to the Greenwich meridian. Latitude is the
angular measure defining the position of point P' on the meridian BP'N.
Latitude is denoted by ϕ. The position of the meridian that contains P is
defined by the longitude, λ. The longitude is the angle AOB, measured in
the equatorial plane, from the Greenwich meridian. The conventions for

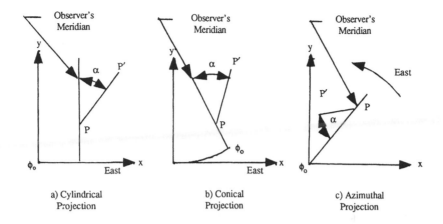

a) Cylindrical
Projection

b) Conical
Projection

c) Azimuthal
Projection

FIGURE 2.   Mapping coordinate systems.

latitude and longitude are as follows: latitude is measured positive to the north, and negative to the south. Longitude is measured positive to the east, and negative to the west.

The circles of parallel are generated by cutting planes parallel to the equatorial plane which intersect the Earth. All points on a circle of parallel have the same latitude.

The coordinate system for the map (Figure 2) is a two-dimensional Cartesian system. The interpretation of the system differs for the projection schemes of Chapters 4, 5, and 6 that are portrayed.

For the cylindrical projections, and the world maps of Chapter 4, the line of the equator coincides with the x axis, where $\phi_o = 0°$. East is in the direction of positive x. The line of the central meridian defines the y axis, with north in the positive y direction. The central meridian, λ_o, is that reference meridian chosen by the cartographer to be at the center of the map. The intersection of the central meridian and equator defines the origin of the map.

Consider next a conical projection. In this case, the central meridian, λ_o, again coincides with the y axis, with y positive to the north. The origin is defined by the intersection of the central meridian with a circular arc representing a standard parallel of latitude, ϕ_o or ϕ, East is roughly in the plus x direction, as shown in Figure 2b.

Figure 2c indicates the convention for azimuthal projections. The y axis contains the central meridian. The origin coincides with the North Pole, or $\phi_o = 90°$. East is defined as a counterclockwise rotation.

The object of the practice of map projection is to transform from the terrestial angular coordinates, that is, the geographic coordinates, to a Cartesian map system in the direct transformation. The reverse is also introduced: transformation from Cartesian to geographic coordinates by the inverse transformation. Before this is done in Chapters 4, 5, and 6, more mathematical conventions and fundamentals must be introduced.

Azimuth is the means of measuring the direction of a second point from a given point and is an important quantity on both the model of the Earth and on the map. The azimuth, α, of point P′ in relation to point P is given in Figures 1 and 2 for both the Earth and the map. For the Earth, azimuth is the angle between the meridian containing point P, and the line between P and P′. In a like manner on the map, azimuth is the angle between the representation of the meridian containing P, and the line between P and P′. The convention used in this text is that azimuth is measured from the north in a clockwise manner. Azimuth ranges from 0° to 360°.

III. GRID SYSTEMS[1]

The parallels and meridians form a grid system on the model of the Earth. The grid serves as a reference system for locating points on the surface of the Earth.

In the map projection transformation process, this grid system is transferred to the map. Each projection scheme results in its own characteristic grid arrangement on the map. In Chapters 4, 5, and 6, after each set of plotting equations is derived, an empty grid system is given as an example for that particular projection. These grids give a clear indication of the geometric properties of each map projection.

On the map, the grids perform the same function as on the model of the Earth. They serve as a reference system for the location of points. In this text, the words "grid" and "graticule" are taken as equivalent in meaning.

IV. DIFFERENTIAL GEOMETRY OF SPACE CURVES[3]

It is first necessary to define a coordinate system at any point along a curve in space. The derivatives of the unit vectors are then required. With this information, we are able to define two important characteristics of the space curve: the torsion and the curvature.

The first step is to define a coordinate system located at any point along any space curve. Consider the space curve of Figure 3. Let ξ be an arbitrary parameter. Let the vector[7] to any point P on the curve in the Cartesian coordinate system be

$$\vec{r} = x(\xi)\hat{i} + y(\xi)\hat{j} + z(\xi)\hat{k} \tag{1}$$

Let $|\Delta \mathbf{r}| = \Delta s$. The unit tangent vector at point P is

$$\lim_{\Delta s \to 0} \frac{\Delta \vec{r}}{\Delta s} = \frac{d\vec{r}}{ds} = \hat{t} \tag{2}$$

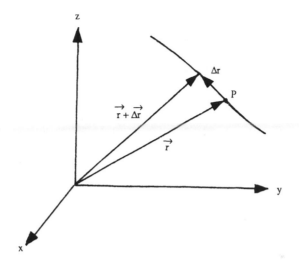

FIGURE 3. Geometry of a space curve.

Applying the chain rule to Equation 2, one finds

$$\hat{t} = \frac{d\vec{r}}{d\xi} \frac{d\xi}{ds} \tag{3}$$

Taking the total differential of Equation 1, we note

$$\hat{t} = \left(\frac{\partial x}{\partial \xi} \hat{i} + \frac{\partial y}{\partial \xi} \hat{j} + \frac{\partial z}{\partial \xi} \hat{k} \right) \left(\frac{d\xi}{ds} \right) \tag{4}$$

Upon taking the dot product of $\hat{t}$ with itself, we have

$$\hat{t} \cdot \hat{t} = 1 = \left[\left(\frac{\partial x}{\partial \xi} \right)^2 + \left(\frac{\partial y}{\partial \xi} \right)^2 + \left(\frac{\partial z}{\partial \xi} \right)^2 \right] \left(\frac{d\xi}{ds} \right)^2$$

$$\left(\frac{d\xi}{ds} \right)^2 = \frac{1}{\left(\frac{\partial x}{\partial \xi} \right)^2 + \left(\frac{\partial y}{\partial \xi} \right)^2 + \left(\frac{\partial z}{\partial \xi} \right)^2}$$

$$\frac{d\xi}{ds} = \frac{1}{\sqrt{\left(\frac{\partial x}{\partial \xi} \right)^2 + \left(\frac{\partial y}{\partial \xi} \right)^2 + \left(\frac{\partial z}{\partial \xi} \right)^2}}$$

$$= \frac{1}{\left| \frac{\partial \vec{r}}{\partial \xi} \right|} \tag{5}$$

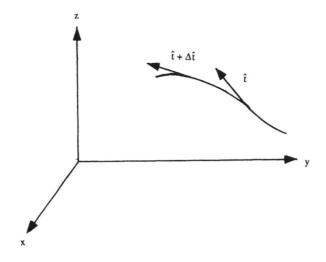

FIGURE 4.   Consecutive unit tangents on a space curve.

Next, we look at two consecutive tangent vectors, as shown in Figure 4.

$$\underset{\Delta s \to 0}{\text{Lim}} \frac{\Delta \hat{t}}{\Delta s} = \frac{d\hat{t}}{ds}$$

Let

$$\frac{d\hat{t}}{ds} = -k\hat{n} \tag{6}$$

where k is defined as the curvature, and $\hat{n}$ is the principal unit normal. Note the relationship of the curvature and the normal to the curve.

Upon dotting $\hat{t}$ with itself, and differentiating, we find

$$\hat{t} \cdot \hat{t} = 1$$

$$2\hat{t} \cdot \frac{d\hat{t}}{ds} = 0$$

$$\hat{t} \cdot \hat{n} = 0$$

This means that $\hat{t}$ is perpendicular to $d\hat{t}/ds$, and from Equation 6, $\hat{n}$ is perpendicular to $\hat{t}$.

In order to obtain a right-handed triad, define the binormal vector.

$$\hat{b} = \hat{t} \times \hat{n} \tag{7}$$

Thus, we have the fundamental set of unit vectors $\hat{t}$, $\hat{n}$, and $\hat{b}$ associated with a continuous space curve at point P.

It is useful now to obtain the derivatives of the unit vectors as a function of distance along the curve. This is needed to obtain the torsion. From the definition of the unit vectors, we have the relations

$$
\left.
\begin{aligned}
\hat{b} &= \hat{t} \times \hat{n} \\
\hat{t} &= \hat{n} \times \hat{b} \\
\hat{n} &= \hat{b} \times \hat{t}
\end{aligned}
\right\}
\tag{8}
$$

From the first of Equations 8,

$$
\begin{aligned}
\frac{d\hat{b}}{ds} &= \frac{d}{ds} (\hat{t} \times \hat{n}) \\
&= \frac{d\hat{t}}{ds} \times \hat{n} + \hat{t} \times \frac{d\hat{n}}{ds}
\end{aligned}
\tag{9}
$$

Substitute Equation 6 into Equation 9 to obtain

$$
\begin{aligned}
\frac{d\hat{b}}{ds} &= -k\hat{n} \times \hat{n} + \hat{t} \times \frac{d\hat{n}}{ds} \\
&= \hat{t} \times \frac{d\hat{n}}{ds}
\end{aligned}
\tag{10}
$$

Dot $\hat{n}$ with itself, and differentiate, so that

$$
\hat{n} \cdot \hat{n} = 1
$$

$$
2\hat{n} \cdot \frac{d\hat{n}}{ds} = 0
$$

Thus, $d\hat{n}/ds$ is perpendicular to $\hat{n}$ and must lie in the plane defined by $\hat{t}$ and $\hat{b}$ and has the components

$$
\frac{d\hat{n}}{ds} = \psi\hat{t} + \tau\hat{b}
\tag{11}
$$

Substituting Equation 11 into Equation 10, we find

$$\frac{d\hat{b}}{ds} = \hat{t} \times (\psi\hat{t} + \tau\hat{b})$$

$$= \tau\hat{t} \times \hat{b}$$

$$= -\tau\hat{n} \tag{12}$$

The variable τ is called the torsion. It is essentially a measure of the twist of the curve out of the plane defined by $\hat{n}$ and $\hat{t}$.

From the last of Equations 8,

$$\frac{d\hat{n}}{ds} = \frac{d}{ds}(\hat{b} \times \hat{t})$$

$$= \frac{d\hat{b}}{ds} \times \hat{t} + \hat{b} \times \frac{d\hat{t}}{ds} \tag{13}$$

Substitute Equations 6 and 12 into Equation 13; this gives

$$\frac{d\hat{n}}{ds} = -\tau\hat{n} \times \hat{t} + \hat{b} \times (-k\hat{n}) \tag{14}$$

$$= \tau\hat{b} + k\hat{t}$$

Equations 6, 13, and 14 can be arranged in matrix form.

$$\begin{Bmatrix} d\hat{t}/ds \\ d\hat{n}/ds \\ d\hat{b}/ds \end{Bmatrix} = \begin{bmatrix} 0 & -k & 0 \\ k & 0 & \tau \\ 0 & -\tau & 0 \end{bmatrix} \begin{Bmatrix} \hat{t} \\ \hat{n} \\ \hat{b} \end{Bmatrix} \tag{15}$$

These are the Frenet-Serret formulas.[7]

We now have the torsion and curvature as coefficients in vectorial relationships. The next step is to obtain the mathematical relations for the curvature and the torsion so they may be numerically evaluated.

The curvature, in general parametric form, is obtained from Equations 2 and 6.

$$-k\hat{n} = \frac{d\hat{t}}{ds}$$

$$= \frac{d}{ds}\left(\frac{d\vec{r}}{ds}\right)$$

$$= \frac{d^2\vec{r}}{ds^2} \tag{16}$$

Taking the cross product of $\hat{t}$ with Equation 16, we have

$$\hat{t} \times (-k\hat{n}) = \hat{t} \times \left(\frac{d^2\vec{r}}{ds^2}\right) \tag{17}$$

Using the first of Equations 8 and Equation 2 in Equation 17,

$$-k\hat{b} = \frac{d\vec{r}}{ds} \times \frac{d^2\vec{r}}{ds^2}$$

$$k = \left|\left(\frac{d\vec{r}}{ds} \times \frac{d^2\vec{r}}{ds^2}\right)\right| \tag{18}$$

For the general parametrization, we have the $\vec{r}$ derivatives

$$\frac{d\vec{r}}{ds} = \frac{d\vec{r}}{d\xi}\frac{d\xi}{ds} \tag{19}$$

$$\frac{d^2\vec{r}}{ds^2} = \frac{d^2\vec{r}}{d\xi^2}\left(\frac{d\xi}{ds}\right)^2 + \frac{d\vec{r}}{d\xi}\left(\frac{d^2\xi}{ds^2}\right) \tag{20}$$

Note, substituting Equations 19 and 20 into Equation 18, that

$$k = \left|\frac{d\vec{r}}{d\xi}\frac{d\xi}{ds} \times \left[\frac{d^2\vec{r}}{d\xi^2}\left(\frac{d\xi}{ds}\right)^2 + \frac{d\vec{r}}{d\xi}\left(\frac{d^2\xi}{ds^2}\right)\right]\right|$$

$$= \left|\frac{d\vec{r}}{d\xi} \times \frac{d^2\vec{r}}{d\xi^2}\right|\left(\frac{d\xi}{ds}\right)^3 \tag{21}$$

Upon substituting Equation 5 into Equation 21, we arrive at

$$k = \frac{\left| \dfrac{d\vec{r}}{d\xi} \times \dfrac{d^2\vec{r}}{d\xi} \right|}{\left| \dfrac{d\vec{r}}{d\xi} \right|^3} \tag{22}$$

We now have a relationship that can be used to evaluate curvature. Of more use in map projections is the radius of curvature. The radius of curvature, R_c, is the reciprocal of the curvature.

$$R_c = \frac{1}{k} \tag{23}$$

Its utility is demonstrated in Chapter 3 with respect to curvature along a meridian, and perpendicular to a meridian.

A similar procedure can be followed to obtain the torsion. Upon taking the dot product of Equation 12 with n.

$$\tau = -\frac{d\hat{b}}{ds} \cdot \hat{n} \tag{24}$$

Substitute the first of Equation 8 into Equation 24; this gives

$$\tau = -\frac{d}{ds}(\hat{t} \times \hat{n}) \cdot \hat{n}$$

$$= -\left(\frac{d\hat{t}}{ds} \times \hat{n} + \hat{t} \times \frac{d\hat{n}}{ds} \right) \cdot \hat{n}$$

$$= (\hat{t} \times \hat{n}) \cdot \frac{d\hat{n}}{ds} \tag{25}$$

From Equation 16

$$\hat{n} = \frac{-\dfrac{d^2\vec{r}}{ds^2}}{k} \tag{26}$$

$$\frac{d\hat{n}}{ds} = \frac{-\dfrac{d^3\vec{r}}{ds^3}}{k} + \frac{\dfrac{dk}{ds}}{k^2} \frac{d^2\vec{r}}{ds^2} \tag{27}$$

Substitute Equations 2, 26, and 27 into Equation 25 to obtain

$$
\tau = -\frac{d\vec{r}}{ds} \times \frac{\dfrac{d^2\vec{r}}{ds^2}}{k} \cdot \left(\frac{\dfrac{d^3\vec{r}}{ds^3}}{k} + \frac{dk}{ds}\frac{d^2\vec{r}}{k^2}\frac{d^2\vec{r}}{ds^2} \right)
$$

$$
= \frac{1}{k^2} \left(\frac{d\vec{r}}{ds} \times \frac{d^2\vec{r}}{ds^2} \right) \cdot \frac{d^3\vec{r}}{ds^3}
\tag{28}
$$

For the general parameterization, differentiate Equation 20

$$
\frac{d^3\vec{r}}{ds^3} = \frac{d^3\vec{r}}{d\xi^3}\left(\frac{d\xi}{ds}\right)^3 + \frac{3d^2\vec{r}}{d\xi^2}\left(\frac{d\xi}{ds}\right)\left(\frac{d^2\xi}{ds^2}\right) + \frac{d\vec{r}}{d\xi}\frac{d^3\xi}{ds^3}
\tag{29}
$$

Substitute Equations 19, 20, and 29 into Equation 28 to obtain

$$
\tau = \frac{\left[\left(\dfrac{d\vec{r}}{d\xi} \times \dfrac{d^2\vec{r}}{d\xi^2}\right) \cdot \left(\dfrac{d^3\vec{r}}{d\xi^3}\right)\right]\left(\dfrac{d\xi}{ds}\right)^6}{k^2}
\tag{30}
$$

Substituting Equation 5 into Equation 30 simplifies Equation 30 to

$$
\tau = \frac{\left[\left(\dfrac{d\vec{r}}{d\xi} \times \dfrac{d^2\vec{r}}{d\xi}\right) \cdot \left(\dfrac{d^3\vec{r}}{d\xi^3}\right)\right]}{k^2 \left|\dfrac{d\vec{r}}{d\xi}\right|^6}
\tag{31}
$$

We now have a numerical relation for torsion. The torsion is important for such curves as the geodesic (Chapter 3). For plane curves, such as the meridian curve, and the equator, $\tau = 0$.

A. Example 1

As an example, consider a plane curve, and let $\xi = x$, and $y = y(x)$. The radius vector is

$$
\mathbf{r} = x\hat{i} + y(x)\hat{j}
$$

Find the radius of curvature and the torsion. Differentiate twice to obtain

$$\frac{d\vec{r}}{dx} = \hat{i} + \frac{dy}{dx}\hat{j}$$

$$\frac{d^2\vec{r}}{dx} = \frac{d^2y}{dx^2}\hat{j}$$

Substituting into Equation 22 to obtain

$$k = \frac{\left|\left(\hat{i} + \frac{dy}{dx}\hat{j}\right) \times \left(\frac{d^2y}{dx^2}\hat{j}\right)\right|}{\left[\left(\hat{i} + \frac{dy}{dx}\hat{j}\right) \cdot \left(\hat{i} + \frac{dy}{dx}\hat{j}\right)\right]^{3/2}}$$

$$= \frac{\left|\frac{d^2y}{dx^2}\hat{k}\right|}{\left[1 + \left(\frac{dy}{dx}\right)^2\right]^{3/2}}$$

The radius of curvature is the reciprocal of the curvature. Thus,

$$R_c = \frac{\left[1 + \left(\frac{dy}{dx}\right)^2\right]^{3/2}}{\frac{d^2y}{dx^2}}$$

Taking the third derivative of the given equation,

$$\frac{d^3\vec{r}}{dx^3} = \frac{d^3y}{dx^3}\hat{j}$$

Substituting these relations into Equation 31,

$$\tau = \frac{\begin{vmatrix} 1 & \frac{dy}{dx} & 0 \\ 0 & \frac{d^2y}{dx^2} & 0 \\ 0 & \frac{d^3y}{dx^3} & 0 \end{vmatrix}}{k^2\left|\hat{i} + \frac{dx}{dy}\hat{j}\right|^6} = 0$$

This is no surprise, since we have a plane curve.

B. Example 2

Given: the plane curve $y = x^3 + 2x^2 + 3$.

Find: k, R_c, τ at $x = 3$.

$$\frac{dy}{dx} = (3x^2 + 4x)|_{x=3} = 39$$

$$\frac{d^2y}{dx^2} = (6x + 4)|_{x=3} = 22$$

$$k = \frac{\left|\dfrac{d^2y}{dx^2}\,\hat{k}\right|}{\left[1 + \left(\dfrac{dy}{dx}\right)^2\right]^{3/2}} = \frac{22}{[1 + 39^2]^{3/2}}$$

$$= 0.0003705$$

$$R_c = \frac{1}{k} = 2{,}699$$

$$\tau = \frac{\begin{vmatrix} 1 & \dfrac{dy}{dx} & 0 \\[2mm] 0 & \dfrac{d^2y}{dx^2} & 0 \\[2mm] 0 & \dfrac{d^3y}{dx^3} & 0 \end{vmatrix}}{k^2\left|\dfrac{d\vec{r}}{dx}\right|^6} = 0$$

V. DIFFERENTIAL GEOMETRY OF A GENERAL SURFACE[3]

Parametric curves are defined for a general surface. Next, the tangents to these parametric curves are obtained. The tangent plane at a point follows from the tangents to the curves. Finally, a total differential is found as a function of the arbitrary parameters.

The parametric representation of a surface requires two parameters. Let the two arbitrary parameters for the parametric representation of the general surface be α_1 and α_2. Later, for the special surfaces of interest to mapping, α_1 and α_2 are identified with ϕ and λ, or latitude and longitude, respectively.

The vector to any point P on the surface is given by

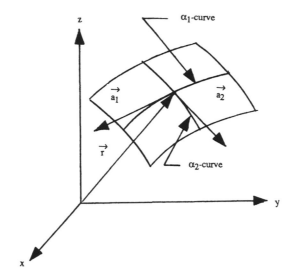

FIGURE 5.   Parametric curves.

$$\vec{r} = \vec{r}(\alpha_1, \alpha_2) \tag{32}$$

If either of the two parameters is held constant, and the other one is varied, a space curve results. This space curve is the parametric curve. Figure 5 gives the parametric representation of space curves on a surface. The α_1 curve is the parametric curve along which α_2 is constant, and the α_2 curve is the parametric curve along which α_1 is constant.

The next step is to obtain the tangents to the parametric curves at point P. The tangent vector to the α_1 curve is

$$\vec{a}_1 = \frac{\partial \vec{r}}{\partial \alpha_1} \tag{33}$$

The tangent to the α_2 curve is

$$\vec{a}_2 = \frac{\partial \vec{r}}{\partial \alpha_2} \tag{34}$$

The plane spanned by the vectors $\vec{a}_1$ and $\vec{a}_2$ is the tangent plane to the surface at point P.

The total differential of Equation 32 is

$$d\vec{r} = \frac{\partial \vec{r}}{\partial \alpha_1} d\alpha_1 + \frac{\partial \vec{r}}{\partial \alpha_2} d\alpha_2 \tag{35}$$

Substituting Equations 33 and 34 into Equation 35, we have

$$d\vec{r} = \vec{a}_1 d\alpha_1 + \vec{a}_2 d\alpha_2 \tag{36}$$

Armed with Equation 36, we are now ready to introduce the first fundamental form.

VI. FIRST FUNDAMENTAL FORM[3]

The first fundamental form of a surface is now to be derived. The first fundamental form is useful in dealing with arc length, area, angular measure on a surface, and the normal to the surface. Take the dot product of Equation 36 with itself to obtain a scalar.

$$(ds)^2 = d\vec{r} \cdot d\vec{r}$$

$$= (\vec{a}_1 d\alpha_1 + \vec{a}_2 d\alpha_2) \cdot (\vec{a}_1 d\alpha_1 + \vec{a}_2 d\alpha_2)$$

$$= \vec{a}_1 \cdot \vec{a}_1 (d\alpha_1)^2 + \vec{a}_2 \cdot \vec{a}_2 (d\alpha_2)^2 + 2(\vec{a}_1 \cdot \vec{a}_2) d\alpha_1 d\alpha_2 \tag{37}$$

Define new variables.

$$\left.\begin{aligned} E &= \vec{a}_1 \cdot \vec{a}_1 \\ F &= \vec{a}_1 \cdot \vec{a}_2 \\ G &= \vec{a}_2 \cdot \vec{a}_2 \end{aligned}\right\} \tag{38}$$

These new variables are known as the Gaussian fundamental quantities. These are defined for particular surfaces of interest to map projection. They are used in the derivation of the projections themselves and in the mathematical development of an estimate of distortion.

Substituting Equations 38 into Equation 37, we obtain the important differential form

$$(ds)^2 = E(d\alpha_1)^2 + 2F d\alpha_1 d\alpha_2 + G(d\alpha_2)^2 \tag{39}$$

Equation 39 is the first fundamental form of a surface, and this is very useful through the whole process of map projection. The first fundamental form is now applied to linear measure on any surface.

Arc length can be found immediately from the integration of Equation 39. The distance between two arbitrary points P_1 and P_2 on the surface is given by

$$s = \int_{P_1}^{P_2} \sqrt{E(d\alpha_1)^2 + 2Fd\alpha_1 d\alpha_2 + G(d\alpha_2)^2}$$

$$= \int_{P_1}^{P_2} \left\{ \sqrt{E + 2F\left(\frac{d\alpha_2}{d\alpha_1}\right) + G\left(\frac{d\alpha_1}{d\alpha_2}\right)^2} \right\} d\alpha_1 \qquad (40)$$

Equation 40 is useful as soon as $d\alpha_2/d\alpha_1$ is defined, and it is used in Chapter 3 for distance along the spheroid.

Angles between two unit tangents $\hat{a}_1$ and $\hat{a}_2$ on the surface can be found by taking the dot product of Equations 33 and 34 and applying Equation 38.

$$\cos\theta = \frac{\vec{a}_1}{|\vec{a}_1|} \cdot \frac{\vec{a}_2}{|\vec{a}_2|} \qquad (41)$$

$$= \frac{F}{\sqrt{EG}} \qquad (42)$$

$$\sin\theta = \sqrt{1 - \cos^2\theta}$$

$$= \sqrt{1 - \frac{F^2}{EG}}$$

$$= \sqrt{\frac{EG - F^2}{EG}} \qquad (43)$$

Define

$$H = EG - F^2 \qquad (44)$$

and substitute Equation 44 into Equation 43 to obtain

$$\sin\theta = \sqrt{\frac{H}{EG}} \qquad (45)$$

The normal to the surface at point P is

$$\hat{n} = \frac{\vec{a}_1 \times \vec{a}_2}{|\vec{a}_1 \times \vec{a}_2|}$$

$$= \frac{\vec{a}_1 \times \vec{a}_2}{|\vec{a}_1||\vec{a}_2|\sin\theta} \qquad (46)$$

Substituting Equations 38 and 45 into Equation 46, we have

$$\hat{n} = \frac{\vec{a}_1 \times \vec{a}_2}{\sqrt{E} \sqrt{G} \sqrt{\dfrac{H}{EG}}}$$

$$= \frac{\vec{a}_1 \times \vec{a}_2}{\sqrt{H}} \tag{47}$$

Incremental area can be obtained by a consideration of incremental distance along the parametric curves. Along the α_2 curve and the α_1 curve, respectively,

$$ds_1 = \sqrt{E} \, d\alpha_1$$

$$ds_2 = \sqrt{G} \, d\alpha_2 \tag{48}$$

The differential area is then

$$dA = ds_1 ds_2 \sin\theta$$

$$= \sqrt{EG} \, \sin\theta d\alpha_1 d\alpha_2 \tag{49}$$

Substituting Equation 45 into Equation 49, we have

$$dA = \sqrt{EG} \sqrt{\frac{H}{EG}} \, d\alpha_1 d\alpha_2$$

$$= \sqrt{H} \, d\alpha_1 d\alpha_2 \tag{50}$$

Thus, the first fundamental form has given a means to derive the arc length, the unit normal to the surface at every point, and incremental area. In conjunction with the second fundamental form of the next section, it is useful in determining the radii of curvature of the surface.

Assume now that $\vec{a}_1$, and $\vec{a}_2$ are chosen to be orthogonal. Thus $\theta = 90°$. If this is substituted in Equation 42

$$\cos 90° = \frac{F}{\sqrt{EG}} = 0$$

$$F = 0$$

When orthogonality is present, Equation 50 is simplified. The first fundamental form is then

$$(ds)^2 = E(d\alpha_1)^2 + G(d\alpha_2)^2 \qquad (51)$$

This is the case for the practical, parametric systems of map projection. Thus, the form of Equation 51 always applies.

A. Example 3

The first fundamental form for a planar surface in Cartesian coordinates or a cylindrical surface is

$$(ds)^2 = (dx)^2 + (dy)^2$$

Thus, the fundamental quantities are

$$E = 1$$

$$G = 1$$

B. Example 4

The first fundamental form for a plane surface in polar coordinates or a conical surface is

$$(ds)^2 = (d\rho)^2 + \rho^2(d\theta)^2$$

The first fundamental quantities are

$$E = 1$$

$$G = \rho^2$$

C. Example 5

The first fundamental form for a sphere in geographic coordinates is

$$(ds)^2 = R^2(d\phi)^2 + R^2\cos^2\theta(d\lambda)^2$$

Thus, the fundamental quantities are

$$E = R^2$$

$$G = R^2\cos^2\phi$$

D. Example 6

The first fundamental form for a spheroid in geographic coordinates is

$$(ds)^2 = R_m^2(d\phi)^2 + R_p^2\cos^2\phi(d\lambda)^2$$

Thus, the fundamental quantities are

$$E = R_m^2$$

$$G = R_p^2 \cos^2\phi$$

E. Example 7

Given: the first fundamental form for the sphere.
Find: distance between P_1 and P_2 and incremental area.

$$s = \int_{P_1}^{P_2} \sqrt{R^2(d\phi)^2 + R^2\cos^2\phi(d\lambda)^2}$$

$$= R\int_{P_1}^{P_2} \sqrt{1 + \cos^2\phi\left(\frac{d\lambda}{d\phi}\right)^2}\, d\phi$$

$$dA = \sqrt{R^2 R^2 \cos^2\phi}\, d\phi d\lambda$$

$$= R^2\cos\phi d\phi d\lambda$$

VII. SECOND FUNDAMENTAL FORM

The second fundamental form provides the means for evaluating the curvature of a surface and determining the principal directions on the surface. The section begins by defining the second fundamental forms for a general surface. The second fundamental form, in conjunction with the first fundamental form, defines the curvature. The second fundamental quantities are then defined. The principal directions on the surface are then obtained by an optimization process. Since the two principal directions are found to be orthogonal, the curvatures in the two principal directions are given by simplified equations.

We begin the derivation for the curvature of a surface with a consideration of normal sections through the surface. A normal section implies that the normal to the parametric curve and the surface coincide. We start with the normal at an arbitrary point. For the parametric curve, take the dot product of $\hat{n}$ with Equation 6.

$$\hat{n} \cdot \frac{d\hat{t}}{ds} = -k\hat{n} \cdot \hat{n}$$

$$= -k \tag{51a}$$

Substitute the derivative of Equation 2 into Equation 51 to obtain

$$k = -\frac{d^2\vec{r}}{ds^2} \cdot \hat{n} \tag{52}$$

Since $\hat{t}$ and $\hat{n}$ are orthogonal,

$$\hat{t} \cdot \hat{n} = 0 \tag{53}$$

Then, substituting Equation 2 into Equation 53, we see

$$\frac{d\vec{r}}{ds} \cdot \hat{n} = 0 \tag{54}$$

Upon taking the derivative of Equation 54 we obtain

$$\frac{d}{ds}\left(\frac{d\vec{r}}{ds} \cdot \hat{n}\right) = \frac{d^2\vec{r}}{ds^2} \cdot \hat{n} + \frac{d\vec{r}}{ds} \cdot \frac{d\hat{n}}{ds} = 0$$

$$\frac{d\vec{r}}{ds} \cdot \frac{d\hat{n}}{ds} = -\frac{d^2\vec{r}}{ds} \cdot \hat{n} \tag{55}$$

Substitute Equation 52 into Equation 55; this gives

$$\frac{d\vec{r}}{ds} \cdot \frac{d\hat{n}}{ds} = k \tag{56}$$

From the total differential of **r** and $\hat{n}$, we have

$$d\vec{r} = \frac{\partial\vec{r}}{\partial\alpha_1} d\alpha_1 + \frac{\partial\vec{r}}{\partial\alpha_2} d\alpha_2 \tag{57}$$

$$d\hat{n} = \frac{\partial\hat{n}}{\partial\alpha_1} d\alpha_1 + \frac{\partial\hat{n}}{\partial\alpha_2} d\alpha_2 \tag{58}$$

Substituting Equations 57 and 58 into Equation 56 gives

$$k = \frac{\left(\dfrac{\partial\vec{r}}{\partial\alpha_1} d\alpha_1 + \dfrac{\partial\vec{r}}{\partial\alpha_2} d\alpha_2\right) \cdot \left(\dfrac{\partial\hat{n}}{\partial\alpha_1} d\alpha_1 + \dfrac{\partial\hat{n}}{\partial\alpha_2} d\alpha_2\right)}{(ds)^2} \tag{59}$$

Finally, substituting Equations 33, 34, and 39 into Equation 59, we obtain

$$k = \frac{\vec{a}_1 \cdot \dfrac{\partial \hat{n}}{\partial \alpha_1} (d\alpha_1)^2 + \vec{a}_2 \cdot \dfrac{\partial \hat{n}}{\partial \alpha_2} (d\alpha_2)^2 + \left(\vec{a}_1 \cdot \dfrac{\partial \hat{n}}{\partial \alpha_2} + \vec{a}_2 \cdot \dfrac{\partial \hat{n}}{\partial \alpha_1} \right) d\alpha_1 d\alpha_2}{E(d\alpha_1)^2 + 2F d\alpha_1 d\alpha_2 + G(d\alpha_2)^2}$$

(60)

The second fundamental form is defined as

$$\vec{a}_1 \cdot \frac{\partial \hat{n}}{\partial \alpha_1} (d\alpha_1)^2 + \vec{a}_2 \cdot \frac{\partial \hat{n}}{\partial \alpha_2} (d\alpha_2)^2 + \left(\vec{a}_1 \cdot \frac{\partial \hat{n}}{\partial \alpha_2} + \vec{a}_2 \cdot \frac{\partial \hat{n}}{\partial \alpha_1} \right) d\alpha_1 d\alpha_2 \quad (61)$$

Thus, Equation 60 is the ratio

$$k = \frac{\text{Second fundamental form}}{\text{First fundamental form}}$$

We now have curvature in terms of the first and second fundamental forms for a general surface.

It remains to define the coefficients of the differentials in the second fundamental form. These coefficients are the second fundamental quantities. From the definitions of the tangent and normal vectors,

$$\hat{n} \cdot \vec{a}_i = \hat{n} \cdot \frac{\partial \vec{r}}{\partial \alpha_i}$$

$$= 0 \qquad (62)$$

Taking the derivative of Equation 62, we have

$$\frac{\partial}{\partial \alpha_j} \left(\hat{n} \cdot \frac{\partial \vec{r}}{\partial \alpha_i} \right) = 0$$

$$\frac{\partial \hat{n}}{\partial \alpha_j} \cdot \frac{\partial \vec{r}}{\partial \alpha_i} + \hat{n} \cdot \frac{\partial^2 \vec{r}}{\partial \alpha_i \partial \alpha_j} = 0 \qquad (63)$$

By definition, the second fundamental quantities are

$$b_{ij} = \frac{\partial \hat{n}}{\partial \alpha_j} \cdot \frac{\partial \vec{r}}{\partial \alpha_i} \qquad (64)$$

Substitute Equation 64 into Equation 63 to obtain

$$b_{ij} = -\hat{n} \cdot \frac{\partial^2 \vec{r}}{\partial \alpha_i \partial \alpha_j} \qquad (65)$$

Substituting Equations 45 and 46 into Equation 65, we have

$$
\begin{aligned}
b_{ij} &= -\frac{\vec{a}_1 \times \vec{a}_2}{\sqrt{H}} \cdot \frac{\partial^2 \vec{r}}{\partial \alpha_i \partial \alpha_j} \\
&= -\frac{1}{\sqrt{H}} \left(\frac{\partial^2 \vec{r}}{\partial \alpha_i \partial \alpha_j} \times \vec{a}_1 \right) \cdot \vec{a}_2
\end{aligned}
$$

$$
= -\frac{1}{\sqrt{H}}
\begin{vmatrix}
\dfrac{\partial^2 x}{\partial \alpha_i \partial \alpha_j} & \dfrac{\partial^2 y}{\partial \alpha_i \partial \alpha_j} & \dfrac{\partial^2 z}{\partial \alpha_i \partial \alpha_j} \\[2ex]
\dfrac{\partial x}{\partial \alpha_1} & \dfrac{\partial y}{\partial \alpha_1} & \dfrac{\partial z}{\partial \alpha_1} \\[2ex]
\dfrac{\partial x}{\partial \alpha_2} & \dfrac{\partial y}{\partial \alpha_2} & \dfrac{\partial z}{\partial \alpha_2}
\end{vmatrix}
\qquad (66)
$$

Defining the second fundamental quantities as

$$
\left.
\begin{aligned}
L &= b_{11} \\
M &= b_{12} = b_{21} \\
N &= b_{22}
\end{aligned}
\right\}
\qquad (67)
$$

and substituting Equation 67 into Equation 60, we arrive at the result,

$$
k = \frac{L(d\alpha_1)^2 + 2M d\alpha_1 d\alpha_2 + N(d\alpha_2)^2}{E(d\alpha_1)^2 + 2F d\alpha_1 d\alpha_2 + G(d\alpha_2)^2}
\qquad (68)
$$

We now have the curvature in terms of the first and second fundamental quantities for a general surface.

For any surface, at a point, there will be distinct curvature in two intersecting normal sections of the surface. The next step is to maximize Equation 68 to obtain the principal directions in which these curvature occur. Consult Figure 6 for a visualization of the curvature. On the general surface are the intersecting parametric curves defined by α_1 and α_2. The normal $\hat{n}$ is located at the intersection of the curves. Since the curvature is generally distinct at any point, P, two radii of curvature are needed. One of these is R_1, centered on O_1, located on the extension of the normal. The second is R_2, centered at O_2, also located on the extension of the normal.

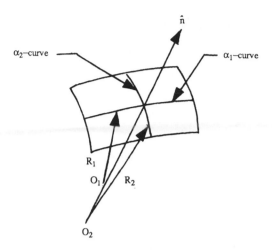

FIGURE 6. Radii of curvature in the parametric directions.

In order to determine the principal parametric directions, we begin by letting $\alpha_2 = \alpha_2(\alpha_1)$ and $\lambda = d\alpha_2/d\alpha_1$ where λ is an unspecified parametric direction. From Equation 68,

$$
\begin{aligned}
k &= \frac{L + 2M\left(\dfrac{d\alpha_2}{d\alpha_1}\right) + N\left(\dfrac{d\alpha_2}{d\alpha_1}\right)^2}{E + 2F\left(\dfrac{d\alpha_2}{d\alpha_1}\right) + G\left(\dfrac{d\alpha_2}{d\alpha_1}\right)^2} \\[2mm]
&= \frac{L + 2M\lambda + N\lambda^2}{E + 2F\lambda + G\lambda^2}
\end{aligned}
\tag{69}
$$

To find the directions for which k is an extremum, take the derivative of Equation 69 with respect to λ, and set this equal to zero.

$$
\begin{aligned}
\frac{dk}{d\lambda} &= \frac{(2M + 2N\lambda)}{E + 2F\lambda + G\lambda^2} - \frac{(L + 2M\lambda + N\lambda^2)(2F + 2G\lambda)}{(E + 2F\lambda + G\lambda^2)^2} \\[2mm]
&= 0
\end{aligned}
\tag{70}
$$

Substitute Equation 69 into Equation 70 to obtain

$$
\begin{aligned}
\frac{dk}{d\lambda} &= \frac{2M + 2N\lambda}{E + 2F\lambda + G\lambda^2} - \frac{k(2F + 2G\lambda)}{E + 2F\lambda + G\lambda^2} \\[2mm]
&= 0
\end{aligned}
$$

Since the denominator will never be zero, we have the necessary condition that

$$2M + 2N\lambda - k(2F + 2G\lambda) = 0$$

$$k = \frac{M + N\lambda}{F + G\lambda} \tag{71}$$

Write Equation 69 as

$$k = \frac{(L + M\lambda) + \lambda(M + N\lambda)}{(E + F\lambda) + \lambda(F + G\lambda)}$$

$$k[(E + F\lambda) + \lambda(F + Gk)] = (L + M\lambda) + \lambda(M + N\lambda) \tag{72}$$

and substitute Equation 71 into Equation 72; this gives

$$k[(E + F\lambda) + (F + G)\lambda] = (L + M\lambda) + \lambda(F + G\lambda)k$$

$$k(E + F\lambda) = L + M\lambda$$

$$k = \frac{L + M\lambda}{E + F\lambda} \tag{73}$$

Now, cross multiplying Equations 71 and 73 to form a quadratic in λ. This is solved for the principal directions.

$$(L + M\lambda)(F + G\lambda) = (M + N\lambda)(E + F\lambda)$$

$$LF + FM\lambda + GL\lambda + MG\lambda^2 = ME + NF\lambda + NE\lambda + NF\lambda^2$$

$$\lambda^2(MG - NF) + (LG - NE)\lambda + (LF - ME) = 0 \tag{74}$$

The solutions of Equation 74 are

$$\begin{Bmatrix} \lambda_1 \\ \lambda_2 \end{Bmatrix} = \frac{-(LG - NE) \pm \sqrt{(LG - NE)^2 - 4(MG - NF)(LF - ME)}}{2(MG - MF)} \tag{75}$$

Apply the theory of equations[9] for a quadratic to Equation 74 to obtain

$$\lambda_1 + \lambda_2 = -\frac{LG - NE}{MG - NF} \tag{76}$$

$$\lambda_1 \lambda_2 = \frac{LF - ME}{MG - NF} \tag{77}$$

Now it is necessary to show that the principal directions defined by these roots are orthogonal. This is done by investigating the angle between the principal directions. Let

$$\lambda_1 = \frac{\partial \alpha_2}{\partial \alpha_1} \tag{78}$$

$$\lambda_2 = \frac{\delta \alpha_2}{\delta \alpha_1} \tag{79}$$

Let θ be the angle between these two directions, and let $d\mathbf{r}$ and $\delta\mathbf{r}$ be infinitesimal vectors along λ_1 and λ_2. The cosine of the angle between the vectors is

$$\cos\theta = \frac{d\vec{r}}{|d\vec{r}|} \cdot \frac{\delta\vec{r}}{|\delta\vec{r}|}$$

$$= \frac{d\vec{r}}{ds} \cdot \frac{\delta\vec{r}}{\delta s} \tag{80}$$

The total differentials can be developed as follows.

$$d\vec{r} = \frac{\partial\vec{r}}{\partial\alpha_1}\,d\alpha_1 + \frac{\partial\vec{r}}{\partial\alpha_2}\,d\alpha_2 \tag{81}$$

$$\delta\vec{r} = \frac{\partial\vec{r}}{\partial\alpha_1}\,\delta\alpha_1 + \frac{\partial\vec{r}}{\partial\alpha_2}\,\delta\alpha_2 \tag{82}$$

Substitute Equations 81 and 82 into Equation 80 and carry out the dot product to obtain

$$\cos\theta = \left[\left(\frac{\partial\vec{r}}{\partial\alpha_1} \cdot \frac{\partial\vec{r}}{\partial\alpha_1}\right)d\alpha_1\delta\alpha_1 + \left(\frac{\partial\vec{r}}{\partial\alpha_1} \cdot \frac{\partial\vec{r}}{\partial\alpha_2}\right)d\alpha_1\delta\alpha_2 \right.$$

$$\left. + \left(\frac{\partial\vec{r}}{\partial\alpha_2} \cdot \frac{\partial\vec{r}}{\partial\alpha_1}\right)d\alpha_2\delta\alpha_1 + \left(\frac{\partial\vec{r}}{\partial\alpha_2} \cdot \frac{\partial\vec{r}}{\partial\alpha_2}\right)d\alpha_2\delta\alpha_2 \right]$$

$$\times \frac{1}{ds\delta s} \tag{83}$$

Substitute Equation 38 into Equation 83; this yields

$$\cos\theta = \frac{1}{ds\delta s}[Ed\alpha_1\delta\alpha_1 + F(d\alpha_1\delta\alpha_2 + d\alpha_2\delta\alpha_1) + G(d\alpha_2\delta\alpha_2)]$$

$$\frac{\cos\theta}{\delta\alpha_1 d\alpha_1} = \frac{1}{ds\delta s}\left[E + F\left(\frac{\delta\alpha_2}{\delta\alpha_1} + \frac{d\alpha_2}{d\alpha_1}\right) + G\left(\frac{\delta\alpha_2}{\delta\alpha_1}\right)\left(\frac{d\alpha_2}{d\alpha_1}\right)\right] \qquad (84)$$

Substitution of Equations 78 and 79 into Equation 84 gives

$$\frac{\cos\theta}{\delta\alpha_1 d\alpha_1} = \frac{1}{ds\delta s}[E + F(\lambda_1 + \lambda_2) + G(\lambda_1\lambda_2)] \qquad (85)$$

Finally, substituting Equations 76 and 77 into Equation 85 results in

$$\frac{\cos\theta}{\delta\alpha_1 d\alpha_1} = \frac{1}{ds\delta s}\left[E + \frac{F(LG - NE)}{MG - NF} + \frac{G(LF - ME)}{MG - NF}\right]$$

$$= \frac{1}{ds\delta s}\left[\frac{EMG - ENF - FLG + ENF + FLG - EMG}{MG - NF}\right]$$

$$= 0$$

Thus, $\theta = 90°$, and the principal directions are orthogonal. Since the principal directions are orthogonal, it is useful to choose the parametric curves to coincide with the directions of principal curvature. This provides additional simplification.

The next step is to find the simplified relations for the curvature along the principal directions.

The equations of the lines of curvature are

$$\lambda_1 = \lambda_2 = 0 \qquad (86)$$

If the lines of principal curvature coincide with the parametric lines, then from Equations 86 and 77,

$$LF - ME = 0 \qquad (87)$$

From orthogonality, $F = 0$. This means, from Equation 87, that

$$ME = 0 \qquad (88)$$

For an actual surface, neither E nor G can be zero. Thus, from Equation 88, $M = 0$. Upon substituting $F = M = 0$ into Equations 71 and 73, we find

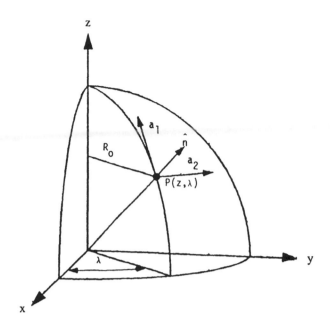

FIGURE 7.   Geometry of an arbitrary surface of revolution.

$$k_1 = \frac{L}{E} \qquad \left.\right\} \qquad (89)$$

$$k_2 = \frac{N}{G} \qquad (90)$$

Equations 89 and 90 give the means for obtaining the principal curvature of a surface from the first and second fundamental quantities.

So far, we have considered curves on general surfaces. The next step is to particularize the discussion to curves on surfaces of revolution. Following that, we deal with those surfaces of revolution of particular interest to mapping: the sphere or spheroid for the model of the Earth, and the plane, cone, or cylinder for the map.

VIII. SURFACES OF REVOLUTION[3]

Except for the planar map, all the surfaces of interest to map projections are surfaces of revolution. This section makes the step from a general surface in space to the surface of revolution. The basic geometry of a surface of revolution is discussed. The first and second fundamental quantities of a surface of revolution are defined. Once this is done, we have the means to define the curvatures and then the radii of curvature.

Surfaces of revolution are formed when a space curve is rotated about an arbitrary axis. In particular, the space curves we consider are planar curves, that is the torsion is zero. Generating curves are meridians. In this development, the two parameters needed to define a position on the surface of revolution are z and λ. Figure 7 gives the geometry for the development.

Let $R_o = R_o(z)$. The position of point P in Cartesian coordinates is

$$\vec{r} = R_o\cos\lambda\hat{i} + R_o\sin\lambda\hat{j} + z\hat{k} \tag{91}$$

Develop the vectors $\vec{a}_1$ and $\vec{a}_2$ at the point P (z, λ), and the unit normal vector there. From Equations 33 and 34, we have

$$\vec{a}_1 = \frac{\partial R_o}{\partial z}\cos\lambda\hat{i} + \frac{\partial R_o}{\partial z}\sin\lambda\hat{j} + \hat{k} \tag{92}$$

$$\vec{a}_2 = -R_o\sin\lambda\hat{i} + R_o\cos\lambda\hat{j} \tag{93}$$

From Equations 44 and 47, the normal to the surface is

$$\hat{n} = -\frac{\vec{a}_1 \times \vec{a}_2}{\sqrt{EG - F^2}} \tag{94}$$

It is now necessary to define the quantities of the first fundamental form in terms of z and λ. The first fundamental quantities are, from Equations 38, 92, and 93,

$$E = \left(\frac{\partial R_o}{\partial z}\right)^2\cos^2\lambda + \left(\frac{\partial R_o}{\partial z^2}\right)^2\sin^2\lambda + 1$$

$$= 1 + \left(\frac{\partial R_o}{\partial z}\right)^2 \tag{95}$$

$$F = -\frac{\partial R_o}{\partial z}(\cos\lambda)(R_o\sin\lambda) + \frac{\partial R_o}{\partial z}(\sin\lambda)(R_o\cos\lambda)$$

$$= 0 \tag{96}$$

$$G = R_o^2\sin^2\lambda + R_o^2\cos^2\lambda$$

$$= R_o^2 \tag{97}$$

The next step is to define the second fundamental quantities for the surface of revolution. Also, from Equations 92 and 93,

$$\vec{a}_1 \times \vec{a}_2 = \begin{vmatrix} \hat{i} & \hat{j} & \hat{k} \\ \dfrac{\partial R_o}{\partial z}\cos\lambda & \partial R_o\sin\lambda & 1 \\ -R_o\sin\lambda & R_o\cos\lambda & 0 \end{vmatrix}$$

$$= -R_o\cos\lambda\hat{i} - R_o\sin\lambda\hat{j} + \left(R_o\frac{\partial R_o}{\partial z}\cos^2\lambda + R_o\frac{\partial R_o}{\partial z}\sin^2\lambda \right)\hat{k}$$

$$= -R_o\left(\cos\lambda\hat{i} + \sin\lambda\hat{j} - \frac{\partial R_o}{\partial z}\hat{k} \right) \tag{98}$$

Substitute Equations 95, 96, 97, and 98 into Equation 94 to obtain

$$\hat{n} = -\frac{R_o\left(\cos\lambda\hat{i} + \sin\lambda\hat{j} - \dfrac{\partial R_o}{\partial z}\hat{k} \right)}{R_o\sqrt{1 + \left(\dfrac{\partial R_o}{\partial z}\right)^2}}$$

$$= -\frac{\cos\lambda\hat{i} + \sin\lambda\hat{j} - \dfrac{\partial R_o}{\partial z}\hat{k}}{\sqrt{1 + \left(\dfrac{\partial R_o}{\partial z}\right)^2}} \tag{99}$$

The quantities of the second fundamental form, in terms of z and λ, are given by Equations 66, 67, and 99 as

$$L = -\frac{\partial^2 \vec{r}}{\partial z^2} \cdot \hat{n}$$

$$= -\left(\frac{\partial^2 R_o}{\partial z^2}\cos\lambda\hat{i} + \frac{\partial^2 R_o}{\partial z^2}\sin\lambda\hat{j} \right)$$

$$\cdot -\frac{\left(\cos\lambda\hat{i} + \sin\lambda\hat{j} - \dfrac{\partial R_o}{\partial z}\hat{k} \right)}{\sqrt{1 + \left(\dfrac{\partial R_o}{\partial z}\right)^2}}$$

$$= \frac{\dfrac{\partial^2 R_o}{\partial z^2}\cos^2\lambda + \dfrac{\partial^2 R_o}{\partial z^2}\sin^2\lambda}{\sqrt{1 + \left(\dfrac{\partial R_o}{\partial z}\right)^2}}$$

$$= \frac{\partial^2 R_o}{\partial z^2} \bigg/ \sqrt{1 + (\partial z)^2} \tag{100}$$

$$M = -\frac{\partial^2 \vec{r}}{\partial z \partial \lambda} \cdot \hat{n}$$

$$= -\left(-\frac{\partial R_o}{\partial z} \sin\lambda \hat{i} + \frac{\partial R_o}{\partial z} \cos\lambda \hat{j}\right)$$

$$\cdot \left(-\frac{\cos\lambda \hat{i} + \sin\lambda \hat{j} - \frac{\partial R_o}{\partial z} \hat{k}}{\sqrt{1 + \left(\frac{\partial R_o}{\partial z}\right)^2}}\right)$$

$$= \frac{-\frac{\partial R_o}{\partial z} \sin\lambda\cos\lambda + \frac{\partial R_o}{\partial z} \sin\lambda\cos\lambda}{\sqrt{1 + \left(\frac{\partial R_o}{\partial z}\right)^2}}$$

$$= 0 \tag{101}$$

$$N = -\frac{\partial^2 \vec{r}}{\partial \lambda^2} \cdot \hat{n}$$

$$= -(-R_o\cos\lambda \hat{i} - R_o\sin\lambda \hat{j})$$

$$\cdot \left(-\frac{\cos\lambda \hat{i} + \sin\lambda \hat{j} - \frac{\partial R_o}{\partial z} \hat{k}}{\sqrt{1 + \left(\frac{\partial R_o}{\partial z}\right)^2}}\right)$$

$$= \frac{R_o\cos^2\lambda + R_o\sin^2\lambda}{\sqrt{1 + \left(\frac{\partial R_o}{\partial z}\right)^2}}$$

$$= \frac{R_o}{\sqrt{1 + \left(\frac{\partial R_o}{\partial z}\right)^2}} \tag{102}$$

Note that both F and M are zero. Thus, our choice of parameters has gained us both orthogonality and coincidence with the principal directions.

The first and second fundamental quantities are now used to obtain the curvatures in the two principal directions. Consider first the curvature in a plane perpendicular to a meridian, k_p. Substitute Equations 97 and 102 into Equation 90.

$$k_p = \cfrac{\cfrac{R_o}{\sqrt{1 + \left(\cfrac{\partial R_o}{\partial z}\right)^2}}}{R_o^2}$$

$$= \cfrac{1}{R_o\sqrt{1 + \left(\cfrac{\partial R_o}{\partial z}\right)^2}} \qquad (103)$$

The radius of curvature in a plane perpendicular to a meridian is then

$$R_p = R_o\sqrt{1 + \left(\frac{\partial R_o}{\partial z}\right)^2} \qquad (104)$$

Consider next the curvature in the plane of the meridian, k_m. Substitute Equations 95 and 100 into Equation 89.

$$k_m = \cfrac{\cfrac{\partial^2 R_o}{\partial z^2}}{\sqrt{1 + \left(\cfrac{\partial R_o}{\partial z}\right)^2}}{1 + \left(\cfrac{\partial R_o}{\partial z}\right)^2}$$

$$= \cfrac{\cfrac{\partial^2 R_o}{\partial z^2}}{\left[1 + \left(\cfrac{\partial R_o}{\partial z}\right)^2\right]^{3/2}} \qquad (105)$$

The radius of curvature in the meridian plane is then

$$R_m = \cfrac{\left[1 + \left(\cfrac{\partial R_o}{\partial z}\right)^2\right]^{3/2}}{\cfrac{\partial^2 R_o}{\partial z^2}} \qquad (106)$$

With the general results for the surface of revolution, we now have the tools to deal with the specialized surfaces of revolution of Chapter 3, that is, the sphere and the spheroid.

IX. DEVELOPABLE SURFACES

We mentioned in Chapter 1 that there are two types of surfaces of interest to map projections: developable and nondevelopable. One way to make the distinction between the two is to consider the principal radii of curvature. Nondevelopable surfaces have two finite radii of curvature. Developable surfaces have one finite and one infinite radius of curvature. Once a surface has been developed into a plane, the surface is defined by two infinite radii of curvature in two orthogonal directions.

This section expands on the differences between the two types of surfaces. The two developable surfaces of interest to mapping, the cone and cylinder, and the nondevelopable surface, the sphere, are considered. A mathematical criterion is given which differentiates between developable and nondevelopable surfaces.

The surfaces which are envelopes of one-parameter families of planes are called developable surfaces. Every cone or cylinder is an envelope of a one-parameter family of tangent planes. Moreover, every tangent plane has a contact with the surface along a straight line. Consequently, a developable surface is swept out by a family of rectilinear generators.

It is necessary, in the present section, to consider the tangent planes for cones, cylinders, and spheres and to note their characteristics. This requires the investigation of the tangent plane defined by $\vec{a}_1$ and $\vec{a}_2$.

For the cone, consider arbitrary parameters u and v. Let the origin of the coordinate system be at the vertex of the cone. The parametric equation of the cone is

$$\vec{r} = v\vec{q}(u) \tag{107}$$

From Equations 33 and 34,

$$\vec{a}_1 = v\dot{\vec{q}}(u) \tag{108}$$

$$\vec{a}_2 = \vec{q}(u) \tag{109}$$

Take the cross product of Equations 108 and 109,

$$\vec{a}_1 \times \vec{a}_2 = v\dot{\vec{q}}(u) \times \vec{q}(u) \tag{110}$$

If $q(u)$ and $\dot{q}(u)$ are not collinear, the point (u, v) is regular, and the tangent plane has the following equation, after the substitution of Equations 107 and 110:

$$[\vec{r} - v\vec{q}(u)] \cdot v\dot{\vec{q}}(u) \times \vec{q}(u) = 0$$

$$\vec{r} \cdot \dot{\vec{q}}(u) \times \vec{q}(u) = 0 \tag{111}$$

Thus, $\vec{r}$ from Equation 111, depends only u, and family of tangent planes is a one-parameter family.

For a cylinder with elements parallel to a constant vector $\vec{c}$, the parametric equation is

$$\vec{r} = \vec{q}(u) + v\vec{c} \tag{112}$$

Applying Equations 33 and 34 to Equation 112,

$$\vec{a}_1 = \dot{\vec{q}}(u) \tag{113}$$

$$\vec{a}_2 = \vec{c} \tag{114}$$

Taking the cross product of Equations 113 and 114,

$$\vec{a}_1 \times \vec{a}_2 = \dot{\vec{q}}(u) \times \vec{c} \tag{115}$$

From Equations 112 and 115, the equation of the tangent plane is

$$[\vec{r} - \vec{q}(u) - v\vec{c}] \cdot \dot{\vec{q}}(u) \times \vec{c} = 0$$

$$\vec{r} \cdot \dot{\vec{q}}(u) \times \vec{c} = \vec{q}(u) \cdot \dot{\vec{q}}(u) \times \vec{c} \tag{116}$$

Again, we have an equation which depends only on the parameter u.

A different situation occurs when a nondevelopable surface, such as the sphere of radius R is considered. Let the two parameters be ϕ and λ. The parametric equation of the surface is

$$\vec{r} = R(\cos\lambda\cos\phi\hat{i} + \sin\lambda\cos\phi\hat{j} + \sin\phi\hat{k}) \tag{117}$$

Using Equations 33 and 34,

$$\vec{a}_1 = R(-\cos\lambda\sin\phi\hat{i} - \sin\lambda\sin\phi\hat{j} + \cos\phi\hat{k}) \tag{118}$$

$$\vec{a}_2 = R(-\sin\lambda\cos\phi\hat{i} + \cos\lambda\sin\phi\hat{j}) \tag{119}$$

Taking the cross product of Equations 118 and 119

$$\vec{a}_1 \times \vec{a}_2 = R^2 \begin{vmatrix} \hat{i} & \hat{j} & \hat{k} \\ -\cos\lambda\sin\phi & -\sin\lambda\cos\phi & \cos\phi \\ -\sin\lambda\cos\phi & \cos\lambda\cos\phi & 0 \end{vmatrix}$$

$$= R^2(-\cos\lambda\cos^2\phi\hat{i} - \sin\lambda\cos^2\phi\hat{j}$$

$$- \hat{k}(\cos^2\lambda\sin\phi\cos\phi + \sin^2\lambda\sin\phi\cos\phi)]$$

$$= -R^2(\cos\lambda\cos^2\phi\hat{i} + \sin\lambda\cos^2\phi\hat{j} + \cos\phi\sin\phi\hat{k}) \quad (120)$$

From Equations 117 and 120, the tangent plane to the sphere has the equation

$$[\vec{r} - R(\cos\lambda\cos\phi\hat{i} + \sin\lambda\cos\phi\hat{j} + \sin\phi\hat{k})]$$

$$\cdot [-R^2(\cos\lambda\cos^2\phi\hat{i} + \sin\lambda\cos^2\phi\hat{j} + \cos\phi\sin\phi\hat{k})] = 0 \quad (121)$$

Equation 121 depends on two parameters, and, thus, the sphere is a nondevelopable surface.

Thus, the basic criterion for a developable surface is that the tangent plane must depend on only one parameter.

X. TRANSFORMATION MATRICES[8]

It would be hoped that we could find a relatively simple relationship between geographic and mapping variables such as that given in Equation 122, where T is the transformation.

$$\begin{Bmatrix} x \\ y \end{Bmatrix} = [T] \begin{Bmatrix} \phi \\ \lambda \end{Bmatrix} \quad (122)$$

We are frustrated in this hope since we are dealing with partial derivatives and fundamental quantities couched in transcendental terms. However, the derivation below gives relations that can be usefully applied to the map projections of Chapters 4 and 5, where the differential geometry approach is most useful. This section relates the first fundamental quantities and parameters for the model of the Earth and the mapping surface. The two conditions for a useful transformation are defined. Following this, one form of the transformation matrix useful in Chapter 4 and a second form useful in Chapter 5 are derived. Finally, the Jacobian determinant is defined.

We begin with a consideration of the model of the Earth. On this model, let the parametric curves be defined by the parameters ϕ and λ. The fundamental quantities are to be e, f, and g.

Consider next a two-dimensional projection surface with parametric curves defined by the parameters u and v. The corresponding fundamental quantities are E', F', and G'. Consider also on the plotting surface a second set of parameters, X and Y, with fundamental quantities E, F, and G. The relationship between the two sets of parameters on the plane is given by

$$X = X(u, v) \\ Y = Y(u, v) \Big\}$$

(123)

The relationship between the two sets of fundamental quantities on the plotting surface are to be defined.

The parametric curves on the Earth are related to those on the projection surface by

$$u = u(\phi, \lambda) \\ v = v(\phi, \lambda) \Big\}$$

(124)

For the Earth, and for any plotting surface, only two conditions are to be satisfied. The projection must be (1) unique, and (2) reversible. A point on the Earth must correspond to only one point on the map and vice versa. This requires that

$$\phi = \phi(u, v) \\ \lambda = \lambda(u, v) \Big\}$$

(125)

Substituting Equation 124 into Equation 123, we have

$$X = X[u(\phi, \lambda), \quad v(\phi, \lambda)] \\ Y = Y[u(\phi, \lambda), \quad v(\phi, \lambda)] \Big\}$$

(126)

Differentiate Equations 126 with respect to ϕ and λ; this gives

$$\frac{\partial x}{\partial \phi} = \frac{\partial x}{\partial u}\frac{\partial u}{\partial \phi} + \frac{\partial x}{\partial v}\frac{\partial v}{\partial \phi} \\[2mm] \frac{\partial x}{\partial \lambda} = \frac{\partial x}{\partial u}\frac{\partial u}{\partial \lambda} + \frac{\partial x}{\partial v}\frac{\partial v}{\partial \lambda} \\[2mm] \frac{\partial y}{\partial \phi} = \frac{\partial y}{\partial u}\frac{\partial u}{\partial \phi} + \frac{\partial y}{\partial v}\frac{\partial v}{\partial \phi} \\[2mm] \frac{\partial y}{\partial \lambda} = \frac{\partial y}{\partial u}\frac{\partial u}{\partial \lambda} + \frac{\partial y}{\partial v}\frac{\partial v}{\partial y} \Bigg\}$$

(127)

The procedure for the derivation is to find a relation between E, F, and G, and E', F', and G'. This entails the elimination of the partial derivatives of x and y.

From the definition of the first fundamental quantities in Section VI, we know

$$
\left.
\begin{aligned}
E &= \vec{a}_1 \cdot \vec{a}_1 = \left(\frac{\partial x}{\partial \phi}\right)^2 + \left(\frac{\partial y}{\partial \phi}\right)^2 \\[2mm]
F &= \vec{a}_1 \cdot \vec{a}_2 = \frac{\partial x}{\partial \phi}\frac{\partial x}{\partial \lambda} + \frac{\partial y}{\partial \phi}\frac{\partial y}{\partial \lambda} \\[2mm]
G &= \vec{a}_1 \cdot \vec{a}_2 = \left(\frac{\partial x}{\partial \lambda}\right)^2 + \left(\frac{\partial y}{\partial \lambda}\right)^2
\end{aligned}
\right\}
\tag{128}
$$

Upon substituting Equation 127 into Equation 128, the latter become

$$
\begin{aligned}
E &= \left(\frac{\partial x}{\partial u}\frac{\partial u}{\partial \phi}\right)^2 + 2\frac{\partial x}{\partial u}\frac{\partial u}{\partial \phi}\frac{\partial x}{\partial v}\frac{\partial v}{\partial \phi} + \left(\frac{\partial x}{\partial v}\frac{\partial v}{\partial \phi}\right)^2 \\[2mm]
&\quad + \left(\frac{\partial y}{\partial u}\frac{\partial u}{\partial \phi}\right)^2 + 2\frac{\partial y}{\partial u}\frac{\partial u}{\partial \phi}\frac{\partial y}{\partial v}\frac{\partial v}{\partial \phi} + \left(\frac{\partial y}{\partial v}\frac{\partial v}{\partial \phi}\right)^2 \\[2mm]
G &= \left(\frac{\partial x}{\partial u}\frac{\partial u}{\partial \lambda}\right)^2 + 2\frac{\partial x}{\partial u}\frac{\partial u}{\partial \lambda}\frac{\partial x}{\partial v}\frac{\partial v}{\partial \lambda} + \left(\frac{\partial x}{\partial v}\frac{\partial v}{\partial \lambda}\right)^2 \\[2mm]
&\quad + \left(\frac{\partial y}{\partial u}\frac{\partial u}{\partial \lambda}\right)^2 + 2\frac{\partial y}{\partial u}\frac{\partial u}{\partial \lambda}\frac{\partial y}{\partial v}\frac{\partial v}{\partial \lambda} + \left(\frac{\partial y}{\partial v}\frac{\partial v}{\partial \lambda}\right)^2 \\[2mm]
F &= \left(\frac{\partial x}{\partial u}\right)^2\frac{\partial u}{\partial \phi}\frac{\partial u}{\partial \lambda} + \frac{\partial x}{\partial u}\frac{\partial x}{\partial v}\left(\frac{\partial u}{\partial \phi}\frac{\partial v}{\partial \lambda} + \frac{\partial u}{\partial \lambda}\frac{\partial v}{\partial \phi}\right) \\[2mm]
&\quad + \left(\frac{\partial x}{\partial v}\right)^2\frac{\partial v}{\partial \phi}\frac{\partial v}{\partial \lambda} + \left(\frac{\partial y}{\partial u}\right)^2\frac{\partial u}{\partial \phi}\frac{\partial u}{\partial \lambda} \\[2mm]
&\quad + \frac{\partial y}{\partial u}\frac{\partial y}{\partial v}\left(\frac{\partial u}{\partial \phi}\frac{\partial v}{\partial \lambda} + \frac{\partial u}{\partial \lambda}\frac{\partial v}{\partial \phi}\right) + \left(\frac{\partial y}{\partial v}\right)^2\frac{\partial v}{\partial \phi}\frac{\partial v}{\partial \lambda}
\end{aligned}
\tag{129}
$$

In a similar manner, relying on the relations of Section VI, we have

$$
\begin{aligned}
E' &= \left(\frac{\partial x}{\partial u}\right)^2 + \left(\frac{\partial y}{\partial u}\right)^2 \\[2mm]
F' &= \frac{\partial x}{\partial u}\frac{\partial x}{\partial v} + \frac{\partial y}{\partial u}\frac{\partial y}{\partial v} \\[2mm]
G' &= \left(\frac{\partial x}{\partial v}\right)^2 + \left(\frac{\partial y}{\partial v}\right)^2
\end{aligned}
\tag{130}
$$

Substituting Equation 130 into Equation 129, we find

$$E = \left(\frac{\partial u}{\partial \phi}\right)^2 E' + 2\frac{\partial u}{\partial \phi}\frac{\partial v}{\partial \phi} F' + \left(\frac{\partial v}{\partial \phi}\right)^2 G'$$

$$F = \left(\frac{\partial u}{\partial \phi}\frac{\partial u}{\partial \lambda}\right)E' + \left(\frac{\partial u}{\partial \phi}\frac{\partial v}{\partial \lambda} + \frac{\partial u}{\partial \lambda}\frac{\partial v}{\partial \phi}\right)F' + \left(\frac{\partial v}{\partial \phi}\frac{\partial v}{\partial \phi}\right)G'$$

$$G = \left(\frac{\partial u}{\partial \lambda}\right)^2 E' + 2\frac{\partial u}{\partial \lambda}\frac{\partial v}{\partial \lambda} F' + \left(\frac{\partial v}{\partial \lambda}\right)^2 G' \tag{131}$$

Equations 131 may be written in matrix notation as

$$\begin{Bmatrix} E \\ F \\ G \end{Bmatrix} = \begin{bmatrix} \left(\dfrac{\partial u}{\partial \phi}\right)^2 & 2\dfrac{\partial u}{\partial \phi}\dfrac{\partial v}{\partial \phi} & \left(\dfrac{\partial v}{\partial \phi}\right)^2 \\ \dfrac{\partial u}{\partial \phi}\dfrac{\partial u}{\partial \lambda} & \dfrac{\partial u}{\partial \phi}\dfrac{\partial v}{\partial \lambda} + \dfrac{\partial u}{\partial \lambda}\dfrac{\partial v}{\partial \phi} & \dfrac{\partial v}{\partial \phi}\dfrac{\partial v}{\partial \lambda} \\ \left(\dfrac{\partial u}{\partial \lambda}\right)^2 & 2\dfrac{\partial u}{\partial \lambda}\dfrac{\partial v}{\partial \lambda} & \left(\dfrac{\partial v}{\partial \lambda}\right)^2 \end{bmatrix} \begin{Bmatrix} E' \\ F' \\ G' \end{Bmatrix} \tag{132}$$

The transformation matrix in Equation 132 is the fundamental matrix for mapping transformations. This form of the transformation relation is of particular use in Chapter 5 for conformal projections. Note that the fundamental quantities E, F, and G are functions of ϕ and λ, and the fundamental quantities E', F', and G' are functions of the basic parameters chosen for the mapping surface.

It is now necessary to expand the relations in order to find a form of the transformation matrix of use for equal area projections. This is done by developing the term $EG - F^2$.
From Equation 131,

$$EG - F^2 = \left[\left(\frac{\partial u}{\partial \phi}\right)^2 E' + 2\frac{\partial u}{\partial \phi}\frac{\partial v}{\partial \phi} F' + \left(\frac{\partial u}{\partial \phi}\right)^2 G'\right]$$

$$\times \left[\left(\frac{\partial u}{\partial \lambda}\right)^2 E' + 2\frac{\partial u}{\partial \lambda}\frac{\partial v}{\partial \lambda} F' + \left(\frac{\partial v}{\partial \phi}\right)^2 G'\right]$$

$$- \left[\frac{\partial u}{\partial \phi}\frac{\partial u}{\partial \lambda} E' + \left(\frac{\partial u}{\partial \phi}\frac{\partial v}{\partial \lambda} + \frac{\partial u}{\partial \lambda}\frac{\partial v}{\partial u}\right)F' + \frac{\partial v}{\partial \phi}\frac{\partial v}{\partial \lambda} G'\right]^2$$

$$= (E')^2 \left(\frac{\partial u}{\partial \phi}\right)^2\left(\frac{\partial u}{\partial \lambda}\right)^2 + 2\frac{\partial u}{\partial \phi}\frac{\partial v}{\partial \phi}\left(\frac{\partial u}{\partial \lambda}\right)^2 E'F'$$

$$+ \left(\frac{\partial v}{\partial \phi}\right)^2 \left(\frac{\partial u}{\partial \lambda}\right)^2 E'G' + 2\frac{\partial u}{\partial \lambda}\frac{\partial v}{\partial \lambda}\left(\frac{\partial u}{\partial \phi}\right)^2 E'F'$$

$$+ 4\frac{\partial u}{\partial \phi}\frac{\partial v}{\partial \phi}\frac{\partial u}{\partial \lambda}\frac{\partial v}{\partial \lambda}(F')^2 + 2\frac{\partial u}{\partial \lambda}\frac{\partial v}{\partial \lambda}\left(\frac{\partial v}{\partial \phi}\right)^2 F'G'$$

$$+ \left(\frac{\partial u}{\partial \phi}\right)^2\left(\frac{\partial v}{\partial \lambda}\right)^2 E'G' + 2\frac{\partial u}{\partial \phi}\frac{\partial v}{\partial \phi}\left(\frac{\partial v}{\partial \lambda}\right)^2 E'G'$$

$$+ \left(\frac{\partial u}{\partial \phi}\right)^2\left(\frac{\partial v}{\partial \lambda}\right)^2 (G')^2 - \left(\frac{\partial u}{\partial \phi}\frac{\partial u}{\partial \lambda}\right)^2 (E')^2$$

$$- 2\left(\frac{\partial u}{\partial \phi}\frac{\partial v}{\partial \lambda} + \frac{\partial u}{\partial \lambda}\frac{\partial v}{\partial \phi}\right)\frac{\partial u}{\partial \phi}\frac{\partial u}{\partial \lambda} E'F'$$

$$- 2\frac{\partial u}{\partial \phi}\frac{\partial u}{\partial \lambda}\frac{\partial v}{\partial \phi}\frac{\partial v}{\partial \lambda} E'G' - \left(\frac{\partial u}{\partial \phi}\frac{\partial v}{\partial \lambda} + \frac{\partial u}{\partial \lambda}\frac{\partial v}{\partial \phi}\right)^2 (F')$$

$$- \left(\frac{\partial v}{\partial \phi}\frac{\partial v}{\partial \lambda}\right)^2 (G')^2 - 2\left(\frac{\partial u}{\partial \phi}\frac{\partial v}{\partial \lambda} + \frac{\partial u}{\partial \lambda}\frac{\partial v}{\partial \phi}\right)^2 F'G'$$

$$EG - F^2 = E'G'\left[\left(\frac{\partial u}{\partial \phi}\frac{\partial v}{\partial \lambda}\right)^2 + \left(\frac{\partial v}{\partial \phi}\frac{\partial u}{\partial \lambda}\right)^2 - 2\left(\frac{\partial u}{\partial \phi}\frac{\partial v}{\partial \lambda}\frac{\partial v}{\partial \phi}\frac{\partial u}{\partial \lambda}\right)\right]$$

$$+ \left[-\left(\frac{\partial u}{\partial \phi}\frac{\partial v}{\partial \lambda}\right)^2 - \left(\frac{\partial v}{\partial \phi}\frac{\partial u}{\partial \lambda}\right)^2 + 2\left(\frac{\partial u}{\partial \phi}\frac{\partial v}{\partial \lambda}\frac{\partial v}{\partial \phi}\frac{\partial u}{\partial \lambda}\right)\right](F')^2$$

$$= [E'G' - (F')^2]$$

$$\times \left[\left(\frac{\partial u}{\partial \phi}\frac{\partial v}{\partial \lambda}\right)^2 + \left(\frac{\partial v}{\partial \phi}\frac{\partial u}{\partial \lambda}\right)^2 - 2\left(\frac{\partial u}{\partial \phi}\frac{\partial v}{\partial \lambda}\frac{\partial v}{\partial \phi}\frac{\partial u}{\partial \lambda}\right)\right]$$

$$= [E'G' - (F')^2]\left(\frac{\partial u}{\partial \phi}\frac{\partial v}{\partial \lambda} - \frac{\partial v}{\partial \phi}\frac{\partial u}{\partial \lambda}\right)^2$$

$$= [E'G' - (F')^2]\begin{vmatrix} \dfrac{\partial u}{\partial \phi} & \dfrac{\partial u}{\partial \lambda} \\[2mm] \dfrac{\partial v}{\partial \phi} & \dfrac{\partial v}{\partial \lambda} \end{vmatrix}^2 \tag{133}$$

The determinant

$$J = \begin{vmatrix} \dfrac{\partial u}{\partial \phi} & \dfrac{\partial u}{\partial \lambda} \\[2mm] \dfrac{\partial v}{\partial \phi} & \dfrac{\partial v}{\partial \lambda} \end{vmatrix}$$

is the Jacobian determinant[4] of the transformation from the coordinate set ϕ and λ to the coordinate set u and v.

A further simplification will be introduced, since we are dealing with orthogonal curves: $f = F = F' = 0$. Substituting this into Equation 133, we obtain

$$EG = E'G' \quad \begin{vmatrix} \dfrac{\partial u}{\partial \phi} & \dfrac{\partial u}{\partial \lambda} \\[2mm] \dfrac{\partial v}{\partial \phi} & \dfrac{\partial v}{\partial \lambda} \end{vmatrix}^2 \tag{134}$$

Again, note that E and G are in geographic coordinates and E' and G' are the mapping coordinates.

In the transformation process, the particular conditions of equality of area or conformality are used to relate e and g on the model of the Earth to E and G on the map. Then, Equation 134 is evaluated to obtain the mapping coordinates in the direct transformation.

It should be obvious that a similar procedure can be applied to obtain the inverse transformation. However, in Chapters 4, 5, and 6, the inverse transformation is obtained by a manipulation of the direct transformation equations. This is often done through the aid of trigonometric identities.

A. Example 8

Given: $x = C_1\lambda + C_2$, where C_1 and C_2 are constants, and $y = y(\phi)$.
Find: the Jacobean Determinant.

$$\frac{\partial x}{\partial \phi} = 0 \qquad\qquad \frac{\partial x}{\partial \lambda} = C_1$$

$$\frac{\partial y}{\partial \phi} = \frac{\partial y(\phi)}{\partial \phi} \qquad\qquad \frac{\partial y}{\partial \lambda} = 0$$

$$J = \begin{vmatrix} \dfrac{\partial x}{\partial \phi} & \dfrac{\partial x}{\partial \lambda} \\[2mm] \dfrac{\partial y}{\partial \phi} & \dfrac{\partial y}{\partial \lambda} \end{vmatrix} = \begin{vmatrix} 0 & C_1 \\[2mm] \dfrac{\partial y(\phi)}{\partial \phi} & 0 \end{vmatrix}$$

$$= -C_1 \frac{\partial y(\phi)}{\partial \phi}$$

XI. MATHEMATICAL DEFINITION OF EQUALITY OF AREA AND CONFORMALITY[3]

In Chapter 1, the concepts of equality of area and conformality were introduced in a qualitative sense. In this section, the qualities of equality of area and conformality are defined mathematically. For our purposes, this is done with respect to the first fundamental forms of the model of the Earth and the chosen mapping surface. In the definitions which follow, the fundamental quantities e and g refer to the model of the Earth, and E and G refer to the plotting surface. Only orthogonal systems on the model of the Earth and the plotting surface are considered.

An equal area map is one in which the areas of domains are preserved as they are transformed from the Earth to the map. A theorem of differential geometry requires that a mapping from the Earth to the plotting surface is locally equal area if, and only if,

$$eg = EG \tag{135}$$

This relation is substituted into Equation 134 to obtain the equal area transformation. This is done in Chapter 4 to transform from the model of the Earth to the cylinder, plane, cone, and authalic sphere.

A mapping of the surface of the Earth onto the plane, or a developable surface, is called conformal (or isogonal) if it preserves the angle between intersecting curves on the surface. From a theorem of differential geometry, a mapping is called conformal if, and only if, the first fundamental forms of the Earth and the mapping surface, in compatible coordinates, are proportional at every point. This requires that, for the orthogonal case,

$$\frac{E}{e} = \frac{G}{g} \tag{136}$$

The transformations of Chapter 5 apply the relation of Equation 136 between the model of the Earth and the plane, cylinder, cone, and conformal sphere.

The differential geometry approach, in the general form, can be applied only to the equal area and conformal projections. For the conventional projections, each one has its own special requirements, and no general relations can be defined. Thus, for the conventional projections, the differential geometry approach is not applicable.

XII. ROTATION OF THE COORDINATE SYSTEM[6]

Often it is desirable to derive a map projection for the simplest case, and then rotate the coordinate system to obtain complex cases. This is the alternative to a derivation for the general case, and then a simplification for particular orientations of the mapping surface.

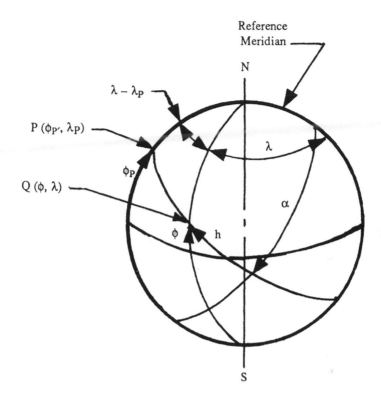

FIGURE 8. Geometry for the rotational transformation.

In this section, the spherical trigonometric formulas are given for rotations of the coordinate system. Obviously, a spherical model of the Earth is required to implement these transformations.

Figure 8 provides the basic geometry for the rotational transformation. Let Q be any arbitrary point with coordinates ϕ and λ on the Earth. Let P be the pole of the auxiliary spherical coordinate system. In the standard equatorial coordinate system, P has the coordinates ϕ_p and λ_p. Let h be the latitude of Q in the auxiliary system, and α, the longitude in that same system. A reference meridian is chosen for the origin of the measurement of α.

The intention is to derive the projection in the (h, α) system, and then transform to the (ϕ, λ) system for the plotting of the coordinates.

The relations between the angles of interest can be found from the spherical triangle PNQ.

From the law of cosines,

$$\cos(90° - \phi) = \cos(90° - \phi_p)\cos(90° - h)$$
$$+ \sin(90° - \phi_p)\sin(90° - h)\cos\alpha \qquad (137)$$

$$\sin\phi = \sin\phi_p\sin h + \cos\phi_p\cos h \cos\alpha$$

From the law of sines,

$$\frac{\sin(\lambda - \lambda_p)}{\sin(90° - h)} = \frac{\sin\alpha}{\sin(90° - \phi)}$$

$$\sin(\lambda - \lambda_p) = \frac{\sin\alpha\cos h}{\cos\phi} \tag{138}$$

Also, applying the four-parts formula,

$$\cos(90° - \phi_p)\cos\alpha = \sin(90° - \phi_p)\cot(90° - h) - \sin\alpha\cot(\lambda - \lambda_p)$$

$$\sin\phi_p\cos\alpha = \cos\phi_p\tan h - \sin\alpha\cot(\lambda - \lambda_p)$$

$$\cot(\lambda - \lambda_p) = \frac{\cos\phi_p\tan h - \sin\phi_p\cos\alpha}{\sin\alpha} \tag{139}$$

The inverse relationships are also of use. From the law of cosines,

$$\cos(90° - h) = \cos(90° - \phi)\cos(90° - \phi_p)$$

$$+ \sin(90° - \phi)\sin(90° - \phi_p)\cos(\lambda - \lambda_p)$$

$$\sin h = \sin\phi\sin\phi_p + \cos\phi\cos\phi_p\cos(\lambda - \lambda_p) \tag{140}$$

From the four-parts formula,

$$\cos(90° - \phi_p)\cos(\lambda - \lambda_p) = \sin(90° - \phi_p)\cot(90° - \phi)$$

$$- \sin(\lambda - \lambda_p)\cot\alpha$$

$$\sin\phi_p\cos(\lambda - \lambda_p) = \cos\phi_p\tan\phi - \sin(\lambda - \lambda_p)\cot\alpha$$

$$\sin(\lambda - \lambda_p)\cot\alpha = \cos\phi_p\tan\phi - \sin\phi_p\cos(\lambda - \lambda_p)$$

$$\tan\alpha = \frac{\sin(\lambda - \lambda_p)}{\cos\phi_p\tan\phi - \sin\phi_p\cos(\lambda - \lambda_p)} \tag{141}$$

A final useful equation is needed for unique quadrant determination. From a consideration of Figure 8,

$$\cos\alpha\cos h = \sin\phi\cos\phi_p - \cos\phi\sin\phi_p\cos(\lambda - \lambda_p) \tag{142}$$

The set of Equations 137 through 142 permit the transformation from polar and regular projections to oblique and transverse cases, and in the reverse order.

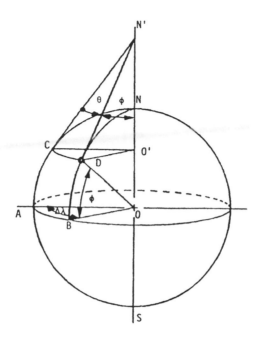

FIGURE 9. Geometry for the angular convergence of the meridians.

XIII. CONVERGENCY OF THE MERIDIANS[2]

Refer back to Figure 1, Chapter 1. It is evident from this figure that as one goes poleward from the equator, the meridians steadily converge. At the pole, the meridians intersect. It is seen that the degree of convergence is a function of latitude. This section gives an estimate of the amount of this convergency. Both angular and linear convergency are considered.

We begin with angular convergence. In Figure 9, ACN and BDN are two meridians separated by a longitude difference of $\Delta\lambda$. Let CD be an arc of the circle of parallel of latitude ϕ. Let the Earth be considered as spherical.

From the figure, with $\Delta\lambda$ in radians,

$$CD = CO'\Delta\lambda \tag{143}$$

$$DN' = \frac{DO'}{\sin\phi} \tag{144}$$

Approximately, the angle of convergency is

$$\theta = \frac{CD}{DN'} \tag{145}$$

Substituting Equations 143 and 144 into Equation 145, and noting that CO′ = DO′,

$$\theta = \frac{CO'\Delta\lambda}{DO'/\sin\phi}$$

$$= \Delta\lambda\sin\phi \qquad (146)$$

Let the distance between the meridians, measured along a parallel of latitude, be d, and let the radius of the Earth be R. From the figure,

$$\Delta\lambda = \frac{d}{R\cos\phi} \qquad (147)$$

Substitute Equation 147 in Equation 146 to obtain

$$\theta = \frac{d\sin\phi}{R\cos\phi}$$

$$= \frac{\tan\phi}{R} \qquad (148)$$

where θ is measured in radians.

Note that θ varies from 0° at the equator to infinity at the pole. The angle θ on the model of the Earth may be compared with the comparable angle on the map. For the three basic types of projections, θ is invariant over the entire projection. For a cylindrical projection, there is no convergence of the meridians, even though the projection may extend to latitudes where convergence is actually quite noticeable. For a conical projection, the convergence may be accurate at the true scale line in a one-standard parallel projection, or halfway between the true scale lines on the two-standard parallel projection. However, convergence on the map will not be the same as convergence on the model of the Earth at other latitudes. Likewise, on the azimuthal projection, convergence will be true at only one latitude.

The next step is to obtain the linear convergency. From Figure 10, let ℓ be the length of the meridian between two parallels ϕ_1 and ϕ_2. Let θ be the mean angular convergency at a mean latitude

$$\phi = \frac{\phi_1 + \phi_2}{2} \qquad (149)$$

The mean distance at the mean latitude is d. Define the linear convergency of the two meridians to be c. Then, as an approximation,

$$\theta = c/\ell \qquad (150)$$

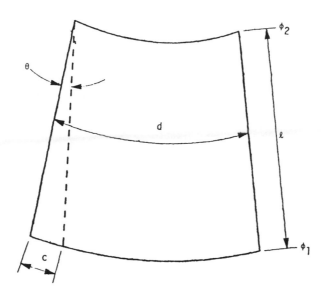

FIGURE 10. Geometry for the linear convergence of the meridians.

Substitute Equation 127 into Equation 129; this gives

$$c = \frac{d \cdot \ell\tan\phi}{R} \qquad (151)$$

This length, ℓ, can also be compared to equivalent lengths on the cylindrical, conical, and azimuthal projections.

A. Example 9

Let $R = 6,378,000$ m, and $d = 2000$ m. Find θ at $\phi = 45°$.

$$\theta = \frac{d \cdot \tan\phi}{R} = \frac{(2000)\tan45°}{R}$$

$$= 0.000314 \text{ rad}$$

$$= 0.0180°$$

If $\ell = 3,000$ m

$$c = \frac{d \cdot \ell\tan\phi}{R}$$

$$= (3,000)(0.000314) = 1.2 \text{ m}$$

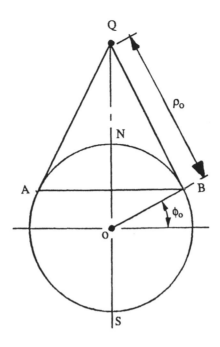

FIGURE 11.   Geometry for the cone tangent to the Earth.

XIV. CONSTANT OF THE CONE AND SLANT HEIGHT[5]

At this point, it is convenient to consider the geometry of a cone and introduce the concepts of the constant of the cone and slant height for a cone. This is done for cones both tangent and secant to the model of the Earth. The constant of a cone is a number which relates angle along a circle of tangency or secancy on the Earth to angle on that same circle on the cone. Slant height of the cone is a polar coordinate from apex of the cone to the circle of tangency or secancy. The slant height is used as the linear coordinate in polar coordinate systems for the map.

Consider first the tangent case, as shown in Figure 11. Let R be the radius of the model for the Earth. the slant height, ρ, corresponds to the distance QB. From trigonometry we find that the slant height of the cone tangent to the Earth at latitude ϕ_o is

$$\rho = R\cot\phi_o \tag{152}$$

The circumference, d, of the parallel circle through AB of latitude ϕ_o, which defines the circle of tangency of the cone, is

$$d = 2\pi R\cos\phi_o \qquad (153)$$

The constant of the cone, c, is defined from the relation between lengths on the developed cone on the Earth. Let the total angle on the cone, θ_T, corresponding to 2π on the Earth be

$$\theta_T = d/\rho \qquad (154)$$

Substitute Equations 152 and 153 into Equation 154 and simplify.

$$\theta_T = \frac{2\pi R\cos\phi_o}{R\cot\phi_o}$$

$$= 2\pi\sin\phi_o \qquad (155)$$

The constant of the cone is defined as

$$c = \sin\phi_o \qquad (156)$$

It is the multiplicative factor that relates longitude on the Earth to the representation of longitude on the map. Since this is a linear function, in general,

$$\theta = \lambda\sin\phi_o \qquad (157)$$

The second polar coordinate on the map is θ. Equations 152, 156, and 157 are needed in Chapters 4, 5, and 6 in the development of the various conical projections.

In the special case of $\phi_o = 90°$, we have an azimuthal projection. Then $\sin\phi_o = 1$ and $\theta = \lambda$.

In the second case to consider, the cone may be secant to the Earth. This is shown in Figure 12, where the circles of secancy are at the latitudes ϕ_1 and ϕ_2. In all the derivations that follow, $\phi_2 > \phi_1$. Let slant height ρ_2 be the distance QC, and slant height ρ_1 be the distance QB. From the similar triangles, the ratio of the slant heights to the these respective latitudes is

$$\frac{\rho_1}{\rho_2} = \frac{R\cos\phi_1}{R\cos\phi_2} \qquad (158)$$

For the secant case, define the constant of the cone, based on Equation 158, as

$$c = \frac{R\cos\phi_1}{\rho_1} = \frac{R\cos\phi_2}{\rho_2} \qquad (159)$$

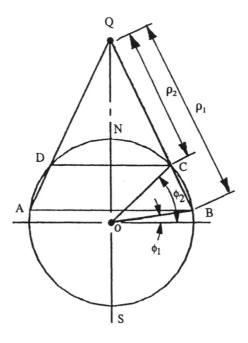

FIGURE 12.   Geometry for the cone secant to the
Earth.

Equation 159 is the basic relation used in Chapters 4 and 5 to obtain the
constant of the cone for conical projections with two standard parallels. In
each case that it is used, the form of ρ_1, and ρ_2 depends on the particular
projection being developed. However, it must be noted that secancy is conceptual for the Albers and Lambert conformal projections.

A. Example 10
Let the cone be tangent to the sphere at $\phi_o = 45°$.

The constant of the cone is

$$c = \sin45° = 0.70711$$

If a change in longitude is given as 30°, the polar coordinate on the map
is

$$\theta = \lambda\sin45°$$
$$= (30°)(0.70711)$$
$$= 21°2133$$

B. Example 11

Give: $\phi = 30°$ and $\phi_2 = 40°$.

Find: the ratio of the polar coordinates.

$$\frac{\rho_1}{\rho_2} = \frac{\cos\phi_1}{\cos\phi_2} = \frac{\cos 30°}{\cos 40°} = 1.130$$

REFERENCES

1. **Adler, C. F.**, *Modern Geometry*, McGraw-Hill, New York, 1958.
2. **Davies, R. E., Foote, F. S., and Kelly, J. E.**, *Surveying: Theory and Practice*, McGraw-Hill, New York, 1966.
3. **Goetz, A.**, *Introduction to Differential Geometry*, Addison-Wesley, Reading, MA, 1958.
4. **Mathews, J. and Walker, R. L.**, *Mathematical Methods of Physics*, Benjamin, Menlo Park, CA, 1970.
5. **Middlemiss, R. R., Marks, J. L., and Smart, J. R.**, *Analytic Geometry*, McGraw-Hill, New York, 1965.
6. **Palmer, C. I., Leigh, C. W., and Kimball, S. H.**, *Plane and Spherical Trigonometry*, McGraw-Hill, New York, 1950.
7. **Phillips, H. B.**, *Vector Analysis*, John Wiley & Sons, New York, 1967.
8. **Richardus, P. and Adler, R. K.**, *Map Projections for Geodesists, Cartographers, and Geographers*, North Holland, Amsterdam, 1972.
9. **Uspensky, J. V.**, *Theory of Equations*, McGraw-Hill, New York, 1948.

Chapter 3

FIGURE OF THE EARTH

I. INTRODUCTION

The basic geometrical surface taken as the model of the Earth is an oblate spheroid generated by revolving an ellipse about its minor axis. This chapter is concerned with the geometry of the spheroid and the simplification of the geometry to the spherical case.

The chapter begins with a brief treatment of the geodetic considerations in defining the size and shape of the model of the Earth. Then, the geometry of the ellipse and the spheroid are investigated, the coordinate system impressed on the spheroid is introduced, and angles and distances on the spheroid are considered.

Many of the projections of Chapters 4, 5, and 6 are based on a transformation from a spherical model of the Earth. Thus, it is necessary to present the increased simplification in equations for coordinates, angles, and distances if a spherical model is used. The accepted radius for the spherical model of the Earth is discussed. Finally, the characteristics of a triaxial ellipsoid model are given.

II. GEODETIC CONSIDERATIONS[1,4]

Geodesy is the science of the measurement and mapping of the surface of the Earth. The problem of geodesy is to determine the figure and the external gravity of the Earth. Map projection theory does not concern itself with the gravity field. Map projection theory needs to rely only on that part of geodesy relating to the determination of the mean Earth ellipsoid.

The figure of the Earth refers to the physical and mathematical surface of the Earth. The irregular solid surfaces and ocean surfaces of the Earth cannot be modeled in a simple manner. Thus, it was necessary to define a fictitious surface which approximates the total shape of the earth. The surfaces of revolution of this section are the geodesists' answer to the problem. They are the most convenient surfaces which best fit the true figure of the Earth. The relation between the irregular Earth and the approximating surfaces is given in Figure 1.

The geodesist's first approximation to the shape of the Earth is the equipotential surface at the mean sea level called the geoid. The geoid is smooth and continuous and extends under the continents at mean sea level. By definition, the perpendicular at any point of the geoid is in the direction of the gravity vector. This surface, however, is not symmetrical about the axis of revolution, since the distribution of matter within the Earth is not uniform.

Ellipsoid

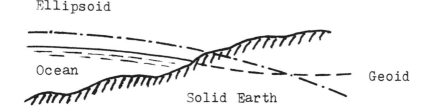

FIGURE 1.   Real and defined surfaces of the Earth.

TABLE 1[1,4]
Geodetic Reference Spheroids

Spheroid	a (m)	b (m)	e	1/f
Everest	6377304	6356103	0.081473	300.8
Bessel	6377397	6356082	0.081690	299.2
Airy	6377563	6356300	0.081591	299.93
Clarke 1858	6378294	6356621	0.082366	294.3
Clarke 1866	6378206	6356585	0.082269	295
Clarke 1880	6378249	6356517	0.082478	293.5
Hayford	6378388	6356912	0.081992	297
Krasovski	6378245	6356863	0.081813	298.3
Hough	6378270	6356794	0.081992	297
Fischer 60	6378166	6356784	0.081813	298.3
Kaula	6378165	6356345	0.082647	292.3
I.U.G.G. 67	6378160	6356775	0.081820	298.25
Fischer 68	6378150	6356330	0.082647	292.3
WGS-72	6378135	6356751	0.081819	298.26
I.U.G.G. 75	6378140	6356755	0.081819	298.257
WGS-84	6378137	6356752	0.081819	298.257

The geoid is the intermediate projection surface between the irregular Earth and the mathematically manageable surface of revolution.

The mean Earth ellipsoid is the best approximation to the geoid. It is a surface of revolution defined by revolving an ellipse about its minor axis. The task of geodesy has been to determine values for the semimajor axis and eccentricity that best represents either a special segment of the Earth or the entire Earth. To accomplish this end, a number of reference spheroids have been defined. In this text, ellipsoid and spheroid are taken as synonymous.

The estimates of the values of the semimajor axis and the flattening have changed over the last 145 years. Progress has meant better instruments, better methods of their use, and better methods of choosing the best fit between the geoid and spheroid. The instruments, their use, and the statistical methods of reducing and fitting the data are beyond the scope of this text. Nevertheless, we must be aware of their existence and their contribution to mapping.

Table 1 lists the important reference spheroids from the Everest to the

WGS-84. Historically, the first of these spheroids, from the Everest through the Clarke 1880, were intended to fit local areas of the world. An example of this is the Clarke 1866, which is still the standard reference spheroid for the U.S. A problem arose in any attempt the "fit" spheroids together at discontinuities. Beginning with the Hayford spheroid, an attempt was made to obtain an internationally acceptable representation for the entire Earth. The WGS-84 spheroid is used in this text as the best representation available today.

In Table 1 are given the semimajor axis, a, semiminor axis, b, flattening, f, and eccentricity, e, for the various reference spheroids. As is indicated in the following section, these are not independent quantities in an ellipse. In usual practice, the semimajor axis and eccentricity are the most important for map protection computations. However, consideration of the semimajor axis compared to the semiminor axis gives an indication of how closely the spheroid approaches a sphere. For the WGS-84 spheroid, the difference between the two is about 0.33%.

The small value for the eccentricity permits a simplification in a number of the derivations in this chapter and in subsequent chapters. A useful technique is to consider a series expansion in powers of the eccentricity. Due to the small value for the eccentricity, these expansions are rapidly convergent. This is demonstrated in Example 1 for the WGS-84 spheroid. If the required accuracy is to be seven figures, it is not necessary to carry any expansion beyond e^6.

Now that the reference spheroids have been introduced, it is necessary to consider the mathematical formulas which describe them.

A. Example 1

Given: the WGS-84 spheroid.

Find: the powers of e, through e^6, to the sixth decimal place.

n	e^n
1	0.081819
2	0.006694
3	0.000548
4	0.000045
5	0.000004
6	0.000000

III. GEOMETRY OF THE ELLIPSE

The ellipse is the planar-generating curve which produces the spheroid of revolution which is identical to the chosen reference spheroid. This section explores the basic geometry of the ellipse. On the planar figure, we define

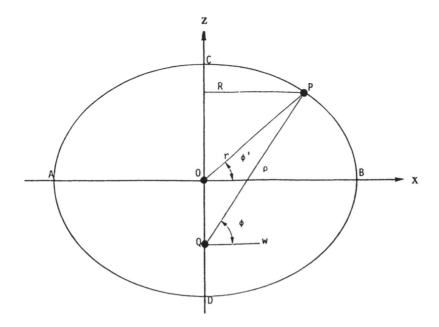

FIGURE 2.   Geometry of the ellipse.

the first independent coordinate for locating positions of the Earth: the latitude.
We then see that two types of latitude may be defined, and the relation between
them is derived.

The nomenclature of the ellipse is best described with reference to Figure
2. The semimajor axis, a, is the length of the line AO or the line OB. The
semiminor axis, b, is the length of the line DO or the line CO. The equation
for the ellipse, for a Cartesian coordinate system with origin at O, is

$$\frac{x^2}{a^2} + \frac{z^2}{b^2} = 1 \tag{1}$$

In the figure, the semimajor axis is along the x axis. The semiminor axis is
along the z axis. The z axis is taken as the axis of rotation in generating the
surface of revolution. The line enclosing the ellipse is one specific meridian.

The degree of departure for circularity is described by the eccentricity,
e, or the flattening, f. The eccentricity, flattening, semimajor axis, and semiminor axis are related as follows:

$$e^2 = \frac{a^2 - b^2}{a^2} \tag{2}$$

$$f = \frac{a - b}{a} \tag{3}$$

$$e^2 = 2f - f^2 \tag{4}$$

At this point we introduce one of the two angular coordinates which uniquely locate a position on the spheroid. This first coordinate is the latitude. Two types of latitude will be noted: the geodetic and the geocentric. The relation between the two is now derived.

The geocentric latitude is the angle between a vector from a center of the ellipse to a point P on the ellipse, or meridian, and the semimajor axis. The geodetic latitude is the angle between a line through the given point, normal to the ellipse, and the semimajor axis. The normal to the ellipse is the line defined by a surveyor's plumb line if all gravity anomalies are ignored. Thus, geocentric latitude is defined by angle POB. Geodetic latitude is given by angle PQW, where line OB is parallel to line QW.

First, consider a polar coordinate system with the origin at O. Let the magnitude of the vector between O and P be r. The relation between the Cartesian and polar coordinates is

$$x = r\cos\phi' \tag{5}$$

$$z = r\sin\phi' \tag{6}$$

Equations 5 and 6 can be combined to form

$$\frac{z}{x} = \tan\phi' \tag{7}$$

Of greater interest is the geodetic latitude, ϕ. This is the angle, PQW, which defines the inclination of line QP, which is normal to the ellipse at point P.

$$\tan\phi = -\frac{dx}{dz} \tag{8}$$

This is the standard calculus definition of a tangent.

The next step is to derive a relation between geocentric and geodetic latitude. This is done by taking the differential of Equation 1.

$$\frac{2x \, dx}{a^2} + \frac{2z \, dz}{b^2} = 0$$

$$\frac{dx}{dz} = -\frac{a^2}{b^2}\frac{z}{x} \tag{9}$$

Substitute Equation 9 into Equation 8 to obtain

$$\tan\phi = \frac{a^2}{b^2} \cdot \frac{z}{x} \tag{10}$$

Substitute Equation 7 into Equation 10; this gives

$$\tan\phi = \frac{a^2}{b^2} \tan\phi'$$

$$\tan\phi' = \frac{b^2}{a^2} \tan\phi \tag{11}$$

Finally, substitute Equation 2 into Equation 11 to arrive at

$$\tan\phi' = (1 - e^2) \tan\phi \tag{12}$$

Because of the small value of the eccentricity for the reference ellipsoids, the difference in values for geodetic and geocentric latitudes is small at any particular point. At latitudes 0 and 90°, the two types are equal. In the vicinity of 45°, the major difference occurs. As an example, for the WGS-84 ellipsoid, the greatest deviation is less than 0.2°.

The convention for measuring geodetic latitude is $+\phi$ in the Northern Hemisphere and $-\phi$ in the Southern Hemisphere.

A. Example 2
Given: the I.U.G.G. spheroid.
Find: ϕ' for $\phi = 45°$, and ϕ for $\phi' = 45°$.

$$f = \frac{1}{298.257} = 0.00335281$$

$$e^2 = 2f - f^2 = 0.00670563 - 0.00001124 = 0.00669439$$

$$1 - e^2 = 1 - 0.00669439 = 0.993305$$

$$\tan\phi' = (1 - e^2) \tan\phi = (0.993305)(i)$$

$$\phi' = 44°.8076$$

$$\tan\phi = \frac{\tan\phi'}{(1 - e^2)} = \frac{1}{0.993305} = 1.00674$$

$$\phi = 45°.1924$$

IV. THE SPHEROID AS A MODEL OF THE EARTH[3]

The spheroidal model of the Earth is obtained by revolving the ellipse of Figure 2 about the z axis. This section explores the geometry of the spheroid.

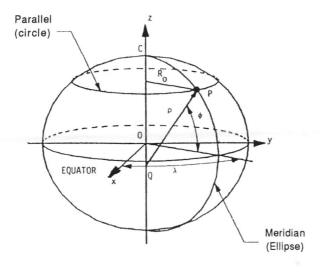

FIGURE 3.   Geometry of the spheroid.

The equation for the surface is given. The coordinate system on the spheroid is illustrated. The second independent coordinates necessary to locate a position on the Earth are introduced. This is the longitude, λ. The two principal radii of curvature and the radius of the parallel circle are derived. The relation of Cartesian coordinates in three-dimensional space to the geographic coordinates is given. The first fundamental form for the spheroid is repeated. Distances along the important lines of the spheroid are developed. These are distances along the meridians, the equator, the parallel circles, the geodesic, and the loxodrome. Finally, the azimuth on a spheroid is derived, and the normal at a point is defined.

The Cartesian three-dimensional coordinate system is shown in Figure 3. In this system, the equation of the spheroidal surface is

$$\frac{x^2}{a^2} + \frac{y^2}{a^2} + \frac{z^2}{b^2} = 1 \tag{13}$$

where a is the semimajor axis of the spheroid of revolution, and b is the semiminor axis.

The nomenclature of the spheroid can be introduced by Figure 3. Each of the infinity of positions of the ellipse as it is rotated about the z axis defines a meridional ellipse, or meridian. The angle λ, measured in the x-y plane, and from the x axis, is the longitude of any and all points on the meridional ellipse. This is the second of the two angular coordinates which uniquely define a position on the spheroid. As a convention, a rotation from +x to +y, or east, will be positive, and the reverse rotation, negative. The meridian containing the positive x axis is the prime meridian.

Consider the point P in Figure 3 to be defined by ϕ and λ. Suppose now that λ is allowed to vary, while ϕ is held constant. The locus on the spheroid traced out by P is a circle of parallel or radius R_o. The circle of parallel for a latitude of zero is the equator.

It remains to derive the equations for several radii of importance in future developments. These are the two principal radii of curvature and the radius of a parallel circle, all as a function of latitude.

Consider the meridional ellipse at any arbitrary λ. From Equation 13,

$$\frac{R_o^2}{a^2} + \frac{z^2}{b^2} = 1 \tag{14}$$

where $R_o^2 = x^2 + y^2$ is the radius of a parallel circle.

Substitute Equation 2 into Equation 14 to obtain

$$\frac{R_o^2}{a^2} + \frac{z^2}{a^2(1 - e^2)} = 1$$

$$R_o^2(1 - e^2) + z^2 = a^2(1 - e^2) \tag{15}$$

Taking the differential of Equation 15 to obtain the slope of the tangent at P, we have

$$2R_o \cdot dR_o(1 - e^2) + 2z\,dz = 0$$

$$\frac{dz}{dR_o} = -\frac{R_o}{z}(1 - e^2) \tag{16}$$

The slope of the normal at P is again

$$-\frac{dR_o}{dz} = \frac{z}{R_o(1 - e^2)} = \tan\phi$$

$$z = R_o(1 - e^2)\tan\phi \tag{17}$$

From Figure 3,

$$\sin\phi = \frac{z}{(1 - e^2)\,\overline{QP}}$$

$$z = \overline{QP}(1 - e^2)\sin\phi \tag{18}$$

Substitute Equation 17 into Equation 15; this gives

$$R_o^2(1 - e^2) + R_o^2(1 - e^2)^2\tan^2\phi = a^2(1 - e^2)$$

$$R_o^2 + R_o^2(1 - e^2)\tan^2\phi = a^2$$

Substitute sin ϕ/cos ϕ for tan ϕ and expand the relation.

$$(\cos^2\phi + \sin^2\phi)R_o^2 - e^2R_o^2\sin^2\phi = a^2\cos^2\phi$$

$$R_o^2(1 - e^2\sin^2\phi) = a^2\cos^2\phi$$

$$R_o = \frac{a \cos\phi}{\sqrt{1 - e^2\sin^2\phi}} \tag{19}$$

The radius of curvature in a plane perpendicular to the plane of the meridian also follows from Figure 3.

$$R_o = \overline{QP} \cos\phi \tag{20}$$

Equating Equations 19 and 20, we find

$$\overline{QP} \cos\phi = \frac{a \cos\phi}{\sqrt{1 - e^2\sin^2\phi}}$$

$$\overline{QP} = \frac{a}{\sqrt{1 - e^2\sin^2\phi}} \tag{20a}$$

$\overline{QP}$ is the radius of curvature of the spheroid in the plane perpendicular to the meridional plane, and will be denoted as R_p.
Thus,

$$R_p = \frac{a}{\sqrt{1 - e^2\sin^2\phi}} \tag{21}$$

The radius of curvature of the meridional ellipse follows from the formula for a plane curve. Consider the magnitude only.

$$R_m = \left| \frac{\left[1 + \left(\dfrac{dz}{dR_o}\right)^2\right]^{3/2}}{\dfrac{d^2z}{dR_o^2}} \right| \tag{22}$$

Take the derivative of Equation 16 to obtain

$$\frac{d^2z}{dR_o^2} = -\frac{1}{z}(1 - e^2) + \frac{R_o}{z^2}(1 - e^2)\frac{dz}{dR} \tag{23}$$

Substituting Equation 16 into Equation 22, we have

$$\frac{d^2z}{dR_o^2} = \frac{1}{z}\left[-(1 - e^2) - \left(\frac{dz}{dR_o}\right)^2 \right]$$

$$= -\frac{1}{z}\left[1 + \left(\frac{dz}{dR_o}\right)^2 - e^2 \right] \tag{24}$$

Since the slope of the normal is $\tan\phi$, that of the tangent is $-\cot\phi$; then

$$\frac{dz}{dR_o} = \cot\phi \tag{25}$$

Substituting Equation 25 into Equation 24 and applying trigonometric relations, we obtain

$$\frac{d^2z}{dR_o^2} = -\frac{1 + \cot^2\phi - e^2}{z}$$

$$= -\frac{\dfrac{1}{\sin^2\phi} - e^2}{z} \tag{26}$$

From Equations 18 and 20a, we have

$$z = \frac{a(1 - e^2)\sin\phi}{\sqrt{1 - e^2\sin^2\phi}} \tag{27}$$

Then, substituting Equations 25, 26, and 27 into Equation 22 and simplifying,

$$R_m = \left| \frac{\left(1 + \dfrac{\cos^2\phi}{\sin^2\phi}\right)^{3/2}}{-\left(\dfrac{1}{\sin^2\phi} - e^2\right)\dfrac{\sqrt{1 - e^2\sin^2\phi}}{a(1 - e^2)\sin\phi}} \right|$$

$$= \left| \frac{\left(\dfrac{\cos^2\phi + \sin^2\phi}{\sin^2\phi}\right)^{3/2}}{-\left(\dfrac{1 - e^2\sin^2\phi}{\sin^2\phi}\right)\dfrac{\sqrt{1 - e^2\sin^2\phi}}{a(1 - e^2)\sin\phi}} \right|$$

$$= \left| \frac{-\dfrac{1}{\sin^3\phi}}{\left(\dfrac{1 - e^2\sin^2\phi}{\sin^3\phi}\right)^{3/2} \cdot \dfrac{1}{a(1 - e^2)}} \right|$$

$$= \frac{a(1 - e^2)}{(1 - e^2\sin^2\phi)^{3/2}} \tag{28}$$

The equations for the radius of the circle of parallel and the radius of curvature are independent of the longitude. This is a characteristic for surfaces of revolution.

Before turning to distances on the spheroid, it is useful to relate the Cartesian coordinates in three-dimensional space for the spheroid to geographic coordinates. This is done as follows:

$$\left.\begin{aligned}
x &= R_p\cos\phi\cos\Delta\lambda \\
y &= R_p\cos\phi\sin\Delta\lambda \\
z &= (1 - e^2)R_p\sin\phi
\end{aligned}\right\} \tag{29}$$

Recall that the Cartesian coordinates are related by Equation 13. Thus, there are only two degrees of freedom. Positions are constrained to be on the spheroidal surface.

The first fundamental form of the spheroid (Section VI, Chapter 2) is

$$(ds)^2 = R_m^2(d\phi)^2 + R_p^2\cos^2\phi(d\lambda)^2 \tag{30}$$

The fundamental quantities are

$$\left.\begin{aligned}
E &= R_m^2 \\
G &= R_p^2\cos^2\phi
\end{aligned}\right\} \tag{31}$$

Equations 31 are found to be useful in the transformations from the spheroid to the sphere in Chapters 4 and 5 and in the discussion of distortions in Chapter 7.

The next step is to consider distance on the spheroid. Three types of distances measured on the spheroid are important. These are distances along a circle of parallel, distances along the meridional ellipse, and distances between two arbitrary points.

We can deal with the distance along a circle of parallel very easily. From Equation 19,

$$d = \frac{a\Delta\lambda\cos\phi}{\sqrt{1 - e^2\sin^2\phi}} \tag{32}$$

where $\Delta\lambda$ is the angular separation of two points on the circle of parallel, in radians.

Distance along the meridional ellipse requires an integration based on

Equation 28. To facilitate this, Equation 28 is expanded by the binomial theorem.

$$R_m = a(1 - e^2)\left(1 + \frac{3}{2} e^2\sin^2\phi + \frac{15}{8} e^4\sin^4\phi + \frac{35}{16} e^6\sin^6\phi + ...\right) \quad (33)$$

Equation 33 is a rapidly converging series, since the values of e are always small.

The distance between positions at latitude ϕ_1 and ϕ_2 on the same meridional ellipse is given by

$$d = \int_{\phi_1}^{\phi_2} R_m d\phi \quad (34)$$

Substitute Equation 33 into Equation 34 and integrate to obtain

$$d = a(1 - e^2)\int_{\phi_1}^{\phi_2}\left(1 + \frac{3}{2} e^2\sin^2\phi + \frac{15}{8} e^4\sin^4\phi + \frac{35}{16} e^6\sin^6\phi + ...\right)d\phi$$

$$= a(1 - e^2)\left\{\phi + \frac{3}{2} e^2\left(\frac{\phi}{2} - \frac{1}{4}\sin2\phi\right) + \frac{15e^4}{8}\left(\frac{3\phi}{8} - \frac{\sin2\phi}{4} + \frac{\sin4\phi}{32}\right)\right.$$

$$\left. + \frac{35e^6}{16}\left[-\frac{\sin^5\phi\cos\phi}{6} + \frac{5}{6}\left(\frac{3\phi}{8} - \frac{\sin2\phi}{4} + \frac{\sin4\phi}{32}\right)\right] + ...\right\}_{\phi_1}^{\phi_2}$$

$$= a(1 - e^2)\left\{\phi\left(1 + \frac{3}{4} e^2 + \frac{45}{64} e^4 + \frac{525}{768} e^6 + ...\right)\right.$$

$$- \sin2\phi\left(\frac{3}{8} e^2 + \frac{15}{32} e^4 + \frac{175}{384} e^6 + ...\right)$$

$$\left. + \sin4\phi\left(\frac{15}{156} e^4 + \frac{175}{3072} e^6 + ...\right) - \frac{35}{96} e^6\sin^5\phi\cos\phi + ...\right\}_{\phi_1}^{\phi_2}$$

$$= a\left\{\left(1 - e^2 + \frac{3}{4} e^2 - \frac{3}{4} e^4 + \frac{45}{64} e^4 - \frac{45}{64} e^6 + \frac{525}{768} e^6 + ...\right)\phi\right.$$

$$- \left(\frac{3}{8} e^2 - \frac{3}{8} e^4 + \frac{15}{32} e^4 - \frac{15}{32} e^6 + \frac{175}{384} e^6 + ...\right)\sin2\phi$$

$$+ \left(\frac{15}{256} e^4 - \frac{15}{256} e^6 + \frac{175}{3072} e^6 + ...\right)\sin4\phi$$

$$\left. - \frac{35}{96} e^6\left(\frac{\sin\phi\cos\phi}{8}\right)(3 - 4\cos2\phi + \cos4\phi)\right\}_{\phi_1}^{\phi_2}$$

$$d = a\left\{\left(1 - \frac{e^2}{4} - \frac{3}{64}e^4 - \frac{5}{256}e^6 - \ldots\right)\phi\right.$$

$$\left(\frac{3}{8}e^2 + \frac{3}{32}e^4 - \frac{5}{384}e^6 + \ldots\right)\sin2\phi$$

$$+ \left(\frac{15}{256}e^4 - \frac{5}{3072}e^6 + \ldots\right)\sin4\phi$$

$$\left. - \frac{35}{96}e^6\left(\frac{\sin2\phi}{16}\right)(3 - 4\cos^2\phi + \cos4\phi)\right\}_{\phi_1}^{\phi_2} \quad (35)$$

Further, expanding the last term in Equation 35,

$$-\frac{35}{96}e^6\left(\frac{\sin2\phi}{16}\right)(3 - 4\cos2\phi + \cos4\phi)$$

$$= -\frac{35}{96}e^6\left(\frac{3}{16}\sin2\phi - \frac{1}{4}\sin2\phi\cos2\phi + \frac{1}{16}\sin2\phi\cos4\phi\right)$$

$$= \frac{35}{96}e^6\left(\frac{3}{16}\sin2\phi - \frac{1}{8}\sin4\phi - \frac{1}{32}\sin2\phi + \frac{1}{32}\sin6\phi\right) \quad (36)$$

Finally, substitute Equation 36 into Equation 35.

$$d = a\left\{\left(1 - \frac{e^2}{4} - \frac{3}{64}e^4 - \frac{5}{256}e^6 - \ldots\right)\phi\right.$$

$$- \left[\frac{3}{8}e^2 + \frac{3}{32}e^4 + \left(-\frac{5}{384} + \frac{35}{96}\left(\frac{3}{16} - \frac{1}{32}\right)\right)e^6 + \ldots\right]\sin2\phi$$

$$+ \left[\frac{15}{256}e^4 + \left(-\frac{5}{3072} + \frac{35}{96} - \frac{1}{8}\right)e^6 + \ldots\right]\sin4\phi$$

$$\left. - \frac{35}{96}\cdot\frac{e^6}{32}\sin6\phi + \ldots\right\}_{\phi_1}^{\phi_2}$$

$$= a\left\{\left(1 - \frac{e^2}{4} - \frac{3}{64}e^4 - \frac{5}{256}e^6 - \ldots\right)\phi\right.$$

$$- \frac{3}{8}e^2 + \frac{3}{32}e^4 + \frac{45}{1024}e^6 + \ldots\bigg)\sin2\phi$$

$$\left. + \left(\frac{15}{256}e^4 + \frac{45}{1024}e^6 + \ldots\right)\sin4\phi - \frac{35}{3072}e^6\sin6\phi + \ldots\right\}_{\phi_1}^{\phi_2} \quad (37)$$

The third type of distance to consider is the shortest distance on the spheroid between two arbitrary points: $P_1(\phi_1, \lambda_2)$ and $P_2(\phi_2, \lambda_2)$. This distance is the geodesic curve. To obtain this distance, we start with

$$d_g = \int_{\phi_1}^{\phi_2} ds \tag{38}$$

Substitute Equation 30 into Equation 38.

$$d_g = \int_{\phi_1}^{\phi_2} \left[R_m^2 + R_p^2 \cos^2\phi \left(\frac{d\lambda}{d\phi} \right)^2 \right]^{1/2} d\phi \tag{39}$$

To obtain the geodesic, it is necessary to minimize Equation 39 by means of the Euler-Lagrange technique. The procedure begins by developing the terms

$$L(\phi, \lambda, \lambda') = R_m^2 + R_p^2 \cos^2\phi (\lambda')^2$$

$$\frac{d}{d\phi} \left(\frac{\partial L}{\partial \lambda'} \right) = \frac{\partial L}{\partial \lambda} = 0$$

$$\frac{\partial L}{\partial \lambda'} = R_p^2 \cos^2\phi \frac{d\lambda}{d\phi} = c \tag{40}$$

where c is a constant.

Substitute Equation 40 into Equation 39 to obtain

$$d_g = \int_{\phi_1}^{\phi_2} \left(R_m^2 + \frac{c^2}{R_p^2 \cos^2\phi} \right)^{1/2} d\phi \tag{41}$$

It remains to evaluate c in Equation 41. Integrate Equation 40; this gives

$$\lambda = c \int \frac{d\phi}{R_p^2 \cos^2\phi} + k \tag{42}$$

Substituting Equation 21 into Equation 42, we have

$$\lambda = c \int \left(\frac{1 - e^2 \sin^2\phi}{a^2 \cos^2\phi} \right) d\phi + k$$

$$= c \int \left[\frac{1 - e^2(1 - \cos^2\phi)}{a^2 \cos^2\phi} \right] d\phi + k$$

$$= c \int \left(\frac{1 - e^2}{a^2 \cos^2\phi} + \frac{e^2}{a^2} \right) d\phi + k$$

$$= c \int \left[\left(\frac{1 - e^2}{a^2} \right) \tan\phi + \frac{e^2}{a^2} \phi \right] + k \tag{43}$$

Evaluate Equation 43 at P_1 and P_2 and substract to eliminate k, thus,

$$\lambda_1 = c\left[\frac{1 - e^2}{a^2} \tan\phi_1 + \frac{e^2}{a^2} \phi_1\right] + k$$

$$\lambda_2 = c\left[\frac{1 - e^2}{a^2} \tan\phi_2 + \frac{e^2}{a^2} \phi_2\right] + k$$

$$\lambda_2 - \lambda_1 = c\left[\left(\frac{1 - e^2}{a^2}\right)(\tan\phi_2 - \tan\phi_1) + \frac{e^2}{a^2}(\phi_2 - \phi_1)\right]$$

$$c = \frac{\lambda_2 - \lambda_1}{\left(\frac{1 + e^2}{a^2}\right)(\tan\phi_2 - \tan\phi_1) + \frac{e^2}{a^2}(\phi_2 - \phi_1)} \tag{44}$$

Substituting Equations 21, 28, and 44 into Equation 41, we have

$$d_g = \int_{\phi_1}^{\phi_2} \left\{ \frac{a^2(1 - e^2)}{(1 - e^2\sin^2\phi)^3} + \frac{(1 - e^2\sin^2\phi)}{a^2\cos^2\phi} \right.$$

$$\left. \times \frac{(\lambda_2 - \lambda_1)^2}{\left[\left(\frac{1 - e^2}{a^2}\right)(\tan\phi_2 - \tan\phi_1) - \frac{e^2}{a^2}(\phi_2 - \phi_1)\right]^2} \right\}^{1/2} d\phi$$

$$= a\int_{\phi_1}^{\phi_2} \left\{ \frac{(1 - e^2)^2}{(1 - e^2\sin^2\phi)^3} \right.$$

$$\left. + \frac{(\lambda_2 - \lambda_1)^2(1 - e^2\sin^2\phi)}{[(1 - e^2)(\tan\phi_2 - \tan\phi_1 - e^2(\phi_2 - \phi_1)]^2\cos^2\phi} \right\}^{1/2} d\phi \tag{45}$$

Equation 45 is completely general. It is of such complexity that it must be integrated numerically.

In general, the geodesic is not a plane curve, that is, the torsion is not zero. The meridional ellipses and the equator are particular geodesic curves. The meridians and the equator are the only geodesics which are plane curves.

At this point, formulas for the azimuth on a spheroid are considered. Three differential formulas are now derived which apply to any point on the geodesic and relate latitude, longitude, distance, and azimuth. These are as follows:

$$\frac{d\phi}{ds} = \frac{\cos\alpha}{R_m} \tag{46}$$

$$\frac{d\lambda}{ds} = \frac{1}{R_p} \frac{\sin\alpha}{\cos\phi} \tag{47}$$

$$\frac{d\alpha}{ds} = \frac{1}{R_p} \tan\phi\sin\alpha \qquad (48)$$

Equations 46 and 47 can be obtained from a consideration of the angle between a curve on the spheroidal surface and one of the parametric curves, the meridional ellipse, where λ is a constant.

Since $\lambda_1 = c$

$$d\lambda_1 = 0 \qquad (49)$$

and

$$ds_1 = \sqrt{E} \, d\phi_1 \qquad (50)$$

$$\cos\alpha = \frac{Ed\phi d\phi_1 + Gd\lambda d\lambda_1}{dsds_1} \qquad (51)$$

Substitute Equations 49 and 50 into Equation 51; this gives

$$\cos\alpha = \frac{E}{\sqrt{E}} \frac{d\phi}{ds} \frac{d\phi_1}{d\phi_1}$$

$$= \sqrt{E} \frac{d\phi}{ds} \qquad (52)$$

Upon substituting the first of Equations 31 into Equation 52, we have

$$\cos\alpha = R_m \frac{d\phi}{ds}$$

This establishes Equation 46.

Consider next

$$\sin\alpha = \sqrt{EG} \left(\frac{d\phi_1 d\lambda - d\phi d\lambda_1}{dsds_1} \right) \qquad (53)$$

Substituting Equations 15 and 50 into Equation 53,

$$\sin\alpha = \frac{\sqrt{EG} \, d\phi_1 d\lambda}{\sqrt{E} \, d\phi_1 ds}$$

$$= \sqrt{G} \frac{d\lambda}{ds} \qquad (54)$$

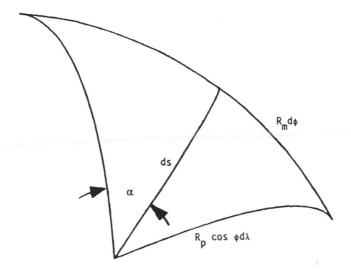

FIGURE 4. Differential element defining a rhumbline on a spheroid.

Then, substituting the second of Equations 31 into Equation 54, we find

$$\sin\alpha = R_p \cos\phi \, \frac{d\lambda}{ds}$$

which establishes Equation 47.

From Equations 46 and 45, the azimuth at P at the initiation of the geodesic can be calculated.

$$\cos\alpha_1 = R_m \frac{d\phi}{ds}$$

$$= (R_m)_1 \left\{ \frac{(1 - e^2)^2}{(1 - e^2\sin^2\phi_1)^3} \right.$$

$$\left. + \frac{(\lambda_2 - \lambda_1)^2(1 - e^2\sin^2\phi_1)}{(1 - e^2)(\tan\phi_2 - \tan\phi_1) - e^2(\phi_2 - \phi_1)^2\cos^2\phi_1} \right\}^{1/2} \quad (55)$$

The final distance of interest which occurs on the spheroid is the lox-odrome. The rhumbline (or loxodrome) is a curve on the spheroid which meets each consecutive meridian at the same azimuth. From Figure 4,

$$\tan\alpha = \frac{R_p}{R_m d\phi} \cos\phi d\lambda$$

$$d\lambda = \tan\alpha \, \frac{R_p}{R_m\cos\phi} d\phi \quad (56)$$

Substitute Equations 21 and 28 into Equation 56 to obtain

$$
d\lambda = \tan\alpha \, \dfrac{\dfrac{a}{\sqrt{1 - e^2\sin^2\phi}}}{\dfrac{a(1 - e^2)\cos\phi}{(1 - e^2\sin^2\phi)^{3/2}}} \, d\phi
$$

$$
= \tan\alpha \left[\dfrac{1 - e^2\sin^2\phi}{(1 - e^2)\cos\phi} \right] d\phi
$$

$$
\Delta\lambda = \tan\alpha \left\{ \ln\left[\tan\left(\dfrac{\pi}{4} + \dfrac{\phi}{2}\right)\left(\dfrac{1 - e\sin\phi}{1 + e\sin\phi}\right)^{e/2} \right] \right\}_{\phi_1}^{\phi_2} \tag{57}
$$

The kernal of Equation 57 arises again when we treat the conformal projections of Chapter 5. The rhumbline, used in conjunction with the Mercator projection (Chapter 5), and the great circle on the gnomonic projection (Chapter 6) are longstanding aids to marine and aerial navigation.

One last mathematical tool is now developed with respect to the spheroid. It is the normal vector to the surface of the spheroid at an arbitrary point in three-dimensional Cartesian space. From Equation 99, Chapter 2, the outward pointing normal at $P(x, y, z)$ is given by

$$
\hat{n} = \dfrac{\cos\lambda\hat{i} + \sin\lambda\hat{j} - \dfrac{\partial R_o}{\partial z}\hat{k}}{\sqrt{1 - \left(\dfrac{\partial R_o}{\partial z}\right)^2}} \tag{58}
$$

After considerable manipulation, it can be shown that

$$
\dfrac{\partial R_o}{\partial z} = -\tan\phi \tag{59}
$$

Substitute Equation 59 into Equation 58 to obtain

$$
\hat{n} = \dfrac{\cos\lambda\hat{i} + \sin\lambda\hat{j} + \tan\phi\hat{k}}{\sqrt{1 + \tan^2\phi}} \tag{60}
$$

If one is in possession of ϕ and λ, the evaluation of Equation 60 is straight forward. If only the Cartesian coordinates are available, Equations 29 must be inverted. From the last of Equations 29, with the inclusion of the defining relation for R_p,

$$z = \frac{a(1 - e^2)\sin\phi}{\sqrt{1 - e^2\sin^2\phi}} \tag{61}$$

Equation 61 is then solved for ϕ.

$$z^2 = \frac{a^2(1 - e^2)\sin^2\phi}{1 - e^2\sin^2\phi}$$

$$z^2(1 - e^2\sin^2\phi) = a^2(1 - e^2)\sin^2\phi$$

$$z^2 - z^2e^2\sin^2\phi = a^2(1 - e^2)\sin^2\phi$$

$$\sin^2\phi[a^2(1 - e^2)^2 + z^2e^2] = z^2$$

$$\sin^2\phi = \frac{z^2}{a^2(1 - e^2)^2 + z^2e^2}$$

$$\sin\phi = \left[\frac{z}{\sqrt{a^2(1 - e^2)^2 + z^2e^2}}\right]$$

$$\phi = \sin^{-1}\left[\frac{z}{\sqrt{a^2(1 - e^2)^2 + z^2e^2}}\right] \tag{62}$$

Equation 62 uniquely defines latitude with respect to quadrant.

The first and second of Equation 29 are then needed to solve for the longitude.

$$\frac{R_p\cos\phi\sin}{R_p\cos\phi\cos} = \frac{y}{x}$$

$$\tan\lambda = \frac{y}{x} \tag{63}$$

$$\lambda = \tan^{-1}\left(\frac{y}{x}\right) \tag{64}$$

By considering the signs of x and y, Equation 64 uniquely defines longitude. Returning to Equation 62, the values of $\sin\lambda$ and $\cos\lambda$ to be used in Equation 60 are

$$\sin\lambda = \frac{y}{\sqrt{x^2 + y^2}}$$

$$\cos\lambda = \frac{x}{\sqrt{x^2 + y^2}} \tag{65}$$

A. Example 3

Given: $\phi = 37°$N on the WGS-72 ellipsoid.

Find: R_m, R_p, and R_o.

$$a = 6,378,135 \text{ meters}, \quad c = 0.081819$$

$$R_o = \frac{a \cos\phi}{\sqrt{1 - e^2\sin^2\phi}} = \frac{(6,378,135)(0.79864)}{\sqrt{1 - (0.081819)^2(0.60182)^2}}$$

$$= \frac{5,093,834}{\sqrt{1 - 0.002425}} = \frac{5,093,834}{0.998787}$$

$$= \underline{5,099,991 \text{ m}}$$

$$R_p = \frac{a}{\sqrt{1 - e^2\sin^2\phi}} = \frac{6,378,135}{0.99878}$$

$$= \underline{6,385,926 \text{ m}}$$

$$R_m = \frac{a(1 - e^2)}{(1 - e^2\sin^2\phi)^{3/2}} = \frac{(6,378,135)(1 - 0.006694)}{(0.997561)^{3/2}}$$

$$= \frac{(6,378,135)(0.993306)}{0.996344}$$

$$= \underline{6,358,684 \text{ m}}$$

B. Example 4

Given: $\phi = 43°$ on the WGS-72 ellipsoid.

Find: the distance along the parallel circle for $\Delta\lambda = 10°$.

$$R_o = \frac{a \cos\phi}{\sqrt{1 - e^2\sin^2\phi}} = (6,378,135)(0.73135)$$

$$= 4,670,347$$

$$d = \Delta\lambda \cdot R_o = \left(\frac{10°}{57.295}\right)(4,670,347)$$

$$= \underline{815,140 \text{ m}}$$

C. Example 5

Given: the WGS-72 spheroid.

Find: the distance along the meridian between $\phi_1 = 23°$ and $\phi_2 = 47°$.

$$d = a \left\{ \left(1 - \frac{e^2}{4} - \frac{3}{64} e^4\right)\Delta\phi \right.$$

$$- \left(\frac{3}{8} e^2 + \frac{3}{32} e^4\right)(\sin 2 \cdot 47 - \sin 2 \cdot 23)$$

$$\left. + \frac{15}{256} e^4(\sin 4.47 - \sin 4.23) \right\}$$

$$= (6,378,135)\left\{ (1 - 0.001674 - 0.000002) \left(\frac{24}{57.295}\right) \right.$$

$$- (0.002510 + 0.000004)(0.997564 - 0.719340)$$

$$\left. + (0.000003)(-0.139173 - 0.999391) \right\}$$

$$= (6,378,135)\{(0.998324)(0.418885)$$

$$- (0.002514)(0.278224) - (0.000003)(1.138564)\}$$

$$= (6,378,135)(0.418182 - 0.000699 - 0.000003)$$

$$= (6,378,135)(0.417480)$$

$$= \underline{2,662,743 \text{ meters}}$$

V. THE SPHERICAL MODEL OF THE EARTH[4]

The use of the sphere as the model of the Earth greatly simplifies the mathematics required. In the spherical case, the generating curve for the surface of revolution is the circle, which is an ellipse of zero eccentricity. Thus, the semimajor and semiminor axes are the same. In general, the spheroidal formulas derived in the previous section can be reduced to the spherical case by a substitution of e = 0. However, for the distance between arbitrary points, and azimuth, it is easier to start with a spherical trigonometry approach.

This section contains the equation of the sphere in Cartesian coordinates and the relation between Cartesian and polar coordinates. The coordinate system is defined, the first fundamental form for the sphere is repeated, the radius of curvature of the surface and the parallel circle are obtained, and distance is derived for the great circle and the loxodrome. The azimuth is also defined. Finally, the radius to best represent the Earth is investigated.

First, the relations of the coordinate systems are given. The equation for the sphere in Cartesian coordinates is given by

$$\frac{x^2}{R^2} + \frac{y^2}{R^2} + \frac{z^2}{R^2} = 1 \tag{66}$$

where R is the radius of the sphere.

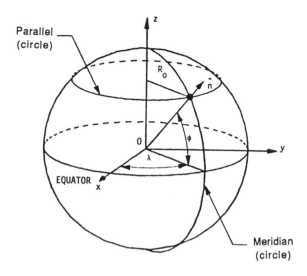

FIGURE 5.   Geometry of the sphere.

Figure 5 gives the geometry of the spherical Earth. Note that the normal to the sphere at a point, P, coincides with the geocentric radius vector. For the sphere, there is only one type of latitude, since geocentric and geodetic latitudes coincide. Longitude is measured in the same way it was for the spheroidal case, that is, east or west of the prime meridian, again taken as coinciding with the x axis. The sign conventions for latitude and longitude in the spheroidal case also holds for the spherical case.

The equations for the important radii in the spherical model simplify as follows: the radius of a circle of parallel becomes

$$R_o = R\cos\phi \qquad\qquad (67)$$

The radii of curvature become

$$R_p = R_m = R \qquad\qquad (68)$$

These radii are no longer dependent on latitude.

The relation between the Cartesian coordinates and latitude and longitude in the three-dimensional space of Figure 5 are

$$\left. \begin{aligned} x &= R\cos\phi\cos\lambda \\[6pt] y &= R\cos\phi\sin\lambda \\[6pt] z &= R\sin\phi \end{aligned} \right\} \qquad (69)$$

Remember that the Cartesian coordinates are not independent, but are related by Equation 66. There are only two degrees of freedom.

The first fundamental form simplifies to

$$(ds)^2 = R^2(d\phi)^2 + R^2\cos^2\phi(d\lambda)^2 \tag{70}$$

Thus, the first fundamental quantities are

$$\left. \begin{aligned} E &= R^2 \\ G &= R^2\cos^2\phi \end{aligned} \right\} \tag{71}$$

Distance along the circle of parallel is, from Equation 32,

$$d = R\Delta\lambda\cos\phi \tag{72}$$

with $\Delta\lambda$ in radians. From Equation 37, distance along the meridian circle is simply

$$d = R\Delta\phi \tag{73}$$

with $\Delta\phi$ in radians.

On the sphere, the shortest curve connecting two arbitrary points is an arc of the great circle. The great circle (also called the orthodrome) corresponds to the geodesic curve on the spheroid, but with many simplifications. The great circle is a planar curve which contains the arbitrary points. The plane of the curve also contains the center of the Earth. Since it is a planar curve, the torsion is zero.

The great circle distance, d, on the surface of the Earth can be determined by consideration of Figure 6. The equations of spherical trigonometry are applied in this derivation. Consider the law of cosines.

$$\begin{aligned} \cos\theta &= \cos(90° - \phi_1)\cos(90° - \phi_2) + \sin(90° - \phi_1)\sin(90° - \phi_2)\cos\Delta\lambda \\ &= \sin\phi_1\sin\phi_2 + \cos\phi_1\cos\phi_2\cos\Delta\lambda \end{aligned} \tag{74}$$

Taking the arc-cosine of Equation 74,

$$d = R\cos^{-1}(\sin\phi_1\sin\phi_2 + \cos\phi_1\cos\phi_2\cos\Delta\lambda) \tag{75}$$

The azimuth can also be determined from Figure 6. The azimuth, α, of point P_2 from P_1 is also obtained from the spherical triangle NP_1P_2. Taking the arc-cosine of Equation 74 again, the angle θ is now available. Then, the law of sines is applied.

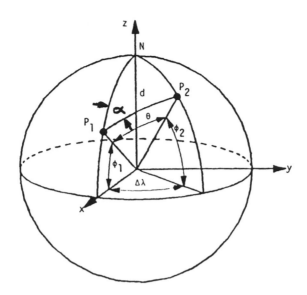

FIGURE 6.   Great circle on the sphere.

$$\frac{\sin\alpha}{\sin(90° - \phi_2)} = \frac{\sin\Delta\lambda}{\sin\theta}$$

$$\sin\alpha = \frac{\cos\phi_2\sin\Delta\lambda}{\sin\theta} \qquad (76)$$

Also,

$$\cos\alpha = \frac{\cos(90° - \phi_2) - \cos\theta\cos(90° - \phi_1)}{\sin\theta\sin(90° - \phi_1)}$$

$$= \frac{\sin\phi_2 - \cos\theta\sin\phi_1}{\sin\theta\cos\phi_1} \qquad (77)$$

From Equations 76 and 77, the quadrant of the azimuth can be seen. As was mentioned in Chapter 1, azimuth is measured from the north, positive to the east.

The rhumbline or loxodrome is obtained from Equation 57 by substituting in e = 0.

$$\Delta\lambda = \tan\alpha\left[\ln\tan\left(\frac{\pi}{4} + \frac{\phi_2}{2}\right) - \ln\tan\left(\frac{\pi}{4} + \frac{\phi_1}{2}\right)\right] \qquad (78)$$

Equation 78 can be investigated. If $\phi_1 = \phi_2$, then $\alpha = 90°$. This is an azimuth along a parallel circle. If $\lambda_1 = \lambda_2$, $\tan\alpha = 0$, $\alpha = 0$, yielding a meridian.

The distance along the rhumbline is found from Figure 4, where $R_m = R_p = R_o$.

$$d = \int_{P_1}^{P_2} ds = \int_{P_1}^{P_2} \frac{R}{\cos\alpha} \, d\phi$$

$$= \frac{R}{\cos\alpha} (\phi_2 - \phi_1) \tag{79}$$

As was mentioned before, the great circle and loxodrome are considered again in conjunction with the Mercator and gnomonic projections.

Note that the azimuth is not an independent variable in Equation 79. It is necessary to solve for the azimuth of P_2 with respect to P_1 before approaching Equation 79. Also note that when, $\Delta\lambda = 0$, $\cos\alpha = 1$, and we have distance along the great circle defined by the meridian. Also, when $\Delta\phi = 0$, we have a distance along a circle of parallel, and Equation 72 applies.

Next, we define the normal to the surface of the sphere at an arbitrary point $P(x, y, z)$. Is is a duplication of Equation 60, since that equation is not an explicit function of the semimajor axis of the eccentricity.

$$\hat{n} = \frac{\cos\lambda\hat{i} + \sin\lambda\hat{j} + \tan\phi\hat{k}}{\sqrt{1 + \tan^2\phi}} \tag{80}$$

The inverse relations for Equations 69 are

$$\tan\phi = \frac{z}{x^2 + y^2}$$

$$\tan\lambda = \frac{y}{z}$$

$$\sin\lambda = \frac{y}{x^2 + y^2}$$

$$\cos\lambda = \frac{x}{x^2 + y^2} \tag{81}$$

Substitute the first, third, and fourth of Equation 81 into Equations 80 and simplify.

$$\hat{n} = \frac{\dfrac{x\hat{i}}{\sqrt{x^2 + y^2}} + \dfrac{y\hat{j}}{\sqrt{x^2 + y^2}} + \dfrac{z\hat{k}}{\sqrt{x^2 + y^2}}}{\sqrt{1 + \dfrac{z^2}{\sqrt{x^2 + y^2}}}}$$

$$= \frac{x\hat{i} + y\hat{j} + z\hat{k}}{\sqrt{x^2 + y^2 + z^2}} \tag{82}$$

The final question to consider in this section is what radius to use for a model of the equivalent spherical Earth. One approach is to take a simple average of the semimajor and semiminor axes of the reference spheroid. A second approach is a weighted average of the semimajor and semiminor axes as given by

$$R = \frac{2a + b}{3} \tag{83}$$

A third approach is to take the radius of a sphere with the same area as the reference spheroid. This equivalent-area sphere is the authalic sphere, and the radius is the authalic radius. The definition of the authalic sphere is given in Chapter 4.

It is the author's intention to use the mean radius given by Equation 83 in the examples in this text.

A. Example 6

For the WGS-72 spheroid, find the simple average radius, the weighted average radius, and the authalic radius.

Simple average

$$R = \frac{a + b}{2} = \frac{6,378,135 + 6,356,750}{2}$$

$$= 6,367,442 \text{ m}$$

Weighted average

$$R = \frac{2a + b}{2} = \frac{(2)(6,378,135) + 6,356,750}{3}$$

$$= 6,371,007 \text{ m}$$

Authalic radius (From Chapter 4)

$$R = 6,371,004 \text{ m}$$

B. Example 7

Given: $\phi = 30°$.

Find: the radius of a parallel circle.

$$R_o = R\cos\phi$$

$$= (6,371,007)\cos 30°$$

$$= (6,371,007)(0.866025)$$

$$= 5,517,450 \text{ m}$$

C. Example 8

Given: $\phi = 30°$, $\lambda = 45°$.

Find: the Cartesian coordinates in three-dimensional space.

$$x = R\cos\phi\cos\lambda$$
$$= 6,371,007\cos30°\cos45°$$
$$= (6,371,007)(0.866025)(0.707107)$$
$$= 3,901,428 \text{ m}$$

$$y = R\cos\phi\cos\lambda$$
$$= 6,371,007\cos30°\sin45°$$
$$= (6,371,007)(0.866025)(0.707107)$$
$$= 3,901,428 \text{ m}$$

$$z = R\sin\phi$$
$$= 6,371,007\sin30°$$
$$= (6,371,007)(0.500000)$$
$$= 3,185,504 \text{ m}$$

D. Example 9

Given: $\phi = 30°$, $\Delta\lambda = 45°$, $R = 6,371,004$ m.

Find: the distance along the circle of parallel.

$$d = R\Delta\lambda\cos\phi$$
$$= (6,371,004)\left(\frac{45}{57.2958}\right)(0.866025)$$
$$= 4,333,393 \text{ m}$$

E. Example 10

Given: $\Delta\phi = 45°$, $R = 6,371,004$ m.

Find: the distance along the meridian.

$$d = R\Delta\phi$$
$$= (6,371,004)\left(\frac{45}{57.2958}\right)$$
$$= 5,003,773 \text{ m}$$

F. Example 11

Given: P_1 at $\phi_1 = 30°$, $\lambda_1 = 10°$ and P_2 at $\phi_2 = 45°$, $\lambda_2 = 70°$.
R = 6,371,004 m.

Find: the great circle distance between P_1 and P_2.

$$d = R\cos^{-1}[\sin\phi_1\sin\phi_2 + \cos\phi_1\cos\phi_2\cos\Delta\lambda]$$

$$\Delta\lambda = \lambda_2 - \lambda_1 = 70 - 10 = 60°$$

$$d = (6,371,004)[\sin30°\sin45° + \cos30°\cos45°\cos60°]$$

$$= (6,371,004)[(0.5)(0.707107)$$

$$+ (0.866025)(0.707107)(0.5)]$$

$$= (6,371,004)[0.353554 + 0.306186]$$

$$= (6,371,004)(0.659740)$$

$$= 4,203,206 \text{ m}$$

VI. THE TRIAXIAL ELLIPSOID

Some astronomical bodies are closer to the shape of a triaxial ellipsoid, as opposed to a spheroid of revolution. The outstanding example of this is the figure of the moon. This section briefly considers methods for modeling a surface by means of a triaxial ellipsoid.

The equation of the surface of a triaxial ellipsoid in three-dimensional Cartesian space is

$$\frac{x^2}{a^2} + \frac{y^2}{b^2} + \frac{z^2}{c^2} = 1 \tag{84}$$

This is illustrated in Figure 7 for a single octant of the triaxial ellipsoid. In the figure, a is along the x axis, b is along the y axis, and c is along the z axis. The model is taken such that a > b in the equatorial plane, and c, along the axis of rotation of the body, is less than both a and b. Table 2 has the value of the semiaxes a, b, and c for the moon.

Consider now the basic geometry in the case of the triaxial ellipsoid, as compared to the case of the spheroid of revolution. Beside the meridians, the parallels and equator become ellipses. All, however, remain as plane curves. Latitude and longitude are measured in the same way as for a spheroid of revolution. Differences occur with regard to the radii of curvature. The radii of curvature at any point become functions of longitude as well as latitude. This introduces considerable complexity in all mathematical relations. If extreme accuracy is required by the user, the general differential geometry formulas can be developed. This produces very complicated equations and seldom would be required for practical work.

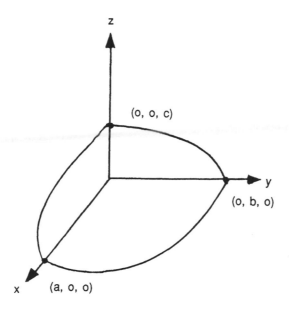

FIGURE 7.   Octant of a triaxial ellipsoid.

TABLE 2
Semiaxes for the Moon

$$a = 1,738,570 \text{ m}$$

$$b = 1,738,210 \text{ m}$$

$$c = 1,737,490 \text{ m}$$

A more tractable procedure is to consider the equation for the equator

$$\frac{x^2}{a^2} + \frac{y^2}{b^2} = 1 \tag{85}$$

Define an auxiliary semimajor axis $a_A(\lambda)$ which is a function of longitude. This may be obtained from Equation 85 from some specific longitude, where $a > a_A(\lambda) > b$. With this value of $a_A(\lambda)$, we can define the equation for the meridional ellipse at longitude λ, given by

$$\frac{x^2}{a_A(\lambda)^2} + \frac{z^2}{c^2} = 1 \tag{86}$$

As an approximation, we can consider that, locally, Equation 86 represents the meridian of a surface of revolution. Then, the spheroid of revolution formulas of Section IV can be applied.

For the usual requirements of accuracy it is sufficient to simply form the average:

$$a_A = \frac{a + b}{2} \qquad (87)$$

This gives an approximate spheroid of revolution, and all applicable formulas of Section IV can be used. This procedure, applied to the figure of the moon, yields $a_A = 1{,}738{,}390$ m, and $f = 1/1031.5$.

REFERENCES

1. **Bomford, G.,** *Geodesy,* Oxford University Press, New York, 1962.
2. **Langhaar, H. L.,** *Energy Methods in Applied Mechanics,* John Wiley & Sons, New York, 1962.
3. **Seppelin, T. O.,** The Department of Defense world geodetic system, 1972, *Can. Surv.,* December, 1974.
4. **Torge, W.,** *Geodesy,* Walter de Gruyter, New York, 1980.

Chapter 4

EQUAL AREA PROJECTIONS

I. INTRODUCTION

All of the projections of this chapter maintain equivalency of area between the map and the model of the Earth. Every section of the resulting map bears a constant ratio to the area of the Earth which it represents. This occurs because the condition of Equation 135, Chapter 2, is imposed.

The chapter begins with an outline of the general procedures to be followed in the transformation process for all of the projections which permit the differential geometry approach. This is followed by the introduction of the authalic sphere, a sphere with the same area as the spheroidal model of the Earth.

Two types of projections are considered. The first type are those which are best applied to a local area of the Earth. These are the Albers with one or two standard parallels, the Bonne, and the azimuthal and cylindrical equal area. The second type are the world maps: the sinusoidal, the Mollweide, the parabolic, the Hammer-Aitoff, the Boggs eumorphic, and Eckert IV. These world maps, with the exception of the Hammer-Aitoff, are also classified as pseudocylindrical. For all of these projections above, plotting equations are derived for the direct transformation from the geographic to the Cartesian coordinates. For a selected few of these projections, the inverse transformation from Cartesian to geographic coordinates is also included.

A quantitative overview of the theory of distortion is delayed until Chapter 7, where representative projections of all types are considered. However, as each projection is introduced, the trends in distortion are indicated. The interrupted projections are introduced as an easy way to minimize distortion. Grids and plotting tables are given for the projections. Uses of the projections are noted in each section.

We are finally at the point of dealing with the projections themselves.

II. GENERAL PROCEDURE

Many of the equal area projections may be derived by applying the processes of differential geometry. This is done in this text for the transformation to the authalic sphere and the derivation of the Albers with one standard parallel, the Bonne, the cylindrical, and the sinusoidal projections. The derivations of all of these direct transformations follow a similar procedure which is outlined here.

The first step is to apply the condition of equality of area of Equation 135, Chapter 2, to the transformation relation of Equation 134, Chapter 2.

This results in

$$eg = E'G' \begin{vmatrix} \dfrac{\partial u}{\partial \phi} & \dfrac{\partial u}{\partial \lambda} \\[2mm] \dfrac{\partial v}{\partial \phi} & \dfrac{\partial v}{\partial \lambda} \end{vmatrix}^2 \tag{1}$$

In Equation 1, the fundamental quantities e and g refer to the spherical or spheroidal model of the Earth and are functions of ϕ and λ. The fundamental quantities E' and G' refer to the surface on which the transformation is to be mapped. In this chapter, these fundamental quantities are functions of the Cartesian coordinates x and y for a plane or cylinder, of the polar coordinates θ and ρ for a plane or cone, and of ϕ and λ for a sphere.

The next step is to develop relations for the Jacobian determinant of Equation 1. The fact that ϕ and λ are independent always leads to one diagonal of the determinant being equal to zero.

Equation 1 is then expanded to a squared partial differential equation. The square root of both sides is then taken. The plus or minus root is taken depending on the coordinates of the mapping surface chosen. For the Cartesian and spherical coordinates, a positive root is needed. For the polar coordinates, the negative root is chosen.

Next, the partial differential equation is converted to an ordinary differential equation. This is justified since ϕ and λ are independent.

The ordinary differential equation is then integrated, and applicable boundary conditions are applied. This gives one of the plotting equations. The other plotting equation generally results from the conditions specified for the evaluation of the Jacobian determinant.

The final result for the transformation to a plotting surface is a set of Cartesian coordinates, x and y. The y axis is the central meridian arbitrarily chosen by the producer of the projection. The x axis intersects the y axis at some latitude, depending on the particular projection chosen. For the world maps and the cylindrical projection, this latitude is $\phi_o = 0°$, or the equator. For the polar azimuthal projections, this latitude is $\phi_o = 90°$, or the pole. For the conical projections with one standard parallel, ϕ_o equals the latitude of the circle of tangency of the projection. In the case of conical projections with two standard parallels, the latitude corresponding to the location of the origin can be at any parallel, depending on the choice of the user. The location of the origin for oblique cases of any projection depends on the point of tangency for the azimuthal projections.

Any of the projections that can be derived by differential geometry can also be derived by other means. This is done for comparison for several projections such as the Albers and the cylindrical. The azimuthal polar projection is most easily derived by considering a plane as the limiting case of a cone, and simplifying the conical relations.

Some of the projections, such as the parabolic, the Eckert IV, and the Mollweide, resist the differential geometry approach and are best derived from the geometry in a plane. The price paid for such a derivation is the lack of information to be applied to the mathematical theory of distortions.

Remember that all projection formulas developed in this chapter are for the Northern Hemisphere. In Chapter 8, the adaptations that are required to make them applicable to the Southern Hemisphere are given.

III. THE AUTHALIC SPHERE[1,2,6]

The authalic sphere is a spherical model of the Earth that has the same surface area as that of the reference ellipsoid. If required by the user, the authalic sphere may be defined as an intermediate step in the transformation from a spheroid to the mapping surface.

In the transformation from the spheroid to the authalic sphere, longitude is invariant. The latitude of a position, however, must be adjusted to maintain the equal area condition. This requires the definition of authalic latitude. It is also necessary to define the radius of the sphere with the same surface area as the reference spheroid. This requires the definition of the authalic radius, R_A.

The derivation of this transformation begins with the particular form of Equation 1 given below.

$$eg = E'G' \begin{vmatrix} \dfrac{\partial \phi_A}{\partial \phi} & \dfrac{\partial \phi_A}{\partial \lambda} \\[2ex] \dfrac{\partial \lambda_A}{\partial \phi} & \dfrac{\partial \lambda_A}{\partial \lambda} \end{vmatrix}^2 \tag{2}$$

In Equation 2, ϕ_A and λ_A are the authalic latitude and longitude, respectively, on the authalic sphere, and ϕ and λ are the geodetic latitude, respectively, on the reference spheroid.

From the reference spheroid, the first fundamental quantities are

$$\left. \begin{aligned} e &= R_m^2 \\ g &= R_p^2 \cos^2\phi \end{aligned} \right\} \tag{3}$$

On the authalic sphere, the first fundamental quantities are

$$E' = R_A^2 \left.\begin{array}{c} \\ \\ \\ \end{array}\right\} \tag{4}$$

$$G' = R_A^2\cos^2\phi_A$$

Substitute Equations 3 and 4 into Equation 2.

$$R_m^2 R_p^2\cos^2\phi = R_A^4\cos^2\phi_A \begin{vmatrix} \dfrac{\partial\phi_A}{\partial\phi} & \dfrac{\partial\phi_A}{\partial\lambda} \\[3mm] \dfrac{\partial\lambda_A}{\partial\phi} & \dfrac{\partial\lambda_A}{\partial\lambda} \end{vmatrix}^2 \tag{5}$$

The next step is to evaluate the Jacobian determinant. As mentioned above, the longitude is invariant under the transformation: $\lambda = \lambda_A$. Thus,

$$\frac{\partial\lambda_A}{\partial\lambda} = 1 \tag{6}$$

$$\frac{\partial\lambda_A}{\partial\phi} = 0 \tag{7}$$

Also, the authalic latitude is independent of λ_A.

$$\frac{\partial\phi_A}{\partial\lambda} = 0 \tag{8}$$

Substitute Equations 6, 7, and 8 into Equation 5 and take the positive square root.

$$R_m^2 R_p^2\cos^2\phi = R_A^4\cos^2\phi_A \begin{vmatrix} \dfrac{\partial\phi_A}{\partial\phi} & 0 \\[3mm] 0 & 1 \end{vmatrix}^2 \tag{9}$$

$$R_m R_p\cos\phi = R_A^2\cos\phi_A\left(\frac{\partial\phi_A}{\partial\phi}\right)$$

Equation 9 can now be converted into an ordinary differential form.

$$R_m R_p\cos\phi \, d\phi = R_A^2\cos\phi_A \, d\phi_A \tag{10}$$

Apply the values of R_m and R_p derived in Chapter 3 (Equations 28 and 21, respectively).

$$R_m = \frac{a(1 - e^2)}{(1 - e^2\sin^2\phi)^{3/2}} \tag{11}$$

$$R_p = \frac{a}{(1 - e^2\sin^2\phi)^{1/2}} \tag{12}$$

Substitute Equations 11 and 12 into Equation 10 to obtain

$$\frac{a^2(1 - e^2)}{(1 - e^2\sin^2\phi)^2} \cos\phi d\phi = R_A^2\cos\phi_A d\phi_A \tag{13}$$

The next step is to integrate Equation 13. The term on the right is easily integrated.

$$\int_0^{\phi_A} R_A^2\cos\phi_A d\phi_A = R_A^2\sin\phi_A \tag{14}$$

The term on the left requires a binomial expansion before integrating it.

$$a^2(1 - e^2)\int_0^{\phi} \frac{\cos\phi}{(1 - e^2\sin^2\phi)^2} d\phi$$

$$= a^2(1 - e^2)\int_0^{\phi} \cos\phi(1 + 2e^2\sin^2\phi + 3e^4\sin^4\phi$$

$$+ 4e^6\sin^6\phi + ...)d\phi$$

$$= a^2(1 - e^2)\left(\sin\phi + \frac{2}{3} e^2\sin^3\phi + \frac{3}{5} e^4\sin^5\phi \right.$$

$$\left. + \frac{4}{7} e^6\sin^7\phi + ... \right) \tag{15}$$

Equate Equations 14 and 15 to obtain

$$R_A^2\sin\phi_A = a(1 - e^2)\left(\sin\phi + \frac{2}{3} e^2\sin^3\phi + \frac{3}{5} e^4\sin^5\phi \right.$$

$$\left. + \frac{4}{7} e^6\sin^7\phi + ... \right) \tag{16}$$

In order to determine R_A, we introduce the condition that $\phi_A = \pi/2$ when $\phi = \pi/2$. Then, Equation 16 becomes

$$R_A^2 = a^2(1 - e^2)\left(1 + \frac{2}{3}e^2 + \frac{3}{5}e^4 + \frac{4}{7}e^6 + ...\right) \qquad (17)$$

Equation 17 gives the radius of an authalic sphere with a surface area equal to that of the reference spheroid.

Substitute Equation 17 into Equation 16 to obtain the relation between authalic latitude and geodetic latitude.

$$\sin\phi_A = \sin\phi\left(\frac{1 + \frac{2}{3}e^2\sin^3\phi + \frac{3}{5}e^4\sin^5\phi + \frac{4}{7}e^6\sin^7\phi + ...}{1 + \frac{2}{3}e^2 + \frac{3}{5}e^4 + \frac{4}{7}e^6 + ...}\right) \qquad (18)$$

As was seen in Chapter 3, the eccentricity, e, is a small number for all of the accepted spheroids. Thus, Equation 18 contains a rapidly converging series. The relation between authalic and geodetic latitudes is identically equal at latitudes 0 and 90°. At latitudes near 45°, the difference between the two is on the order of 0°.1 for the WGS-84 spheroid.

The transformation from the reference spheroid to the authalic sphere distorts distances between given positions on the spheroid. The amount of distortion introduced in this transformation is explored in an extended example in Chapter 8.

A. Example 1

Given: the WGS-72 spheroid with a = 6,378,135 m and e = 0.081818.
Find: the radius of the authalic sphere.

$$R_A = a\sqrt{(1 - e^2)\left(1 + \frac{2}{3}e^2 + \frac{3}{5}e^4\right)}$$

$$= (6,378,135)\sqrt{(0.9933059)(1 + 0.0044627 + 0.0000268)}$$

$$= 6,371,004 \text{ m}$$

B. Example 2

Given: the I.U.G.G. spheroid with f = 1/298.257.
Find: the authalic latitude for geodetic latitude 45°.

$$e^2 = 2f - f^2 = 0.00669438$$

$$e^4 = 0.00004481$$

$$\sin\phi_A = \sin\phi \left[\frac{1 + \frac{2}{3} e^2 \sin^2\phi}{1 + \frac{2}{3} e^2 + \frac{3}{5} e^4} \right]$$

$$= (0.707107) \left[\frac{1 + 0.0022314}{1.0044898} \right]$$

$$= 0.705517$$

$$\phi_A = 44°.8713$$

IV. ALBERS, ONE STANDARD PARALLEL[1,6]

The first projection to consider is the Albers projection with one standard parallel. In this projection, the conical plotting surface is tangent to the model of the Earth. The axis of the cone coincides with the polar axis of the sphere. The direct transformation is derived by two methods. Following this, the inverse transformation is given.

We begin with the differential geometry approach. The first fundamental quantities for the sphere are

$$\left. \begin{array}{l} e = R^2 \\ \\ g = R^2\cos^2\phi \end{array} \right\} \tag{19}$$

The first fundamental quantities for a polar coordinate system in a plane, or for a cone, are

$$\left. \begin{array}{l} E' = 1 \\ \\ G' = \rho^2 \end{array} \right\} \tag{20}$$

Equations 19 and 20 are then substituted in Equation 1, particularized to the coordinates of this derivation.

$$R^4\cos^2\phi = \rho^2 \begin{vmatrix} \dfrac{\partial\rho}{\partial\phi} & \dfrac{\partial\rho}{\partial\lambda} \\ \\ \dfrac{\partial\theta}{\partial\phi} & \dfrac{\partial\theta}{\partial\lambda} \end{vmatrix}^2 \tag{21}$$

Next, impose the conditions that

$$\rho = \rho(\phi) \tag{22}$$

$$\theta = c_1\lambda + c_2 \tag{23}$$

In Equation 23, c_1 is the constant of the cone from Section XIV, Chapter 2. If the further condition is imposed that $\theta = 0$, when $\lambda = \lambda_o$, then $c_2 = -c_1\lambda_o$. Equation 23 becomes

$$\theta = c_1(\lambda - \lambda_o) \tag{24}$$

Now we develop the partial derivatives of the Jacobian matrix. From Equation 22,

$$\frac{\partial\rho}{\partial\lambda} = 0 \tag{25}$$

From Equation 23,

$$\frac{\partial\theta}{\partial\phi} = 0 \tag{26}$$

$$\frac{\partial\theta}{\partial\lambda} = c_1 \tag{27}$$

Substitute Equations 25, 26, and 27 in Equation 21 and take the negative square root. Then, simplify the results.

$$R^4\cos^2\phi = \rho^2 \begin{vmatrix} \dfrac{\partial\rho}{\partial\phi} & 0 \\ 0 & c_1 \end{vmatrix}^2$$

$$R^2\cos\phi = -\rho c_1\left(\frac{\partial\rho}{\partial\phi}\right) \tag{28}$$

The minus sign is chosen since an increase in ϕ corresponds to a decrease in ρ.

Convert Equation 28 into an ordinary differential equation and integrate; this gives

$$\rho d\rho = -\frac{R^2}{c_1} \cos\phi d\phi$$

$$\rho^2 = -\frac{2R^2}{c_1} \sin\phi + c_3 \tag{29}$$

In Equation 29, $c_1 = \sin\phi_o$, and c_3 is obtained from an application of boundary conditions. To this end, recall from Section XIV, Chapter 2, that

$$\rho_o = R\cot\phi_o \tag{30}$$

Substitute the constant of the cone into Equation 29 and evaluate at $\phi = \phi_o$.

$$\rho^2 = -2R^2 \frac{\sin\phi}{\sin\phi_o} + c_3 \tag{31}$$

$$\rho_o^2 = -2R^2 + c_3$$

$$c_3 = \rho_o + 2R^2 \tag{32}$$

Substitute Equation 30 into Equation 32.

$$c_3 = R^2\cot^2\phi_o + 2R^2$$

$$= R^2(2 + \cot^2\phi_o) \tag{33}$$

Substitute Equation 33 into Equation 31, simplify, and take the square root.

$$\rho^2 = R^2\left(2 + \cot^2\phi_o - 2\frac{\sin\phi}{\sin\phi_o}\right)$$

$$= R^2\left(2\frac{\sin^2\phi_o}{\sin^2\phi_o} + \frac{\cos^2\phi_o}{\sin^2\phi_o} - 2\frac{\sin\phi\sin\phi_o}{\sin^2\phi_o}\right)$$

$$= \frac{R^2}{\sin^2\phi_o}(1 + \sin^2\phi_o - 2\sin\phi\sin\phi_o)$$

$$\rho = \frac{R}{\sin\phi_o}\sqrt{1 + \sin^2\phi_o - 2\sin\phi\sin\phi_o} \tag{34}$$

Recall the relation between θ and λ from Section XIV, Chapter 2. In this projection,

$$\theta = \Delta\lambda\sin\phi_o \tag{35}$$

Equations 34 and 35 are used in conjunction with Equation 7, Chapter 1, to obtain the Cartesian plotting equations.

$$\left. \begin{array}{l} x = S[\rho\sin(\Delta\lambda\sin\phi_o)] \\[2mm] y = S[R\cot\phi_o - \rho\cos(\Delta\lambda\sin\phi_o)] \end{array} \right\} \tag{36}$$

where S is the scale factor, and $\Delta\lambda = \lambda - \lambda_o$. The origin of the projection has the coordinates ϕ_o and λ_o, where ϕ_o is the latitude of tangency of the cone. The central meridian has the longitude λ_o.

The plotting equations of Equation 36 were evaluated for a choice of $\phi_o = 45°$ and $\lambda_o = 0°$. The grid that results from this choice of ϕ_o and λ_o is in Figure 1. Note that a different plotting table will result from each choice of ϕ_o. Note also the positions of the Cartesian coordinate axes in this projection. The y axis is along the central meridian. The x axis is perpendicular to the y axis, with origin at ϕ_o.

In Figure 1, the parallels are arcs of concentric circles. The meridians are straight lines. The only true scale line is the parallel arc at $\phi = \phi_o$. All other lengths along the parallels are larger than true scale. This grid has been used in some atlas maps.

A second approach can be followed for the single standard parallel case. This involves equating corresponding areas on the cone and the sphere.

Consider a cone with constant $\sin\phi_o$, and let ρ_o be the radius on the map to the standard parallel ϕ_o. The area on the cone bounded by that parallel is $\pi\rho_o^2 \sin\phi_o$. If ρ is the radius of any other parallel of latitude ϕ, then the area bounded is $\pi\rho^2 \sin\phi$. The area in the strip between these parallels is

$$A = \pi(\rho_o^2 - \rho^2)\sin\phi_o \tag{37}$$

The area on the sphere between the parallels ϕ_o and ϕ is

$$A = 2\pi R^2(\sin\phi - \sin\phi_o) \tag{38}$$

For equal area, equate Equations 37 and 38.

$$\pi(\rho_o^2 - \rho^2)\sin\phi_o = 2\pi R^2(\sin\phi - \sin\phi_o)$$

$$(\rho_o^2 - \rho^2)\sin\phi_o = 2R^2(\sin\phi - \sin\phi_o) \tag{39}$$

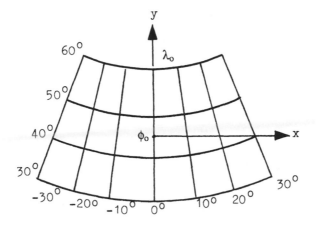

FIGURE 1. Albers projection, one standard parallel.

Substitute Equation 30 into Equation 39.

$$\sin\phi_o(R^2\cot^2\phi_o - \rho^2) = 2R^2(\sin\phi - \sin\phi_o)$$

$$R^2\cot^2\phi_o - \rho^2 = 2R^2\frac{\sin\phi}{\sin\phi_o} - 2R^2$$

$$\rho^2 = R^2\cot^2\phi_o + 2R^2 - 2R^2\frac{\sin\phi}{\sin\phi_o} \qquad (40)$$

Equation 34 can then be reproduced.

The inverse transformation for this projection can be derived From Equations 36. Rewrite Equations 36 in a form conducive to summing the squares of the trigonometric functions.

$$\left.\begin{array}{l} \dfrac{x}{S\rho} = \sin(\Delta\lambda\sin\phi_o) \\[3mm] \dfrac{SR\cot\phi_o - y}{S\rho} = \cos(\Delta\lambda\sin\phi_o) \end{array}\right\} \qquad (41)$$

$$\sin^2(\Delta\lambda\sin\phi_o) + \cos^2(\Delta\lambda\sin\phi_o) = 1$$

$$= \left(\frac{x}{S\rho}\right)^2 + \left(\frac{SR\cot\phi_o - y}{S\rho}\right)^2$$

$$\rho^2 = \frac{1}{S^2}[x^2 + (SR\cot\phi_o - y)^2] \qquad (42)$$

$$\rho = \frac{1}{S} \sqrt{x^2 + (SR\cot\phi_o - y)^2} \tag{43}$$

From the first of Equations 36,

$$\Delta\lambda = \frac{\sin^{-1}\left(\frac{x}{S\rho}\right)}{\sin\phi_o} \tag{44}$$

Equations 43 and 44 may be used with $\lambda = \lambda_o + \Delta\lambda$ to obtain the geographic longitude on the sphere.

Equate the squares of Equations 34 and 43 and solve for ϕ.

$$\frac{1}{S^2}[x^2 + (SR\cot\phi_o - y)^2] = \frac{R^2}{\sin^2\phi_o}(1 + \sin^2\phi_o - 2\sin\phi\sin\phi_o)$$

$$\frac{\sin^2\phi_o}{S^2R^2}\left[x^2 + SR\cot\phi_o - y)^2\right] = 1 + \sin^2\phi_o - 2\sin\phi\sin\phi_o$$

$$\sin\phi = \frac{1}{2\sin\phi_o}\left\{1 + \sin^2\phi_o - \frac{\sin^2\phi_o}{S^2R^2}[x^2 + (SR\cot\phi_o - y)^2]\right\}$$

$$\phi = \sin^{-1}\left\{\frac{1}{2\sin\phi_o} + \frac{\sin\phi_o}{2} - \frac{\sin\phi_o}{2S^2R^2}[x^2 + (SR\cot\phi_o - y)^2]\right\} \tag{45}$$

Thus, latitude and longitude may be obtained from the Cartesian coordinates. This is the inverse transformation.

A. Example 3

Given: $\lambda_o = 45°$, $\phi_o = 30°$, $S = 1:10,000,000$, $\phi = 35°$, $\lambda = 50°$, $R = 6,378,004$ m.

$$\Delta\lambda = \lambda - \lambda_o = 50 - 45 = 5°$$

$$\sin\phi_o = \sin30° = 0.5$$

$$\rho = \left(\frac{6,378,004}{0.5}\right)\sqrt{1 + (0.5)^2 - (2)(0.5)\sin35°}$$

$$= (12,756,008)\sqrt{1 + 0.25 - 0.573578}$$

$$= 10,491,175 \text{ m}$$

$$x = S\rho\sin(\Delta\lambda\sin\phi_o)$$

$$= \frac{10,491,175}{10,000,000} \sin[(0.5)(5)]$$

$$= 0.04576 \text{ m}$$

$$y = S\{R\cot 30° - \rho\cos[(0.5)(5)]\}$$

$$= \frac{1}{10,000,000} [(6,378,004)(1.73205) - (10,491,175)(0.999048)]$$

$$= 0.0565832 \text{ m}$$

B. Example 4

Given: $\lambda_o = 45°$, $\phi_o = 30°$, $S = 1{:}1,000,000$, $x = 0.2$ m, $y = 0.2$ m, $R = 6,371,000$ m.

Find: ϕ and λ for the Albers projection, with one standard parallel.

$$\rho = \frac{1}{S} \sqrt{x^2 + SR\cot\phi_o - y)^2}$$

$$= (1,000,000) \sqrt{(0.2)^2 + \left[\left(\frac{6,371,000}{1,000,000}\right)(1.732051) - 0.2\right]^2}$$

$$= (1,000,000)\sqrt{0.04 + 117.395}$$

$$= (1,000,000)(10.8367) = 10,836,700 \text{ m}$$

$$\Delta\lambda = \frac{\sin^{-1}\left(\dfrac{x}{S\rho}\right)}{\sin\phi_o} = \frac{\sin^{-1}\left(\dfrac{0.2}{10.8367}\right)}{0.5}$$

$$= 2.115°$$

$$\lambda = \lambda_o + \Delta\lambda = 45 + 2.115 = 47°.115$$

$$\phi = \sin^{-1}\left\{\frac{1}{2\sin\phi_o} + \frac{\sin\phi_o}{2} - \frac{\sin\phi_o}{2S^2R^2} [x^2 + (SR\cot\phi_o - y]^2\right\}$$

$$= \sin^{-1}\left\{1 + 0.25 - \frac{0.25}{(6.371)^2} [117.434]\right\}$$

$$= \sin^{-1}\{1.25 - 0.723527\}$$

$$= \sin^{-1}\{0.52647\}$$

$$= 31°.767$$

V. ALBERS, TWO STANDARD PARALLELS[1,6]

The conical equal area projection can also be derived when the mapping cone is secant to the spherical model of the Earth. This is done in two ways: first in reference to the constant of the cone, and second by comparing areas on the map and on the sphere. In addition, an inverse transformation is derived. Note that secancy is a conceptual quality for conical projections.

In the first derivation, consider the cone secant at latitudes ϕ_1 and ϕ_2, with $\phi_2 > \phi_1$. From Equation 159, Chapter 2,

$$\rho_1 = \frac{R\cos\phi_1}{c_1}$$

$$\rho_2 = \frac{R\cos\phi_2}{c_1} \tag{46}$$

Next, substitute Equations 46 into Equation 29 to obtain two equations.

$$\frac{R^2\cos^2\phi_1}{c_1^2} = -\frac{2R^2}{c_1}\sin\phi_1 + c_3$$

$$R^2\cos^2\phi_1 + 2R^2c_1\sin\phi_1 - c_1^2c_3 = 0 \tag{47}$$

$$\frac{R^2\cos^2\phi_2}{c_1^2} = -\frac{2R^2}{c_1}\sin\phi_2 + c_3$$

$$R^2\cos^2\phi_2 + 2R^2c_1\sin\phi_2 - c_1^2c_3 = 0 \tag{48}$$

Equations 47 and 48 can be solved simultaneously for c_1.

$$R^2\cos^2\phi_1 - R^2\cos^2\phi_2 + 2R^2c_1\sin\phi_1 - 2R^2c_1\sin\phi_2 = 0$$

$$\cos^2\phi_1 - \cos^2\phi_2 = 2c_1(\sin\phi_2 - \sin\phi_1)$$

$$c_1 = \frac{\cos^2\phi_1 - \cos^2\phi_2}{2(\sin\phi_2 - \sin\phi_1)}$$

$$= \frac{\sin^2\phi_2 - \sin^2\phi_1}{2(\sin\phi_2 - \sin\phi_1)}$$

$$= \frac{1}{2}(\sin\phi_2 + \sin\phi_1) \tag{49}$$

Substitute Equation 49 into the relation between the polar coordinate θ and the difference in longitude $\Delta\lambda$.

$$\theta = \frac{\Delta\lambda}{2}(\sin\phi_1 + \sin\phi_2) \tag{50}$$

Substitute Equation 49 into Equation 28.

$$\rho^2 = -\frac{4R^2\sin\phi}{\sin\phi_1 + \sin\phi_2} + c_3 \tag{51}$$

Evaluate Equation 51 at ϕ_1.

$$\rho_1^2 = -\frac{4R^2\sin\phi_1}{\sin\phi_1 + \sin\phi_2} + c_3$$

$$c_3 = \rho_1^2 + \frac{4R^2\sin\phi_1}{\sin\phi_1 + \sin\phi_2} \tag{52}$$

Substitute Equation 52 into Equation 51.

$$\rho^2 = \rho_1^2 + \frac{4R^2(\sin\phi_1 - \sin\phi)}{\sin\phi_1 + \sin\phi_2} \tag{53}$$

where, from Equations 46 and 49,

$$\rho_1 = \frac{2R\cos\phi_1}{\sin\phi_1 + \sin\phi_2} \tag{54}$$

A similar development gives

$$\rho^2 = \rho_2^2 + \frac{4R^2(\sin\phi_2 - \sin\phi)}{\sin\phi_1 + \sin\phi_2} \tag{55}$$

$$\rho_2 = \frac{2R\cos\phi_2}{\sin\phi_1 + \sin\phi_2} \tag{56}$$

It only remains to substitute into Equations 7, Chapter 1, to obtain the plotting equations. One form of these is

$$x = S \cdot \sqrt{\rho_1^2 + \frac{4R^2(\sin\phi_1 - \sin\phi)}{\sin\phi_1 + \sin\phi_2}} \sin\left[\frac{\Delta\lambda}{2}(\sin\phi_1 + \sin\phi_2)\right] \tag{57}$$

$$y = S \cdot \left\{ \frac{1}{2} (\rho_1 + \rho_2) \right.$$

$$\left. - \sqrt{\rho_1^2 + \frac{4R^2(\sin\phi_1 - \sin\phi)}{\sin\phi_1 + \sin\phi_2}} \cos\left[\frac{\Delta\lambda}{2} (\sin\phi_1 + \sin\phi_2) \right] \right\} \quad (58)$$

where S is the scale factor, and $\Delta\lambda = \lambda - \lambda_o$.
The second form is

$$x = S \cdot \sqrt{\rho_2^2 + \frac{4R^2(\sin\phi_2 - \sin\phi)}{\sin\phi_1 + \sin\phi_2}} \sin\left[\frac{\Delta\lambda}{2} (\sin\phi_1 + \sin\phi_2) \right] \quad (59)$$

$$y = S \cdot \left\{ \frac{1}{2} (\rho_1 + \rho_2) \right.$$

$$\left. - \sqrt{\rho_2^2 + \frac{4R^2(\sin\phi_2 - \sin\phi)}{\sin\phi_1 + \sin\phi_2}} \cos\left[\frac{\Delta\lambda}{2} (\sin\phi_1 + \sin\phi_2) \right] \right\} \quad (60)$$

The case of Albers projection with two standard parallels can be handled with an alternate approach of equating corresponding areas on the cone and the sphere.

Let ϕ_1 and ϕ_2 be the latitude of the two standard parallels, and ρ_1 and ρ_2 be their respective radii on the projection. Let ϕ_2 be greater than ϕ_1. The constant of the cone is C.

The area of the strip of the cone between these latitudes is

$$A = c\pi(\rho_1^2 - \rho_2^2) \quad (61)$$

The area of a zone of the sphere between the given latitudes is

$$A = 2\pi R^2(\sin\phi_2 - \sin\phi_1) \quad (62)$$

For the equal area projection, equate Equations 61 and 62.

$$c\pi(\rho_1^2 - \rho_2^2) = 2\pi R^2(\sin\phi_2 - \sin\phi_1)$$

$$c(\rho_1^2 - \rho_2^2) = 2R^2(\sin\phi_2 - \sin\phi_1) \quad (63)$$

Since the standard parallels are true length, we can equate these parallels on the map and on the sphere.

$$2\pi\rho_1 c = 2\pi R\cos\phi_1$$

$$\rho_1 = \frac{R\cos\phi_1}{c} \tag{64}$$

$$2\pi\rho_2 c = 2\pi R\cos\phi_2$$

$$\rho_2 = \frac{R\cos\phi_1}{c} \tag{65}$$

Substitute Equations 64 and 65 into Equation 63 and simplify.

$$c\left(\frac{R^2\cos^2\phi_1}{c^2} - \frac{R^2\cos^2\phi_2}{c^2}\right) = 2R^2(\sin\phi_2 - \sin\phi_1)$$

$$\frac{\cos^2\phi_1 - \cos^2\phi_2}{c} = 2(\sin\phi_2 - \sin\phi_1)$$

$$c = \frac{\cos^2\phi_1 - \cos^2\phi_2}{2(\sin\phi_2 - \sin\phi_1)}$$

$$= \frac{\sin^2\phi_2 - \sin^2\phi_1}{2(\sin\phi_2 - \sin\phi_1)}$$

$$= \frac{1}{2}(\sin\phi_1 + \sin\phi_2)$$

We have now reproduced Equation 49.

Substituting the constant of the cone into Equations 64 and 65,

$$\rho_1 = \frac{2R\cos\phi_1}{\sin\phi_1 + \sin\phi_2}$$

$$\rho_2 = \frac{2R\cos\phi_2}{\sin\phi_1 + \sin\phi_2}$$

This has reproduced Equations 54 and 56.

To find the value of ρ for a general latitude ϕ, again, we equate the area on the map and the area on the sphere.

$$C\pi(\rho^2 - \rho_1^2) = 2\pi R^2(\sin\phi_1 - \sin\phi)$$

$$\rho^2 = \rho_1^2 + \frac{2R^2(\sin\phi_1 - \sin\phi)}{C}$$

$$= \rho_1^2 + \frac{4R^2(\sin\phi_1 - \sin\phi)}{\sin\phi_1 + \sin\phi_2}$$

This reproduces Equation 53.

A plotting table for the Albers projection with two standard parallels is given as Table 1. The standard parallels are chosen as $\phi_1 = 35°$, $\phi_2 = 55°$. For this projection, each plotting table is dependent on the user's choice of standard parallels.

The resulting grid is shown in Figure 2. The meridians are all straight lines. The parallels are concentric circles, spaced so that the equal area quality is maintained. The central meridian contains the y axis. The location of the x axis depends on which set of plotting equations is used. As shown in the figure, it can be at either ϕ_1 or ϕ_2.

This projection has been used extensively to portray mid-latitude areas of large east to west extent. It is the standard equal area map for the U.S. In this projection, distortion is a function of latitude and not longitude.

The inverse transformation from Cartesian to geographic coordinates is obtained as follows. Start with relations for ρ_1 and ρ_2.

$$\left.\begin{array}{l} \rho_1 = \dfrac{2R\cos\phi_1}{\sin\phi_1 + \sin\phi_2} \\[3mm] \rho_2 = \dfrac{2R\cos\phi_2}{\sin\phi_1 + \sin\phi_2} \end{array}\right\} \tag{66}$$

Define auxiliary functions

$$f_1 = \frac{\rho_1 + \rho_2}{2}$$

$$f_2 = \frac{\sin\phi_1 + \sin\phi_2}{2}$$

$$f_3 = \sqrt{\rho_1^2 + \frac{4R^2(\sin\phi_1 - \sin\phi)}{\sin\phi_1 + \sin\phi_2}} \tag{67}$$

Substitute Equation 67 into Equations 57 and 58.

$$x = Sf_3\sin(\Delta\lambda f_2)$$

$$y = S[f_1 - f_3\cos(\Delta\lambda f_2)] \tag{68}$$

TABLE 1
Normalized Plotting Coordinates for the Albers Projection,
Two Standard Parallels
($\phi_1 = 35°$, $\phi_2 = 55°$, $\lambda_o = 0°$, $R \cdot S = 1$)

Tabulated Values: x/y

Longitude	Latitude			
	30°	40°	50°	60°
0°	0.000	0.000	0.000	0.000
	−0.263	−0.089	0.088	0.263
10°	0.153	0.132	0.111	0.089
	−0.254	−0.081	0.095	0.268
20°	0.304	0.262	0.219	0.177
	−0.226	−0.057	0.115	0.284
30°	0.450	0.388	0.325	0.263
	−0.180	−0.017	0.148	0.311
40°	0.590	0.509	0.426	0.345
	−0.117	0.038	0.194	0.348
50°	0.721	0.622	0.520	0.421
	−0.037	0.106	0.252	0.395
60°	0.842	0.725	0.607	0.491
	0.058	0.188	0.320	0.450

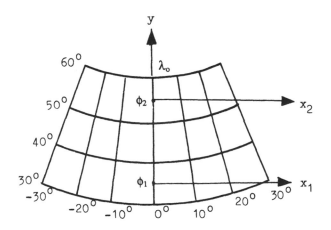

FIGURE 2.   Albers projection, two standard parallels.

Rearrange Equation 68 to have

$$\sin(\Delta\lambda f_2) = \frac{x}{Sf_3}$$

$$\cos(\Delta\lambda f_2) = \frac{Sf_1 - y}{Sf_3} \qquad (69)$$

Develop the tangent function from Equation 69 and solve for the longitude

$$\tan(\Delta\lambda f_2) = \frac{x}{Sf_1 - y}$$

$$\Delta\lambda = \frac{1}{f_2}\tan^{-1}\left(\frac{x}{Sf_1 - y}\right)$$

$$\lambda = \frac{1}{f_2}\tan^{-1}\left(\frac{x}{Sf_1 - y}\right) + \lambda_o \tag{70}$$

To find the latitude, solve Equation 57 for ϕ.

$$x^2 = S^2\left[\rho_1^2 + \frac{2R^2(\sin\phi_1 - \sin\phi)}{f_2}\right]\sin^2(\Delta\lambda f_2)$$

$$\rho_1^2 + \frac{2R^2(\sin\phi_1 - \sin\phi)}{f_2} = \left[\frac{x}{S\sin(\Delta\lambda f_2)}\right]^2$$

$$\frac{2R^2}{f_2}(\sin\phi_1 - \sin\phi) = \left[\frac{x}{S\sin(\Delta\lambda f_2)}\right]^2 - \rho_1^2$$

$$\sin\phi_1 - \sin\phi = \frac{f_2}{2R^2}\left\{\left[\frac{x}{S\sin(\Delta\lambda f_2)}\right]^2 - \rho_1^2\right\}$$

$$\sin\phi = \sin\phi_1 - \frac{f_2}{2R^2}\left\{\left[\frac{x}{S\sin(\Delta\lambda f_2)}\right]^2 - \rho_1^2\right\} \tag{71}$$

Take the arc-sine of Equation 71 to uniquely define ϕ.

A. Example 5

Given: $\lambda_o = 0°, \phi_1 = 30°, \phi_2 = 40°, \phi = 35°, \lambda = 5°, R = 6,371,007$ m, $S = 1/2,000,000$.

Find: the Cartesian plotting coordinates for the Albers projection with two standard parallels.

$$\Delta\lambda = \lambda - \lambda_o$$

$$= 5 - 0 = 5°$$

$$\theta = \frac{\Delta\lambda}{2}(\sin\phi_1 + \sin\phi_2)$$

$$= \frac{5}{2}(\sin 30° + \sin 40°)$$

$$= \left(\frac{5}{2}\right)(0.500000 + 0.642788)$$

$$= 2°.8570$$

$$\rho_1 = \frac{2R\cos\phi_1}{\sin\phi_1 + \sin\phi_2}$$

$$= \frac{(2)(6,371,007)\cos30°}{\sin30° + \sin40°}$$

$$= \frac{(2)(6,371,007)(0.866025)}{1.142788}$$

$$= 9,656,127 \text{ m}$$

$$\rho_2 = \frac{2R\cos\phi_2}{\sin\phi_1 + \sin\phi_2}$$

$$= \frac{(2)(6,371,007)\cos40°}{1.142788}$$

$$= \frac{(2)(6,371,007)(0.766044)}{1.142788}$$

$$= 8,541,350 \text{ m}$$

$$\sqrt{\rho_1^2 + \frac{4R^2(\sin\phi_1 - \sin\phi)}{\sin\phi_1 + \sin\phi_2}}$$

$$= \sqrt{(9.656127)^2 + (4)(6.371007)^2\frac{(0.500000 - 0.573576)}{1.142788}} \times 10^6$$

$$= \sqrt{93.240789 - 10.453137} \times 10^6$$

$$= 9,098,772$$

$$x = \left(\frac{9,098,772}{2,000,000}\right)\sin2.8570$$

$$= 0.2268 \text{ m}$$

$$y = \frac{1}{2,000,000}\left\{\frac{9,656,127 + 8,541,350}{2} - (9,098,772)[\cos(2.8570)]\right\}$$

$$= \frac{1}{2,000,000}[9,098,738 - 9.087463]$$

$$= 0.0056 \text{ m}$$

VI. BONNE[1,2,6]

Bonne's projection is a conical equal area projection in which the defining condition is placed on the constant of the cone. In fact, the constant of the cone is a function of latitude and not a constant at all. A differential geometry approach is used to derive this projection for a spherical Earth model.

For the sphere, the first fundamental quantities are given as

$$\left.\begin{array}{l} e = R^2 \\[2mm] g = R^2\cos^2\phi \end{array}\right\} \tag{72}$$

For the cone, the first fundamental quantities are

$$\left.\begin{array}{l} E' = 1 \\[2mm] G' = \rho^2 \end{array}\right\} \tag{73}$$

Substitute Equations 72 and 73 into Equation 1 to obtain the transformation relation.

$$R^4\cos^2\phi = \rho^2 \begin{vmatrix} \dfrac{\partial\rho}{\partial\phi} & \dfrac{\partial\rho}{\partial\lambda} \\[3mm] \dfrac{\partial\theta}{\partial\phi} & \dfrac{\partial\theta}{\partial\lambda} \end{vmatrix}^2 \tag{74}$$

The condition to be imposed is that the polar coordinate to the parallel circle at latitude ϕ is to be

$$\rho = \rho_o - \int_{\phi_1}^{\phi} R d\phi \tag{75}$$

The next step is to take partial derivatives of Equation 75.

$$\left.\begin{array}{l} \dfrac{\partial\rho}{\partial\phi} = -R \\[4mm] \dfrac{\partial\rho}{\partial\lambda} = 0 \end{array}\right\} \tag{76}$$

Substitute Equation 76 into Equation 74, simplify, and take the positive square root.

$$R^4\cos^2\phi = \rho^2 \begin{vmatrix} -R & 0 \\ \dfrac{\partial\theta}{\partial\phi} & \dfrac{\partial\theta}{\partial\lambda} \end{vmatrix}^2$$

$$= R^2\rho^2\left(\dfrac{\partial\theta}{\partial\lambda}\right)^2$$

$$R\cos\phi = \rho\left(\dfrac{\partial\theta}{\partial\lambda}\right) \tag{77}$$

Convert Equation 77 into an ordinary differential equation and integrate, treating ϕ and ρ as constants.

$$d\theta = \frac{R\cos\phi}{\rho}\,d\lambda$$

$$\theta = \frac{R\cos\phi}{\rho}\,\lambda + k \tag{78}$$

Impose the condition that $\theta = 0$ when $\lambda = \lambda_o$. Then

$$k = -\frac{R\cos\phi}{\rho}\,\lambda_o$$

and Equation 78 becomes

$$\theta = \frac{R\cos\phi}{\rho}\,\Delta\lambda \tag{79}$$

Note that the multiplicative factor of $\Delta\lambda$ is a variable quantity, similar to the constant of the cone in the Albers projections.

The next step is to obtain the relation for ρ. From the definition of the tangent to a sphere in Chapter 2,

$$\rho_o = R\cot\phi_o \tag{80}$$

Integrate Equation 75 to obtain

$$\rho = \rho_o - R(\phi - \phi_o) \tag{81}$$

Substitute Equation 80 into Equation 81.

$$\rho = R\cot\phi_o - R(\phi - \phi_o) \tag{82}$$

We are now in possession of both polar coordinates, and the Cartesian coordinates follow from Equation 7, Chapter 1.

$$\left. \begin{array}{l} x = \rho S \sin\left(\dfrac{\Delta\lambda R\cos\phi}{\rho}\right) \\[2em] y = S\left[R\cot\phi_o - \rho\cos\left(\dfrac{\Delta\lambda R\cos\phi}{\rho}\right)\right] \end{array} \right\} \tag{83}$$

where S is the scale factor, and $\Delta\lambda$, ϕ, and ϕ_o are in radians.

The plotting table for a selection of $\phi_o = 45°$ is given as Table 2. A different plotting table is required for each selection of ϕ_o. There is no universal plotting table.

The grid for the projection is given in Figure 3. The central meridian is the only straight line. The other meridians are characteristic curves. The parallels are concentric circles. Only the parallel at ϕ_o is true scale. The y axis is along the central meridian. The x axis is perpendicular to the y axis at latitude ϕ_o.

Even though the Bonne projection is equal area, its use as a military map by France has been the same as the use of conformal projections by the U.S. Over local areas, it gives an adequate representation.

The inverse transformation from Cartesian to geographic follow from Equation 83. Rearrange the equations to permit the summing of the squares of a sine and cosine.

$$\frac{x}{\rho S} = \sin\left(\frac{\Delta\lambda R\cos\phi}{\rho}\right) \tag{84}$$

$$\frac{SR\cot\phi_o - y}{\rho S} = \cos\left(\frac{\Delta\lambda R\cos\phi}{\rho}\right) \tag{85}$$

Square and sum Equations 84 and 85.

$$\sin^2\left(\frac{\Delta\lambda R\cos\phi}{\rho}\right) + \cos^2\left(\frac{\Delta\lambda R\cos\phi}{\rho}\right) = 1$$

$$= \left(\frac{x}{\rho S}\right)^2 + \left(\frac{SR\cot\phi_o - y}{\rho S}\right)^2$$

$$\rho^2 = \frac{1}{S^2}[x^2 + (SR\cot\phi_o - y)^2]$$

TABLE 2
Normalized Plotting Coordinates for the Bonne Projection
$(\phi_o = 45°, \quad \lambda_o = 0°, \quad R \cdot S = 1)$

Tabulated Values: x/y

Longitude	Latitude			
	30°	40°	50°	60°
0°	0.000	0.000	0.000	0.000
	−0.262	−0.087	0.087	0.262
10°	0.151	0.133	0.112	0.087
	−0.253	−0.079	0.094	0.267
20°	0.299	0.265	0.222	0.173
	−0.226	−0.054	0.115	0.282
30°	0.444	0.392	0.329	0.256
	−0.181	−0.014	0.149	0.808
40°	0.582	0.513	0.431	0.336
	−0.120	0.042	0.195	0.343
50°	0.711	0.627	0.526	0.411
	−0.042	0.112	0.254	0.387
60°	0.831	0.731	0.614	0.481
	0.050	0.195	0.324	0.440

$$\rho = \frac{1}{S} \sqrt{x^2 + (SR\cot\phi_o - y)^2} \tag{86}$$

Rearrange Equation 84 to obtain

$$\Delta\lambda = \frac{\rho}{R\cos\phi} \sin^{-1}\left(\frac{x}{\rho S}\right) \tag{87}$$

Equate Equations 80 and 86 and solve for ϕ.

$$R\cot\phi_o - R(\phi - \phi_o) = \frac{1}{S} \sqrt{x^2 + SR\cot\phi_o - y^2}$$

$$\phi = \phi_o + \cot\phi_o - \frac{1}{RS} \sqrt{x^2 + SR\cot\phi_o - y^2} \tag{88}$$

Equations 86, 87, and 88 are used to implement the inverse transformation.

A. Example 6

Given: $\lambda_o = 0°$, $\phi_o = 30°$, $\lambda = 10°$, $\phi = 35°$, $R = 6,371,007$ m, $S = 1:5,000,000$.

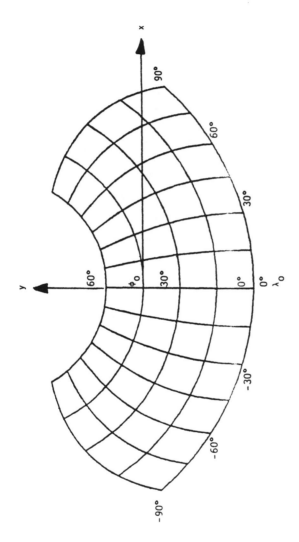

FIGURE 3   Bonne projection.

Find: Cartesian coordinates in the Bonne projection.

$$\Delta\lambda = \lambda - \lambda_o$$

$$= 10 - 0 = 10°$$

$$\rho = R[\cot\phi_o - (\phi - \phi_o)]$$

$$= (6,371,007)\left[\cot 30° - \frac{(35 - 30)}{57.245}\right]$$

$$= (6,371,007)(1.73205 - 0.08727)$$

$$= 10,478,921 \text{ m}$$

$$\theta = \frac{\Delta\lambda R\cos\phi}{\rho}$$

$$= \frac{(10)(6,371,007)\cos 35°}{10,478,921}$$

$$= \frac{(10)(6,371,007)(0.81915)}{10,478,421}$$

$$= 4°.9802$$

$$x = \rho S\sin(4.9802)$$

$$= \left(\frac{10,478,921}{5,000,000}\right)(0.086813)$$

$$= 0.1819 \text{ m}$$

$$y = S[R\cot\phi_o - \rho\cos(4.9802)]$$

$$= \frac{1}{5,000,000}[11,034,896 - 10,439,361]$$

$$= 0.1191 \text{ m}$$

VII. AZIMUTHAL[1,6,7,8]

The azimuthal equal area projection (also called the Lambert zenithal) appears in three variants: the polar, equatorial, and oblique cases. The polar case is derived by two methods. The inverse transformations are then derived. Then, the oblique and equatorial cases are obtained by applying the rotational transformations of Chapter 2.

The azimuthal equidistant projection is obtained by a direct transformation from the spherical model of the Earth onto a plane, without the intermediate step of considering a developable surface.

In the first method of derivation, the azimuthal equal area projection can be obtained from the Albers projection with one standard parallel by setting $\phi_o = 90°$ in that set of plotting equations. By this method, the plane is considered as one limiting case of the cone. When this is done, the constant of the cone equals one, and Equations 34 and 35 reduce to

$$\left. \begin{array}{l} \theta = \Delta\lambda \\[2mm] \rho = R\sqrt{2(1 - \sin\phi)} \end{array} \right\} \tag{89}$$

The Cartesian plotting coordinates, including the scale factor, S, become, from Equation 6, Chapter 1,

$$\left. \begin{array}{l} x = RS\sqrt{2(1 - \sin\phi)}\ \sin\Delta\lambda \\[2mm] y = -RS\sqrt{2(1 - \sin\phi)}\ \cos\Delta\lambda \end{array} \right\} \tag{90}$$

The convention used in this text for all azimuthal projections is that $\Delta\lambda = 0°$ is downward. Thus, the x axis is at $\Delta\lambda = 90°$, and the y axis is at $\Delta\lambda = 180°$.

A plotting table is given as Table 3. A representative grid is given in Figure 4. The meridians are straight lines radiating from the origin. The parallels are concentric circles, unevenly spaced to maintain the equal area quality. Thus, distortion increases as the equator is approached.

The second way to derive the plotting equations is by comparison of the area on the map with the corresponding area on the sphere.[9] The area of the segment of the sphere surrounding the pole, and above the latitude ϕ, is

$$A = 2\pi R(R - R\sin\phi)$$
$$= 2\pi R^2(1 - \sin\phi) \tag{91}$$

This will be transformed into a circle of radius ρ with area

$$A = \pi\rho^2 \tag{92}$$

Equating Equations 91 and 92,

$$\pi\rho^2 = 2\pi R^2(1 - \sin\phi)$$
$$\rho = R\sqrt{2(1 - \sin\phi)}$$

Thus, the second of Equation 89 has been reproduced.

The inverse transformation from Cartesian coordinates to geographic coordinates may be obtained by mathematical manipulation of Equation 90.

TABLE 3
Normalized Plotting Coordinates for the Equal Area
Azimuthal Projection, Polar Case
($\phi_o = 90°$, $\lambda_o = 0°$, $R \cdot S = 1$)

Tabulated Values: x/y

Longitude	Latitude 60°	70°	80°	90°
0°	0.000	0.000	0.000	0.000
	−0.518	−0.347	−0.174	0.000
10°	0.090	0.060	0.030	0.000
	−0.510	−0.342	−0.172	0.000
20°	0.177	0.119	0.060	0.000
	−0.486	−0.326	−0.164	0.000
30°	0.259	0.174	0.087	0.000
	−0.448	−0.301	−0.151	0.000
40°	0.333	0.223	0.112	0.000
	−0.397	−0.266	−0.134	0.000
50°	0.397	0.266	0.134	0.000
	−0.333	−0.233	−0.112	0.000
60°	0.448	0.301	0.151	0.000
	−0.259	−0.174	−0.087	0.000
70°	0.486	0.326	0.164	0.000
	−0.177	−0.119	−0.060	0.000
80°	0.510	0.342	0.172	0.000
	−0.090	−0.060	−0.030	0.000
90°	0.518	0.347	0.174	0.000
	0.000	0.000	0.000	0.000

First, divide the first equation by the second to obtain

$$\frac{\sin\Delta\lambda}{\cos\Delta\lambda} = \frac{x}{-y}$$

$$\Delta\lambda = \tan^{-1}\left(\frac{x}{-y}\right) \tag{93}$$

Equation 93 is used in conjunction with $\lambda = \lambda_o + \Delta\lambda$ to obtain the longitude. The latitude is then obtained from the first of Equation 90. Rearrange this equation to obtain

$$\sqrt{2(1 - \sin\phi)} = \frac{x}{RS\sin\Delta\lambda}$$

$$2(1 - \sin\phi) = \left(\frac{x}{RS\sin\Delta\lambda}\right)^2$$

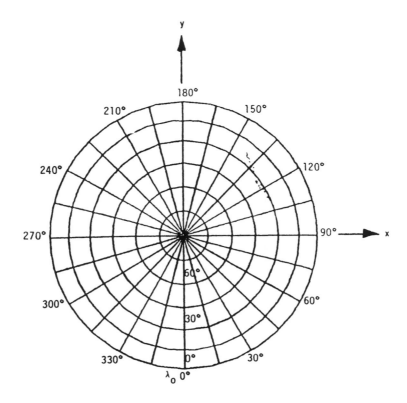

FIGURE 4.   Azimuthal equal area projection, polar case.

$$\sin\phi = 1 - \frac{1}{2}\left(\frac{x}{RS\sin\Delta\lambda}\right)^2$$

$$\phi = \sin^{-1}\left[1 - \frac{1}{2}\left(\frac{x}{RS\sin\Delta\lambda}\right)^2\right] \tag{94}$$

The oblique and equatorial cases can be obtained from the polar case by applying the rotations of Section XII, Chapter 2.

To obtain the oblique variation, write Equations 91 and 92 in the auxiliary coordinate system.

$$\left.\begin{array}{l} x = RS\sqrt{2(1 - \sin h)}\,\sin\alpha \\[2mm] y = RS\sqrt{2(1 - \sin h)}\,\cos\alpha \end{array}\right\} \tag{95}$$

where h corresponds to latitude, and α, to longitude.
From Equations 140 and 141, Chapter 2,

$$\sin h = \sin\phi\sin\phi_p + \cos\phi\cos\phi_p\cos\Delta\lambda \tag{96}$$

$$\tan\alpha = \frac{\sin\Delta\lambda}{\cos\phi_p\tan\phi - \sin\phi_p\cos\Delta\lambda}$$

$$\alpha = \tan^{-1}\left(\frac{\sin\Delta\lambda}{\cos\phi_p\tan\phi - \sin\phi_p\cos\Delta\lambda}\right) \tag{97}$$

Substitute Equations 96 and 97 into Equations 95.

$$\left.\begin{aligned}
x &= R \cdot S \cdot \sqrt{2(1 - \sin\phi\sin\phi_p - \cos\phi\cos\phi_p\cos\Delta\lambda} \\
&\quad \cdot \sin\left[\tan^{-1}\left(\frac{\sin\Delta\lambda}{\cos\phi_p\tan\phi - \sin\phi_p\cos\Delta\lambda}\right)\right] \\
y &= R \cdot S \cdot \sqrt{2(1 - \sin\phi\sin\phi_p - \cos\phi\cos\phi_p\cos\Delta\lambda} \\
&\quad \cdot \cos\left[\tan^{-1}\left(\frac{\sin\Delta\lambda}{\cos\phi_p\tan\phi - \sin\phi_p\cos\Delta\lambda}\right)\right]
\end{aligned}\right\} \tag{98}$$

To obtain the equatorial azimuthal equal area projection, substitute $\phi_p = 0°$ into Equations 98.

$$\left.\begin{aligned}
x &= R \cdot S \cdot \sqrt{1 - \cos\phi\cos\Delta\lambda}\,\sin\left[\tan^{-1}\left(\frac{\sin\Delta\lambda}{\tan\phi}\right)\right] \\
y &= R \cdot S \cdot \sqrt{1 - \cos\phi\cos\Delta\lambda}\,\cos\left[\tan^{-1}\left(\frac{\sin\Delta\lambda}{\tan\phi}\right)\right]
\end{aligned}\right\} \tag{99}$$

Figure 5 gives a representative grid for the oblique case, and Figure 6 gives a grid for the equatorial case. For the oblique case, a tangency of the azimuthal plane was chosen at $\phi_o = 45°$. The y axis is the central meridian. The x axis is perpendicular to the y axis at $\phi_o = 45°$. For the equatorial case, the y axis is again at the central meridian. This time, the x axis is along the equator.

The polar equal area projection is a good means of displaying statistical data when a polar area is a center of interest. The oblique and equatorial projections are useful for the display of statistical data in a region adjacent to the point of tangency of the plane to the sphere.

A. Example 7

Given: $\lambda_o = 0°$, $\lambda = 15°$, $\phi = 80°$ on an azimuthal equal area projection; $R = 6,371,007$ m, $S = 1/5,000,000$.

Find: The Cartesian plotting coordinates.

$$x = R \cdot S\sqrt{2(1 - \sin\phi)}\,\sin\Delta\lambda$$

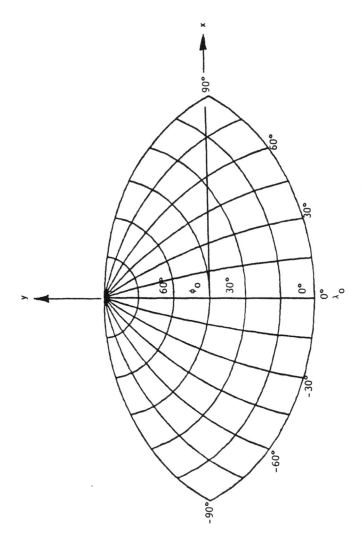

FIGURE 5.   Azimuthal equal area projection, oblique case.

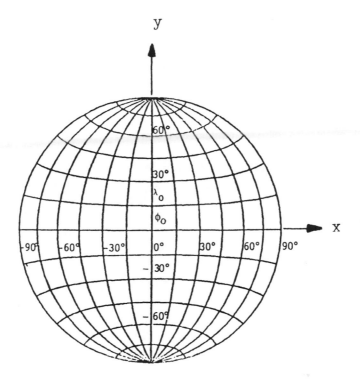

FIGURE 6. Azimuthal equal area projection, equatorial case.

$$= \left(\frac{6,371,007}{5,000,000}\right) \sqrt{2(1 - \sin 80° \, \sin 15°}$$

$$= (1.27420)(0.174312)(0.258819)$$

$$= 0.05749 \text{ m}$$

$$y = R \cdot S \sqrt{2(1 - \sin\phi)} \, \cos\Delta\lambda$$

$$= \left(\frac{6,371,007}{5,000,000}\right)(0.174312)\cos 15°$$

$$= (1.27420)(0.174312)(0.965926)$$

$$= 0.21454 \text{ m}$$

VIII. CYLINDRICAL[1,5,6,7,8]

The cylindrical equal area projection is derived for the regular or equatorial case where the mapping cylinder is parallel to the polar axis of the spherical model of the Earth. In the resulting projection, the meridians and parallels are straight lines, perpendicular to one another. The lines representing the meridians are equally spaced along the equator.

Since the meridians are perpendicular to the equator, and equally spaced, the abscissa is

$$x = R \cdot S \cdot \Delta\lambda \tag{100}$$

where S is the scale factor, and $\Delta\lambda = \lambda - \lambda_o$, in radians. The longitude of the central meridian is λ_o.

It remains to space the parallels so that any area on the projection is equal to the corresponding area on the spherical model of the Earth. This is done by using the differential geometry approach and by a comparison of areas on the sphere and the cylinder. In addition, the inverse transformation is given.

Turning first to the differential geometry approach, the first fundamental quantities for the cylinder are

$$\left.\begin{array}{l} E' = 1 \\[2mm] G' = 1 \end{array}\right\} \tag{101}$$

For the sphere, the first fundamental quantities are

$$\left.\begin{array}{l} e = R^2 \\[2mm] g = R^2\cos^2\phi \end{array}\right\} \tag{102}$$

Substitute Equations 101 and 102 into Equation 100 to obtain the basic transformation relation.

$$R^4\cos^2\phi = (1) \begin{vmatrix} \dfrac{\partial x}{\partial \phi} & \dfrac{\partial x}{\partial \lambda} \\[4mm] \dfrac{\partial y}{\partial \phi} & \dfrac{\partial y}{\partial \lambda} \end{vmatrix}^2 \tag{103}$$

It is now necessary to apply the condition stated in Equation 100. Letting the scale factor equal 1, and taking the partial derivatives,

$$\left.\begin{array}{l} \dfrac{\partial x}{\partial \phi} = 0 \\[4mm] \dfrac{\partial x}{\partial \lambda} = R \end{array}\right\} \tag{104}$$

Substitute Equation 104 into Equation 103 and simplify. Take the positive square root.

$$R^4\cos^2\phi = \begin{vmatrix} 0 & R \\ \dfrac{\partial y}{\partial \phi} & \dfrac{\partial y}{\partial \lambda} \end{vmatrix}^2$$

$$= R^2\left(\frac{\partial y}{\partial \phi}\right)^2$$

$$\frac{\partial y}{\partial \phi} = R\cos\phi \qquad (105)$$

Convert Equation 105 into an ordinary differential equation and integrate.

$$y = R\sin\phi + k \qquad (106)$$

The constant k can be made to be zero by selecting the origin at the equator. Reintroducing the scale factor, the second plotting equation becomes

$$y = R \cdot S\sin\phi \qquad (107)$$

The same result for the y coordinate can be obtained by comparing the area on the sphere with the corresponding area on the cylinder.[9] The area of the zone below latitude ϕ on the sphere is

$$A = 2\pi R^2\sin\phi \qquad (108)$$

The area on a cylinder tangent to this sphere at the equator is

$$A = 2\pi Ry \qquad (109)$$

Equating Equations 108 and 109,

$$2\pi Ry = 2\pi R^2\sin\phi$$

$$y = R\sin\phi$$

This duplicates Equation 106.

The grid resulting from Equations 100 and 107 is given in Figure 7. Table 4 gives the plotting coordinates. On this grid, the central meridian is the y axis. The equator is the x axis. The only true scale line is the equator.

This projection can be used to portray the entire world on a single sheet. However, the distortion becomes intense at higher latitudes, where the pa-

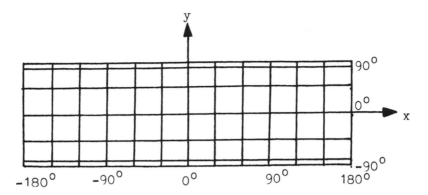

FIGURE 7.   Cylindrical equal area projection.

TABLE 4

**Normalized Plotting Coordinates for the Equal Area Cylindrical
Projection ($\phi_o = 0°$, $\lambda_o = 0°$, $R \cdot S = 1$)**

Latitude or longitude	0°	30°	60°	90°	120°	150°	180°
x	0.000	0.524	1.047	1.571	2.094	2.618	3.142
y	0.000	0.500	0.866	1.000	—	—	—

rallels crowd together. Thus, the projection can be of real service only from
the equator to mid-latitudes. This has limited the usefulness of the projection.

This projection can be made oblique or transverse by applying the rotation
formulas of Chapter 2. If this is done, the area adjacent to the great circle
tangent to the cylinder has a region fairly free of distortion. However, the
oblique and transverse cases of this projection have not been widely used and
are of limited utility.

The inverse transformation from Cartesian to geographic coordinates follows quite easily from Equations 100 and 107.

$$\left.\begin{array}{c} \Delta\lambda = \dfrac{x}{RS} \\[2mm] \phi = \sin^{-1}\left(\dfrac{y}{RS}\right) \end{array}\right\} \tag{110}$$

A. Example 8

Given: $\lambda_o = 10°$, $S = 1/20,000,000$, $\lambda = 35°$, $\phi = 20°$, $R = 6,371,007$
m.

Find: the Cartesian plotting coordinates on the cylindrical equal area
projection.

$$\Delta\lambda = \lambda - \lambda_o$$

$$= 35 - 10 = 25°$$

$$x = RS\Delta\lambda$$

$$= \left(\frac{6,371,007}{20,000,000}\right)\left(\frac{25}{57.295}\right)$$

$$= 0.1390 \text{ m}$$

$$y = RS\sin\phi$$

$$= \left(\frac{6,371,007}{20,000,000}\right)\sin 20°$$

$$= 0.1090 \text{ m}$$

B. Example 9

Given: $\lambda_o = 10°$, $S = 1:20,000,000$, $R = 6,371,007$ m, $x = 0.100$ m, $y = 0.1500$ m.

Find: the geographic coordinates.

$$\Delta\lambda = \frac{x}{RS}$$

$$= \frac{(0.1000)(20,000,000)(57.295)}{6,371,007}$$

$$= 17°.986$$

$$\lambda = \Delta\lambda + \lambda_o$$

$$= 17.986 + 10$$

$$= 27°.986$$

$$\phi = \sin^{-1}\left(\frac{y}{RS}\right)$$

$$= \sin^{-1}\left[\frac{(0.1500)(20,000,000)}{6,371,007}\right]$$

$$= 28°.092$$

IX. SINUSOIDAL[2,5,7,8]

The sinusoidal projection, also called the Sanson-Flamstead projection, is a projection of the entire model of the Earth onto a single map. It gives an adequate whole-world coverage.

The sinusoidal can be derived from the Bonne projection by setting $\phi_o = 0°$. However, we use the differential geometry approach. In this section, both direct and inverse transformations are developed.

From Equation 1, the direct transformation is particularized as

$$eg = E'G' \begin{vmatrix} \dfrac{\partial x}{\partial \phi} & \dfrac{\partial x}{\partial \lambda} \\[2mm] \dfrac{\partial y}{\partial \phi} & \dfrac{\partial y}{\partial \lambda} \end{vmatrix}^2 \tag{111}$$

In this development, the first fundamental quantities for the spherical model for tne Earth are:

$$\left. \begin{aligned} e &= R^2 \\ g &= R^2\cos^2\phi \end{aligned} \right\} \tag{112}$$

On the planar mapping surface, in Cartesian coordinates, the first fundamental quantities are

$$\left. \begin{aligned} E' &= 1 \\ G' &= 1 \end{aligned} \right\} \tag{113}$$

Substitute Equations 112 and 113 into Equation 111.

$$R^2\cos^2\phi = \begin{vmatrix} \dfrac{\partial x}{\partial \phi} & \dfrac{\partial x}{\partial \lambda} \\[2mm] \dfrac{\partial y}{\partial \phi} & \dfrac{\partial y}{\partial \lambda} \end{vmatrix}^2 \tag{114}$$

For this projection, the condition to be imposed is for the y coordinate. Along the central meridian, let

$$y = R\phi \tag{115}$$

Equation 115 is used to develop the necessary partial differential relations.

$$\frac{\partial y}{\partial \phi} = R \left.\begin{array}{c} \\ \\ \end{array}\right\}$$

$$\frac{\partial y}{\partial \lambda} = 0 \qquad \qquad (116)$$

Substitute Equation 116 into Equation 114 and simplify.

$$R^4\cos^2\phi = \begin{vmatrix} \dfrac{\partial x}{\partial \phi} & \dfrac{\partial x}{\partial \lambda} \\[2mm] R & O \end{vmatrix}^2$$

$$= R^2\left(\frac{\partial x}{\partial \lambda}\right) \qquad (117)$$

Taking the positive square root of Equation 117, we have

$$R\cos\phi = \frac{\partial x}{\partial \lambda} \qquad (118)$$

Since ϕ and λ are independent, convert Equation 118 into an ordinary differential equation and integrate.

$$dx = R\cos\phi \, d\lambda$$

$$x = \lambda R\cos\phi + k \qquad (119)$$

Apply the boundary condition that $x = 0$, when $\lambda = \lambda_o$. Then,

$$k = -\lambda_o R\cos\phi$$

Equation 119 becomes, since $\Delta\lambda = \lambda - \lambda_o$

$$x = \Delta\lambda\cos\phi \qquad (120)$$

With the inclusion of the scale factor, S, the plotting equations become, from Equations 115 and 120,

$$x = \Delta\lambda RS\cos\phi \left.\begin{array}{c} \\ \\ \end{array}\right\}$$

$$y = RS\phi \qquad \qquad (121)$$

In Equation 121, $\Delta\lambda$ and ϕ are in radians. A plotting table for the sinusoidal projection is given in Table 5. This is a universal table.

TABLE 5
Normalized Plotting Coordinates for the Sinusoidal Projection
$(\phi_o = 0°, \quad \lambda_o = 0°, \quad R \cdot S = 1)$

Tabulated Values: x/y

Longitude	Latitude			
	0°	30°	60°	90°
0°	0.000	0.000	0.000	0.000
	0.000	0.524	1.047	1.571
30°	0.524	0.453	0.262	0.000
	0.000	0.524	1.047	1.571
60°	1.047	0.907	0.524	0.000
	0.000	0.524	1.047	1.571
90°	1.571	1.360	0.785	0.000
	0.000	0.524	1.047	1.571
120°	2.094	1.814	1.047	0.000
	0.000	0.524	1.047	1.571
150°	2.618	2.267	1.309	0.000
	0.000	0.524	1.047	1.571
180°	3.142	2.721	1.571	0.000
	0.000	0.524	1.047	1.571

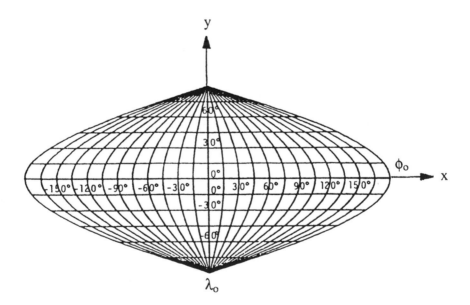

FIGURE 8.   Sinusoidal projection.

Figure 8 is a sinusoidal projection of the Earth. All of the parallels are straight lines. The meridians are sinusoidal curves, with the exception of the central meridian. The central meridian and equator are straight lines of true scale. The y axis is along the central meridian of the projection. The x axis is along the equator, as indicated in the figure. Distortion in this projection is a function of longitude as well as latitude. At the greatest distance from the central meridian, the quadrilaterals become lune shaped. The sinusoidal projection is used in thematic maps. The distortion at extreme latitudes and longitudes is simply ignored.

The inverse transformations from Cartesian to geographical coordinates follow quite simply from Equations 121. These relations are

$$
\left.
\begin{aligned}
\phi &= \frac{y}{RS} \\[2mm]
\Delta\lambda &= \frac{x}{RS\cos\phi}
\end{aligned}
\right\}
\tag{122}
$$

A. Example 10

Given: $\lambda_o = 0°$, $\lambda = 30°$, and $\phi = 30°$; let $S = 1:20,000,000$, $R = 6,371,007$ m.

Find: the Cartesian coordinates in the sinusoidal projection.

$$\Delta\lambda = \lambda - \lambda_o = 30 - 0$$

$$= 30°$$

$$x = \Delta\lambda RS\cos\phi$$

$$= \left(\frac{30}{57.295}\right)\left(\frac{6,371,007}{20,000,000}\right)(0.866025)$$

$$= 0.1444 \text{ m}$$

$$y = RS\phi$$

$$= \left(\frac{6,371,007}{20,000,000}\right)\left(\frac{30}{57.295}\right)$$

$$= 0.1668 \text{ m}$$

$$y = RS\phi$$

$$= \left(\frac{6,371,007}{20,000,000}\right)\left(\frac{30}{57.295}\right)$$

$$= 0.1668 \text{ m}$$

B. Example 11

Given: $\lambda_o = 0°$, $x = 0.1500$ m, $y = 0.1600$ m; $S = 1:20,000,000$, $R = 6,371,007$ m.

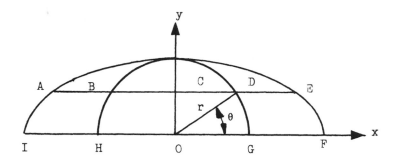

FIGURE 9. Geometry for the Mollweide projection.

Find: ϕ and λ in the sinusoidal projection.

$$\phi = \frac{y}{RS}$$

$$= \frac{(0.1600)(20{,}000{,}000)}{6{,}371{,}007}$$

$$= 0.502275 \text{ radians}$$

$$= 28.778°$$

$$= \frac{x}{RS\cos\phi}$$

$$= \frac{(0.1500)(20{,}000{,}000)}{(6{,}371{,}007)(0.876493)}$$

$$= 0.537236 \text{ radians}$$

$$= 30°.781$$

X. MOLLWEIDE[5,7,8,9]

The Mollweide projection, based on a spherical model of the Earth, is often used in world maps. This projection also goes by the name elliptic or homolographic. The direct and inverse transformations for this projection are produced in the derivations below. The techniques of differential geometry do not apply to the derivation of this projection. The derivation consists of a geometric construction in a plane.

The Mollweide projection of the sphere is derived from the construction in Figure 9. All of the meridians are ellipses. The central meridian is a rectilinear ellipse, or straight line, and the 90° meridians are ellipses of eccentricity zero, or circular arcs. The equator and parallels are straight lines perpendicular to the central meridian.

The main problem in this projection is spacing the parallels so that the property of equivalence of area is maintained. The initial step is to equate the area on the map with radius r, which represents a hemisphere, to the area of a hemisphere on a sphere with radius R.

The area of the circle in Figure 9 centered at 0 is

$$A_1 = \pi r^2 \tag{123}$$

This is to be equal in area to a hemisphere

$$A_1 = 2\pi R^2 \tag{124}$$

Equating Equations 123 and 124,

$$\pi r^2 = 2\pi R^2$$

$$r^2 = 2R^2 \tag{125}$$

$$r = \sqrt{2}\,R \tag{126}$$

The next step is to obtain a variable relationship for latitude ϕ within the context of the geometry given in Figure 9. This requires the introduction of an auxiliary variable θ. The area between latitude ϕ on the sphere and the equator is

$$A = 2\pi R^2 \sin\phi \tag{127}$$

This area is equal to the area AEFI on the figure. For a circle inscribed within an ellipse, where the radius of the circle is one half of the semimajor axis, the area BDGH equals one half of the area AEFI. Consider half of the area BDGH, that is, area CDGO. This area is composed of the triangle OCD and the sector ODG. The area of the triangle is

$$A_A = \frac{1}{2}\,r\sin\phi\, r\cos\theta$$

$$= \frac{r^2}{4}\,\sin2\theta \tag{128}$$

The area of the sector is

$$A_S = \frac{r^2\theta}{2} \tag{129}$$

Equate the spherical area and the map area developed in Equations 127, 128, and 129.

$$2\pi R^2 \sin\phi = 4\left(\frac{1}{2} r^2\theta + \frac{1}{4} r^2\sin2\theta\right)$$

$$\pi R^2 \sin\phi = r^2\theta + \frac{1}{2} r^2\sin2\theta \tag{130}$$

Substitute Equation 125 into Equation 130.

$$\pi R^2 \sin\phi = 2R^2\theta + R^2\sin2\theta$$

$$\pi\sin\phi = 2\theta + \sin2\theta \tag{131}$$

We are now faced with a transcendental equation to be solved for θ. This requires a technique of numerical analysis. We apply the Newton-Raphson method.[3] Write Equation 131 as

$$f(\theta) = \pi\sin\phi - 2\theta - \sin2\theta = 0 \tag{132}$$

Differentiating Equation 132,

$$f'(\theta) = -2 - 2\cos2\theta \tag{133}$$

The iterative solution of Equation 131 for θ as a function of ϕ is

$$\theta_{n+1} = \theta_n - \frac{f(\theta_n)}{f'(\theta_n)} \tag{134}$$

Substitute Equations 132 and 133 into Equation 134.

$$\theta_{n+1} = \theta_n + \frac{\pi\sin\phi - 2\theta_n - \sin2\theta_n}{2 + 2\cos2\theta_n}$$

where θ_n is in radians. This has a rapid convergence if the initial guess for θ is the given value of ϕ. While θ and ϕ are equal at 0 and 90°, the difference between the two is nearly 10° in the vicinity of $\phi = 45°$.

Once θ is found, the mapping equations quickly follow from Figure 9.

$$\left.\begin{array}{l} y = r \cdot S \sin\theta \\[2mm] x = \dfrac{\Delta\lambda}{90} r \cdot S\cos\theta \end{array}\right\} \tag{135}$$

S is the scale factor, and $\Delta\lambda = \lambda - \lambda_o$ is in degrees. The central meridian has longitude λ_o. Substituting Equation 126 into Equation 135, we arrive at the plotting equations

$$\left.\begin{array}{l} y = \sqrt{2}\,R \cdot S\sin\theta \\[2mm] x = \dfrac{\Delta\lambda\sqrt{2}}{90}\,R \cdot S \cdot \cos\theta \end{array}\right\} \qquad (136)$$

A plotting table for the Mollweide projection is given as Table 6. Since the equator is taken as $\phi_o = 0°$ in all uses of this projection, a single, universally applied plotting table is possible.

A representative grid for this projection appears in Figure 10. The y axis is along the central meridian. The x axis is along the equator. The only true scale lines in the projection are along the straight line parallels at latitudes $\pm 40°44'$.

The distortion towards the poles is not as great as in the sinusoidal projection, but it is more noticeable than in the Hammer-Aitoff projection. The chief use of the Mollweide projection is for geographical illustrations relating to area, where shape or other distortions are not disturbing. This projection gives a reliable representation of statistical data.

The inverse transformation from Cartesian coordinates to geographical coordinates on the sphere follows from Equations 131 and 136.

$$\theta = \sin^{-1}\left(\frac{y}{\sqrt{2}\,RS}\right) \qquad (137)$$

$$\phi = \sin^{-1}\left(\frac{2\theta + \sin2\theta}{\pi}\right) \qquad (138)$$

$$\Delta\lambda = \frac{90x}{\sqrt{2}\,RS\cos\theta} \qquad (139)$$

A. Example 12

Given: $\lambda_o = 10°$, $\phi = 20°$, $\lambda = 35°$, $R = 6,371,007$ m, $S = 1/20,000,000$.

Find: The Cartesian coordinates on the Mollweide projection; let the result of the Newton-Raphson iteration be $\theta = 16°$.

$$\Delta\lambda = \lambda - \lambda_o$$

$$= 35 - 10 = 25°$$

$$x = \frac{\Delta\lambda}{90}\sqrt{2}\,R \cdot S\cos\phi$$

TABLE 6
Normalized Plotting Coordinates for the Mollweide Projection
($\phi_o = 0°$, $\lambda_o = 0°$, $R \cdot S = 1$)

Tabulated Values: x/y

Longitude	Latitude 0°	30°	60°	90°
0°	0.000	0.000	0.000	0.000
	0.000	0.571	1.078	1.414
30°	0.471	0.431	0.305	0.000
	0.000	0.571	1.078	1.414
60°	0.943	0.862	0.610	0.000
	0.000	0.571	1.078	1.414
90°	1.414	1.294	0.915	0.000
	0.000	0.571	1.078	1.414
120°	1.886	1.725	1.220	0.000
	0.000	0.571	1.078	1.414
150°	2.357	2.156	1.525	0.000
	0.000	0.571	1.078	1.414
180°	2.828	2.587	1.830	0.000
	0.000	0.571	1.078	1.414

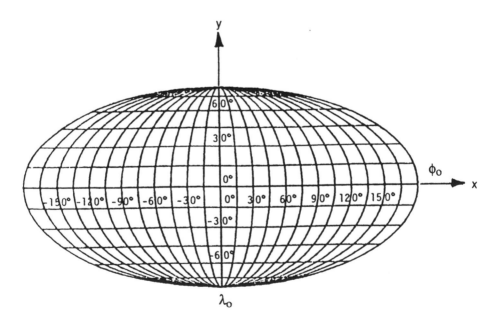

FIGURE 10.   Mollweide projection.

$$= \left(\frac{25}{90}\right) \frac{\sqrt{2}\ (6,371,007)}{20,000,000}\ \cos 16°$$

$$= 0.12029 \text{ m}$$

$$y = \sqrt{2}\ R \cdot S \sin\theta$$

$$= \frac{\sqrt{2}\ (6,371,007)}{20,000,000}\ \sin 16°$$

$$= 0.12417 \text{ m}$$

B. Example 13

Given: a Mollweide projection with $\lambda_o = 10°$, $S = 1 = 20,000,000$, $x = -0.05 \text{ m}$, $y = 0.02 \text{ m}$.

Find: ϕ and λ.

$$\theta = \sin^{-1}\left(\frac{y}{2RS}\right)$$

$$= \sin^{-1}\left[\frac{(0.02)(20,000,000)}{\sqrt{2}\ (6,371,007)}\right]$$

$$= 2°.5445$$

$$\phi = \sin^{-1}\left[\frac{2\theta + \sin 2\theta}{\pi}\right]$$

$$= \sin^{-1}\left[\frac{0.088821 + 0.088703}{\pi}\right]$$

$$= 3°.2394$$

$$\Delta\lambda = \frac{90x}{\sqrt{2}\ RS\cos\theta}$$

$$= \frac{-(90)(0.05)(20,000,000)}{\sqrt{2}\ (6,371,007)\cos(2°.5445)}$$

$$= -9.9988$$

$$\lambda = \Delta\lambda + \lambda_o$$

$$= -9.9988 + 10$$

$$= 0°.0012$$

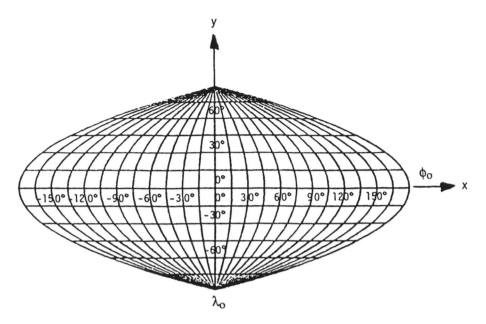

FIGURE 11.   Geometry for the parabolic projection.

XI. PARABOLIC[2,5,9]

The parabolic projection is another equal area projection suitable for world maps. The parabolic projection, also called the Craster projection, is derived by consideration of the geometry on a plane surface. The inverse transformation follows from the direct plotting coordinates.

The parabolic projection is based on the geometry of Figure 11. Let the equator be four arbitrary units of length. Then, the central meridian is two units long. The first step is to obtain the relation for one fourth of the total area on the figure enclosed by the outer meridians. Note the position of the coordinate system in Figure 11 on which the derivation is based.

Consider the area of interest in Figure 11, bounded by an outer meridian, the central meridian, and the equator. The outer meridian is taken to be the parabola, $y^2 = x/2$, with its vertex at $(0,0)$. The mapping criterion requires that one quarter of the area on the sphere is equivalent to the shaded area between $x = 0$ and $x = 2$. Thus, one half of the zone between the equator, and a given parallel, ϕ, is

$$A = \int_0^y (2 - x)dy$$

$$= \int_0^y (2 - 2y^2)dy$$

$$= 2y - \frac{2}{3} y^3 \Big|_0^y$$

$$= 2y - \frac{2}{3} y^3 \tag{140}$$

The next step is to find a relation between the area on the map and the corresponding area on the spherical model of the Earth with radius R. The total area of the sphere is $4\pi R^2$. Substituting this value, and y = 1, into Equation 140,

$$\pi R^2 = 4/3$$

$$R = \sqrt{\frac{4}{3\pi}}$$

$$= 0.651470 \text{ units} \tag{141}$$

Thus, one map unit is equal to 1.53499 R.

Then, relate the map coordinate to latitude. The area of a zone on the sphere from the equator to latitude ϕ is $2\pi R^2 \sin\phi$. Half of this zone is then

$$A = \pi R^2 \sin\phi \tag{142}$$

Equate Equations 140 and 142.

$$2y - \frac{2}{3} y^3 = \pi R^2 \sin\phi \tag{143}$$

Substitute Equation 141 into Equation 143.

$$2y - \frac{2}{3} y^3 = \frac{4}{3} \sin\phi$$

$$y^3 - 3y + 2\sin\phi = 0$$

A solution of this equation is

$$y = 2\sin\phi/3 \tag{144}$$

which can be verified by substitution. A scale factor, S, and radius, R, may be introduced into Equation 144 to obtain the ordinate, recalling also that one map unit equals 1.53499 R.

$$y = 3.06998 \cdot S \cdot R\sin\phi/3 \tag{145}$$

The abscissa is obtained by the following development. The length of a parallel between the central meridian and the outer meridian is given by

$$d = 2 - 2y^2 \qquad (146)$$

Substitute Equation 144 into Equation 146 to obtain the length as a function of latitude.

$$d = 2(1 - 4\sin^2\phi/3)$$

$$= 2\left(1 + 2\cos\frac{2\phi}{3} - 2\right)$$

$$= 2\left(2\cos\frac{2\phi}{3} - 1\right) \qquad (147)$$

The parallels are divided proportionally for the intersections with the meridians. From Equation 147, and including the scale factor, S, the radius, R, and the relation between the map unit and R,

$$x = 1.53499 \, \frac{\Delta\lambda}{180} \cdot d \cdot S \cdot R$$

$$= 1.53499 \, \frac{\Delta\lambda}{90} \, SR\left(2\cos\frac{2\phi}{3} - 1\right) \qquad (148)$$

In Equation 148, $\Delta\lambda = \lambda - \lambda_o$ is the difference in longitude between the given meridian and the central meridian, in degrees.

Equations 145 and 148 have been evaluated to produce Table 7. Since the equator defines ϕ_o, a single table applies to all uses of this projection.

An example of the parabolic grid is given in Figure 12. For this projection, the y axis coincides with the central meridian, and the x axis is along the equator. The parallels in this projection are straight lines parallel to a straight-line equator. The meridians are parabolic arcs. The central meridian can be considered as a rectilinear parabola.

Since this is an equal area projection, its use is for statistical representation. There is still considerable distortion in angles and shapes. However, the distortion is less than in the Mollweide projection because the meridians and parallels do not intersect at such acute angles. The greatest distortion is farthest from the central meridian at high latitudes.

The inverse transformation for the parabolic projection follows from Equations 145 and 148.

TABLE 7
Normalized Plotting Coordinates for the Parabolic Projection
$(\phi_o = 0°, \quad \lambda_o = 0°, \quad R \cdot S = 1)$

Tabulated Values: x/y

Longitude	0°	30°	60°	90°
0°	0.000	0.000	0.000	0.000
	0.000	0.533	1.050	1.535
30°	0.512	0.450	0.272	0.000
	0.000	0.533	1.050	1.535
60°	1.023	0.900	0.545	0.000
	0.000	0.533	1.050	1.535
90°	1.535	1.350	0.817	0.000
	0.000	0.533	1.050	1.535
120°	2.047	1.800	1.089	0.000
	0.000	0.533	1.050	1.535
150°	2.558	2.250	1.361	0.000
	0.000	0.533	1.089	1.535
180°	3.070	2.700	1.634	0.000
	0.000	0.533	1.089	1.535

(Latitude)

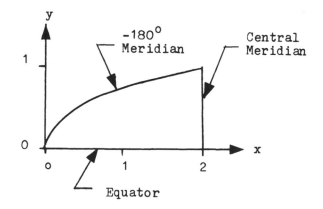

FIGURE 12. Parabolic projection.

$$\phi = 3\sin^{-1}\left(\frac{y}{3.06998 \cdot R \cdot S}\right)$$

$$\Delta\lambda = \frac{90x}{1.53499R \cdot S\left(2\cos\frac{2\phi}{3} - 1\right)} \tag{149}$$

A. Example 14

Given: $\lambda_o = 10°$, $\lambda = 35°$, $\phi = 20°$, R = 6,371,007 m, S = 1:20,000,000.

Find: the Cartesian plotting coordinates in the parabolic projection.

$$\Delta\lambda = \lambda - \lambda_o$$

$$= 35 - 10 = 25°$$

$$x = 1.53499 \frac{\Delta\lambda}{90} S \cdot R\left(2\cos\frac{2\phi}{3} - 1\right)$$

$$= (1.53499)\left(\frac{25}{90}\right)\left(\frac{6,371,007}{20,000,000}\right)(1.94609 - 1)$$

$$= (1.53499)\left(\frac{25}{90}\right)\left(\frac{6,371,007}{20,000,000}\right)(0.94609)$$

$$= 0.1285 \text{ m}$$

$$y = 3.06998S \cdot R\sin\phi/3$$

$$= (3.06998)\left(\frac{6,371,007}{20,000,000}\right)(0.11609)$$

$$= 0.1135 \text{ m}$$

B. Example 15

Given: $\lambda_o = 10°$, S = 1:20,000,000, R = 6,371,007 m, X = -0.05 m, y = 0.02 m

Find: ϕ and λ.

$$\phi = 3\sin^{-1}\left(\frac{y}{3.06998SR}\right)$$

$$= 3\sin^{-1}\left[\frac{(0.02)(20,000,000)}{(3.06998)(6,371,007)}\right]$$

$$= 3\sin^{-1}(0.02045)$$

$$= 3°.515$$

$$\Delta\lambda = \frac{90x}{1.53499RS\left(2\cos\frac{2\phi}{3} - 1\right)}$$

$$= \frac{(90)(-0.05)(20,000,000)}{(1.53499)(6,371,007)(1.998327 - 1)}$$

$$= -9°.2184$$

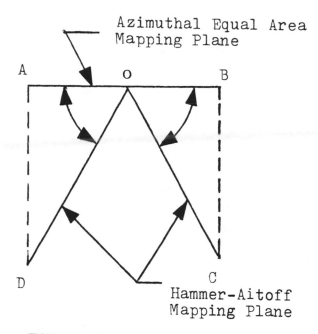

FIGURE 13. Geometry for the Hammer-Aitoff projection.

$$\lambda = \Delta\lambda + \lambda_o$$
$$= -9.2184 + 10$$
$$= 0°.7816$$

XII. HAMMER-AITOFF[7,8]

The Hammer-Aitoff projection is yet another equal area worldwide projection based on a spherical model of the Earth. The direct transformation is developed by a simple mathematical substitution into the plotting equations for the equatorial azimuthal equal area projection of Section VII.

Figure 13 demonstrates the relation of the mapping planes for the Hammer-Aitoff projection and the equatorial azimuthal equal area projection. In this figure, we are looking upon the edges of the mapping planes, which appear as straight lines. Since the angle between the planes is 60°, DO = 2AO, and OB = 2OC. Thus, the total length DO plus OC is the entire equator, as AB is half of the equator. It is assumed that for the Hammer-Aitoff projection, the total map of the sphere is obtained by unfolding DO and OC into a plane DOC, with O as the position of the central meridian. In this projection, the ordinate is not modified from a comparable point on the azimuthal equal area projection. This is easily accomplished by substituting $\lambda/2$ for λ in Equations 99. The resulting plotting equations for the Hammer-Aitoff projection are as follows:

TABLE 8
Normalized Plotting Coordinates for the Hammer-Aitoff Projection
$(\phi_o = 0°, \quad \lambda_o = 0°, \quad R \cdot S = 1)$

Tabulated Values: x/y

Longitude	Latitude			
	0°	30°	60°	90°
0°	0.000	0.000	0.000	0.000
	0.000	0.518	1.000	1.414
30°	0.522	0.478	0.300	0.000
	0.000	0.522	1.006	1.414
60°	1.035	0.926	0.591	0.000
	0.000	0.534	1.023	1.414
90°	1.531	1.364	0.860	0.000
	0.000	0.557	1.053	1.414
120°	2.000	1.772	1.095	0.000
	0.000	0.591	1.095	1.414
150°	2.435	2.138	1.285	0.000
	0.000	0.639	1.152	1.414
180°	2.828	2.449	1.414	0.000
	0.000	0.707	1.225	1.414

$$x = 2RS\sqrt{2(1 - \cos\phi\cos\Delta\lambda/2)} \sin\left[\tan^{-1}\left(\frac{\sin\Delta\lambda/2}{\tan\phi}\right)\right]$$

$$y = RS\sqrt{2(1 - \cos\phi\cos\Delta\lambda/2)} \cos\left[\tan^{-1}\left(\frac{\sin\Delta\lambda/2}{\tan\phi}\right)\right] \quad (149)$$

Evaluation of Equation 149 gives plotting Table 8. Since, $\phi_o = 0$, or the equator always defines the origin, a universal plotting table is possible.

In the Hammer-Aitoff projection, the sphere is represented within an ellipse, with the semimajor axis twice the length of the seminor axis. In this respect, it is similar to the Mollweide projection. However, in the Hammer-Aitoff projection, the parallels are curved lines, rather than straight. This is represented in Figure 14. In the figure, the central meridian and the equator are the only straight lines in the grid. The rest of the meridians and parallels are curves. The curvature of the parallels with respect to the meridians is such that there is less angular distortion than appears at higher latitudes and more distant longitudes in the Mollweide projection. Note in the figure that the y axis coincides with the central meridian, and the x axis is along the equator.

The Hammer-Aitoff projection is a useful projection to display statistical data for thematic mapping.

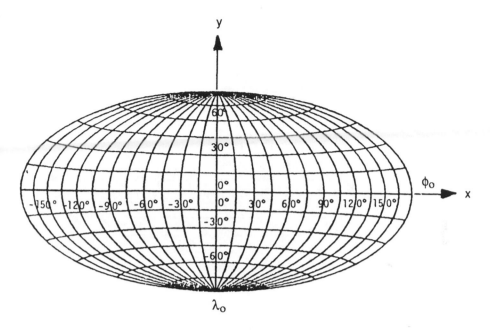

FIGURE 14.   Hammer-Aitoff projection.

A. Example 16

Given: $\lambda_o = 10°$, $\lambda = 35°$, $\phi = 20°$, $R = 6,371,007$ m, $S = 1:20,000,000$.

Find: the Cartesian coordinates in the Hammer-Aitoff projection.

$$x = 2RS\sqrt{2(1 - \cos\phi\cos\Delta\lambda/2)} \, \sin\left[\tan^{-1}\left(\frac{\sin\Delta\lambda/2}{\tan\phi}\right)\right]$$

$$y = RS\sqrt{2(1 - \cos\phi\cos\Delta\lambda/2)} \, \cos\left[\tan^{-1}\left(\frac{\sin\Delta\lambda/2}{\tan\phi}\right)\right]$$

$$\Delta\lambda = \lambda - \lambda_o$$

$$= 35 - 10 = 25°$$

$$\frac{\Delta\lambda}{2} = 12°.5$$

$$\sqrt{2(1 - \cos20°\cos12°.5)} = \sqrt{2[1 - (0.93969)(0.97630)]}$$

$$= \sqrt{2(1 - 0.91742)}$$

$$= \sqrt{0.16517}$$

$$= 0.40641$$

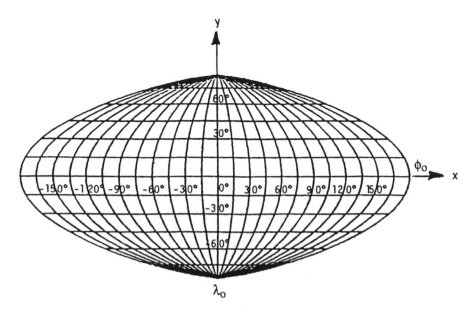

FIGURE 15.   Boggs eumorphic projection.

$$\tan^{-1}\left(\frac{\sin 12°.5}{\tan 20°}\right) = \tan^{-1}\left(\frac{0.21644}{0.36397}\right)$$

$$= 30°.738$$

$$x = (2)\left(\frac{6,371,007}{20,000,000}\right)(0.40641)(0.51112)$$

$$= 0.1322 \text{ m}$$

$$y = \left(\frac{6,371,007}{20,000,000}\right)(0.40641)(0.85951)$$

$$= 0.1113 \text{ m}$$

XIII. BOGGS EUMORPHIC[9]

The Boggs eumorphic projection in Figure 15 is essentially an arithmetic average of the sinusoidal and Mollweide projections. The spacing of the straight line parallels along the central meridian is accomplished by averaging the comparable spacing on the sinusoidal and Mollweide projections. Then, the length of the parallel is needed. This is obtained by requiring that the area between the equator, central meridian, and a selected meridian and parallel on the Boggs eumorphic be the same as the comparable area on the Mollweide or sinusoidal projections. The best approach is to construct a table at a fine enough interval for practical work, and then use a second-order interpolation scheme to obtain needed x coordinates.

TABLE 9
Normalized Plotting Coordinates for the Boggs Eumorphic Projection
($\phi_o = 0°$, $\lambda_o = 0°$, $R \cdot S = 1$)

Tabulated Values: x/y

Longitude	Latitude			
	0°	30°	60°	90°
0°	0.000	0.000	0.000	0.000
	0.000	0.548	1.062	1.492
30°	0.498	0.442	0.284	0.000
	0.000	0.548	1.062	1.492
60°	0.995	0.884	0.567	0.000
	0.000	0.548	1.062	1.492
90°	1.492	1.327	0.850	0.000
	0.000	0.548	1.062	1.492
120°	1.990	1.770	1.134	0.000
	0.000	0.548	1.062	1.492
150°	2.488	2.212	1.417	0.000
	0.000	0.548	1.062	1.492
180°	2.985	2.654	1.700	0.000
	0.000	0.548	1.062	1.492

A widely used alternative is to approximate the projection by averaging both the x coordinate and the y coordinate. In this scheme, to obtain the plotting coordinates for the Boggs eumorphic projection, evaluate the projection equations for the sinusoidal and Mollweide projections at each geographic location. Let $x_m(\phi, \lambda)$ and $y_m(\phi, \lambda)$ be the coordinates calculated for the Mollweide projection. And, let $x_s(\phi, \lambda)$ and $y_s(\phi, \lambda)$ be the coordinates for the sinusoidal projection for the same point. Of course, the same scale must be used. The plotting equations are simply

$$\left. \begin{array}{l} x = \dfrac{x_m + x_s}{2} \\[2mm] y = \dfrac{y_m + y_s}{2} \end{array} \right\} \qquad (150)$$

Of course, in using this method, the strict adherence to the equal area property is lost. However, for most cases, an acceptable approximation results. Table 9 contains the plotting table for the Boggs eumorphic projection, using the approximation. The resulting grid in Figure 15 is less pointed at the north and south poles than is the sinusoidal projection. Angular distortion at higher latitudes is less than in the sinusoidal projection. Along the equator, shape is better preserved in the Boggs eumorphic projection than in the Mollweide projection. In the figure, the x axis for the plotting equations is along the

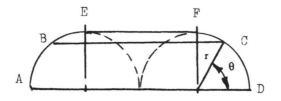

FIGURE 16.   Geometry for the Eckert IV projection.

equator. The y axis is along the central meridian. The eumorphic projection is recommended for thematic mapping.

XIV. ECKERT IV[7,8]

The Eckert IV projection is the last of the equal area world map projections to be considered in this text. Of the six projections developed by Eckert, the Eckert IV projection has been the most popular in the U.S. This projection is defined in a plane and is based on a spherical model of the Earth.

The derivation of the projection follows from the geometric construction of Figure 16. The derivation itself is very similar to that for the Mollweide projection.

The derivation begins by equating areas on the spherical model of the Earth with radius R, and on the map, defined by radius r. The object is to find a relation between R and r, as shown in Figure 16.

The area of the hemisphere is

$$A_1 = 2\pi R^2 \tag{151}$$

The area north of the equator on the Eckert projection is, from Figure 16,

$$A_1 = 2r^2 + \frac{\pi r^2}{2} \tag{152}$$

Equating Equations 151 and 152,

$$2\pi R^2 = 2r^2 + \frac{\pi r^2}{2}$$

$$= r^2 \left(2 + \frac{\pi}{2} \right) \tag{153}$$

Thus, $r = 1.32650$ R.

The next step is to define area as a function of latitude. The area on the hemisphere below latitude ϕ is

$$A_2 = 2\pi R^2 \sin\phi \tag{154}$$

The corresponding area on the projection, including a rectangle, two sectors, and two triangles, is

$$A_2 = 2r \cdot r\sin\theta + \frac{2}{2} r\sin\theta \cdot r\cos\theta + \frac{2}{2} r^2\theta$$

$$= r^2(2\sin\theta + \sin\theta\cos\theta + \theta) \tag{155}$$

Equating Equations 154 and 155,

$$2\pi R^2 \sin\phi = r^2(2\sin\theta + \sin\theta\cos\theta + \theta) \tag{156}$$

Equate Equations 153 and 156 to obtain a relation between ϕ and θ.

$$r^2\left(2 + \frac{\pi}{2}\right)\sin\phi = r^2(2\sin\theta + \sin\theta\cos\theta + \theta)$$

$$\left(2 + \frac{\pi}{2}\right)\sin\phi = 2\sin\theta + \sin\theta\cos\theta + \theta \tag{157}$$

Equation 157 is a transcendental equation to be solved for θ as a function of ϕ. Again, The Newton-Raphson method may be employed to obtain this solution.[3] The numerical procedure is rapidly convergent if $\theta = \phi$ is used to initiate the iteration. The difference between ϕ and θ is not as great as in the Mollweide projection. While θ and ϕ are equal at 0 and 90°, the difference between the two is nearly 5° in the vicinity of $\phi = 45°$.

For use in the Newton-Raphson iteration, write Equation 157 as

$$f(\theta) = \left(2 + \frac{\pi}{2}\right)\sin\phi - 2\sin\theta - \sin\theta\cos\theta - \theta$$

Then,

$$f'(\theta) = -2\cos\theta(1 + \cos\theta)$$

Once a value for θ has been obtained, the Cartesian plotting coordinates follow from the geometry of Figure 16. Recall the necessity of substituting the relation between r and R. The x coordinate is

$$x = 1.32650SR(1 + \cos\theta)\frac{\Delta\lambda}{\pi}$$

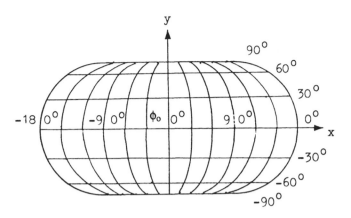

FIGURE 17.   Eckert IV projection.

if $\Delta\lambda$ is in radians. Simplifying, this becomes

$$x = 0.42224 \cdot S \cdot R(1 + \cos\theta)\Delta\lambda \qquad (158)$$

The y coordinate is

$$y = 1.32650 \cdot S \cdot R\sin\theta \qquad (159)$$

Equations 158 and 159 produce the grid of Figure 17. In this grid, the equator, parallels, and central meridian are straight lines. All other meridians are elliptical arcs. The spacing of the parallels decreases with latitude in order to maintain equality of area. The central meridian is one half the length of the equator. The poles are represented by straight lines one half the length of the equator. The only true scale lines are the parallels at $\pm 40°30'$. Again, in this projection, the y axis coincides with the central meridian, and the x axis is along the equator. Table 10 is a universal plotting table for this projection.

XV. INTERRUPTED PROJECTIONS

Interrupted projections of the sphere are a means of reducing maximum distortion at the expense of continuity in a world map. This procedure has been effectively applied to the sinusoidal, Mollweide, parabolic, Boggs eumorphic, and Eckert IV projections. For each projection, the plotting equations developed in the previous sections are applicable.

The procedure to follow in applying the equations in each case is as follows. Certain meridians are chosen by the user as reference meridians. Then other meridians are chosen where the breaks in continuity are to occur. It is not necessary for a reference meridian to continue through both the

TABLE 10
Normalized Plotting Coordinates for the Eckert IV Projection
$$(\phi_o = 0°, \quad \lambda_o = 0°, \quad R \cdot S = 1)$$

Tabulated Values: x/y

Longitude	Latitude			
	0°	30°	60°	90°
0°	0.000	0.000	0.000	0.000
	0.000	0.603	1.097	1.326
30°	0.442	0.418	0.345	0.221
	0.000	0.603	1.097	1.326
60°	0.884	0.836	0.691	0.442
	0.000	0.603	1.097	1.326
90°	1.326	1.254	1.036	0.663
	0.000	0.603	1.097	1.326
120°	1.769	1.672	1.382	0.884
	0.000	0.603	1.097	1.326
150°	2.211	2.090	1.727	1.105
	0.000	0.603	1.097	1.326
180°	2.653	2.508	2.072	1.326
	0.000	0.603	1.097	1.326

Northern and Southern Hemispheres. One can choose a half meridian in either hemisphere.

The parallels are spaced in the same way as in the projection chosen. The difference comes in the method of handling the spacing of the meridians. Each reference meridian becomes the axis for a local coordinate system. A point with a longitude between that reference meridian and the nearest break longitude is mapped with respect to that local coordinate system. If the longitude of the point is greater than the break longitude, the point is mapped in the adjacent coordinate system.

Figure 18 gives the interrupted sinusoidal projection as an example. Note that the reference meridians are straight lines, as is the equator. Distortion, which is always greatest at the farthest longitude from the reference meridian, is significantly decreased.

Maps using interrupted projections are useful for statistical representation of data. Thus, the discontinuities cannot create excessive problems. The breaks are chosen to occur in regions of lesser interest to the user in order to adequately portray regions of greater interest. For example, in oceanography, if a continuous map of the oceans of the world is desired, the breaks would appear in the land masses. Conversely, if land masses were of primary interest, the breaks would appear in the oceans.

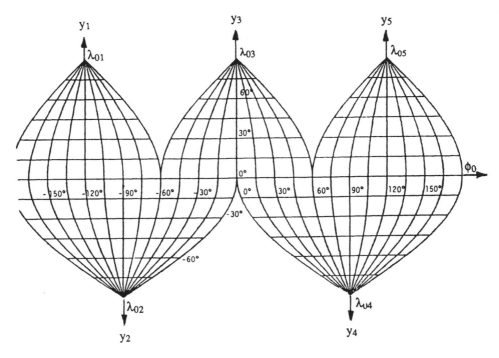

FIGURE 18.   Interrupted sinusoidal projection.

A. Example 17

Consider an interrupted sinusoidal projection in which the reference meridians are defined as follows. (Figure 18)

Southern Hemisphere

Region 1.   Reference meridian at 90°W, breaks at 180°W and 0°.
Region 2.   Reference meridian at 90°E, breaks at 180°E and 0°.

Northern Hemisphere

Region 3.   Reference meridian at 120°W, breaks at 180°W and 60°W.
Region 4.   Reference meridian at 0°, breaks at 60°W and 60°E.
Region 5.   Reference meridian at 120°E, breaks at 60°E and 180°E.

First, consider a point with latitude 30°S and longitude 45°W. This is in Region 1.

$$\Delta\lambda = -45° - 90°$$

$$= 45°$$

The formulas for the sinusoidal projection are to be evaluated for $\phi = -30°$ and $\Delta\lambda = 45°$ to obtain Cartesian coordinates. These coordinates are plotted with reference to the southern axis defined by a straight line at $\lambda = -90°$ and a straight line at $\phi = 0°$.

Second, consider a point at 30°E and 30°N. This is in Region 4.

$$\Delta\lambda = 30° - 0°$$
$$= 30°$$

The formulas for the sinusoidal projection are to be evaluated for $\phi = 30°$ and $\Delta\lambda = 30°$. These coordinates are plotted with reference to the northern axis defined by a straight line at $\lambda = 0°$ and a straight line at $\phi = 0°$.

It remains to locate the reference axes on the projection. This is done by locating the abscissa of the axes of the respective region with reference to the map origin. Taking $\lambda_o = 0°$ as the location of the map origin, the mapping abscissa is the local abscissa minus the abscissa of the particular reference axis.

REFERENCES

1. **Adams, O. S.**, General Theory of Equivalent Projections, Spec. Publ. 236, U.S. Coast and Geodetic Survey, U.S. Gov't. Printing Office, Washington, D.C., 1945.
2. **Deetz, C. H. and Adams, O. S.**, Elements of Map Projection, Spec. Publ. 68, U.S. Coast and Geodetic Survey, U.S. Gov't. Printing Office, Washington, D.C., 1944.
3. **Hildebrand, F. B.**, *Introduction to Numerical Analysis*, McGraw-Hill, New York, 1956.
4. **Maling, D. H.**, *Coordinate Systems and Map Projections*, George Philip & Son, 1973.
5. **McBryde, F. W. and Thomas, P. D.**, Equal Area Projections for World Statistical Maps, Spec. Publ. 245, U.S. Coast and Geodetic Survey, U.S. Gov't. Printing Office, Washington, D.C., 1949.
6. **Richardus, P. and Adler, R. K.**, *Map Projections for Geodesists, Cartographers, and Geographers*, North Holland, Amsterdam, 1972.
7. **Snyder J. P.**, Map Projections used by the U.S. Geological Survey, *U.S. Geol. Surv. Bull.*, U.S. Gov't. Printing Office, Washington, D.C., 1532, 1983.
8. **Snyder, J. P.**, Map projections — a working manual, U.S. Geol. Surv. Prof. Pap., U.S. Gov't. Printing Office, Washington, D.C., 1395, 1987.
9. **Steers, J. A.**, *An Introduction to the Study of Map Projections*, University of London, 1962.

Chapter 5

CONFORMAL PROJECTIONS

I. INTRODUCTION

Conformal projections constitute another important class of map projections. One important characteristic of conformal projections is that the shape of an area is locally maintained during the transformation from the model of the Earth to the mapping surface. Thus, locally, angles are maintained. As a result, the orthogonal system of parallels and meridians on the sphere or spheroid appear as an orthogonal system on the map.

In this chapter, the general procedure for the transformation is discussed. This is followed by an introduction to the conformal sphere, which may be used as intermediate stage in the transformation from the reference spheroid to the map.

The conformal projections to be considered are the Mercator, the Lambert conformal, and the stereographic. Three variations of the Mercator are discussed: the regular (or equatorial), the oblique, and the transverse. The Lambert conformal projection is represented by the one- and two-standard parallel cases. The polar, equatorial, and oblique versions of the stereographic projection are derived. The Mercator, the Lambert conformal, and the stereographic projections are only applicable to limited areas on the model of the Earth for any one map. While there are conformal world maps showing the whole world with or without singular points, these are out of the mainstream of practical use. Thus, no conformal world maps are considered.

The usefulness of the conformal projections is displayed by two important applications: the state plane coordinate system, and the military grid system. A set of approximate equations are given for the direct and inverse transformations using the Lambert conformal and transverse Mercator projections for the state plane coordinate system. This is followed by a development for the universal transverse Mercator grid and the universal polar stereographic grid of the Military Grid System.

II. GENERAL PROCEDURE[12]

The projections of this chapter are predicated on the condition of conformality of Section XI, Chapter 2, which is repeated below.

$$\frac{E}{e} = \frac{G}{g} \tag{1}$$

In this equation, the fundamental quantities e and g refer to the spherical or spheroidal model of the Earth and are functions of ϕ and λ. The fundamental quantities E and G refer to the surface on which the transformation is to be mapped. In this chapter, these fundamental quantities are functions of the Cartesian coordinates x and y for a plane or cylinder, of the polar coordinates θ and ρ for a plane or cone, or of Φ and Λ for a sphere.

The procedure is to use Equation 1 to set up a differential equation to be solved for one of the mapping coordinates for a projection. The basis for this differential equation follows from the transformation theory of Section X, Chapter 2. The second mapping coordinate is defined by a particular condition imposed by the user for a particular mapping surface.

This general procedure is applied to the derivation of the mapping to the conformal sphere and to the derivations of the Lambert conformal projection with one standard parallel and the Mercator projection. The Lambert conformal with two standard parallels and the stereographic projections are obtained by mathematical manipulation of the derivation for the Lambert conformal with one standard parallel.

III. CONFORMAL SPHERE[12]

In a manner similar to the definition of the authalic sphere for equal area transformations, a conformal sphere may be defined for conformal transformations, and, similar to the use of the authalic sphere, the conformal sphere can be used as an intermediate step in the transformation from a reference spheroid to a mapping surface. First, one transforms from the spheroid to the conformal sphere. Next, one transforms from the conformal sphere to a chosen mapping surface. Again, it is necessary to remember that minor changes in scale occur during the two-step transformation. Adjustments in scale, at the user's discretion, may be needed.

The differential geometry approach is used to define the conformal sphere. To this end, it is necessary to find the radius of the conformal sphere for a particular reference spheroid. As is seen below, there is no one radius for a single reference spheroid. Rather, the radius depends on the coordinates of the spheroid about which the derivation is centered. The derivation itself seeks to obtain relations between geodetic latitude and longitude on the reference spheroid and the corresponding latitude and longitude on the conformal sphere.

Consider first the reference spheroid with semimajor axis, a, eccentricity, e, and radii of curvature, R_m and R_p. The geodetic coordinates are φ and λ. The first fundamental quantities of the spheroid are

$$\left. \begin{array}{l} e = R_m^2 \\[2mm] g = R_p^2 \cos^2\phi \end{array} \right\} \tag{2}$$

For the conformal sphere, the conformal latitude and longitude are defined as Φ and Λ, respectively. The radius of the conformal sphere is R_c. The first fundamental quantities of the sphere are

$$
\left.\begin{aligned}
E' &= R_c^2 \\[2mm]
G' &= R_c^2 \cos^2\Phi
\end{aligned}\right\} \tag{3}
$$

The following conditions are then applied: conformal spherical latitude is to be a function of spheroidal geodetic latitude only; conformal longitude is to be a linear function of spheroidal longitude; and conformal longitude and spheroidal longitude are to coincide at zero. These conditions are stated mathematically as

$$
\left.\begin{aligned}
\Phi &= \Phi(\phi) \\[2mm]
\Lambda &= c\lambda
\end{aligned}\right\} \tag{4}
$$

From Equation 4, the following partial derivative relations are obtained

$$
\left.\begin{aligned}
\frac{\partial \Phi}{\partial \lambda} &= 0 \\[3mm]
\frac{\partial \Lambda}{\partial \phi} &= 0 \\[3mm]
\frac{\partial \Lambda}{\partial \lambda} &= c
\end{aligned}\right\} \tag{5}
$$

From Section II, the condition for conformality is

$$
m^2 = \frac{E}{e} = \frac{G}{g} \tag{6}
$$

where m is the distortion at a point, as explained in detail in Chapter 7.

From Section X, Chapter 2, the fundamental quantities on the mapping surface are related as

$$
\left.\begin{aligned}
E &= \left(\frac{\partial \Phi}{\partial \phi}\right)^2 E' + \left(\frac{\partial \Lambda}{\partial \phi}\right)^2 G' \\[3mm]
G &= \left(\frac{\partial \Phi}{\partial \phi}\right)^2 E' + \left(\frac{\partial \Lambda}{\partial \lambda}\right)^2 G'
\end{aligned}\right\} \tag{7}
$$

We are now at a point where we can set up a differential equation to relate Φ and ϕ. Substitute Equations 5 into Equation 7.

$$\left.\begin{array}{l} E = \left(\dfrac{\partial \Phi}{\partial \phi}\right)^2 E' \\[2mm] G = c^2 G' \end{array}\right\} \qquad (8)$$

Substitute Equations 8 into Equation 6.

$$m^2 = \dfrac{\left(\dfrac{\partial \Phi}{\partial \phi}\right)^2 E'}{e} = \dfrac{c^2 G'}{g} \qquad (9)$$

Substitute Equations 2 and 3 into Equation 9.

$$m^2 = \dfrac{1}{R_m^2} \left(\dfrac{\partial \Phi}{\partial \phi}\right)^2 R_c^2 = \dfrac{c^2 R_c^2 \cos^2 \Phi}{R_p^2 \cos^2 \phi} \qquad (10)$$

Convert Equation 10 into an ordinary differential equation and take the square root. Then, by separating the variables,

$$\dfrac{d\Phi}{\cos \Phi} = \dfrac{c R_m}{R_p \cos \phi} \, d\phi \qquad (11)$$

Substitute Equations 21 and 28 of Chapter 3 into Equation 11

$$\begin{aligned} \dfrac{d\Phi}{\cos \Phi} &= \dfrac{\dfrac{ca(1 - e^2)}{(1 - e^2 \sin^2 \phi)^{3/2}}}{\dfrac{a \cos \phi}{(1 - e^2 \sin^2 \phi)^{1/2}}} \, d\phi \\[3mm] &= \dfrac{c(1 - e^2)}{(1 - e^2 \sin^2 \phi)\cos \phi} \, d\phi \end{aligned} \qquad (12)$$

The solution of Equation 12 is

$$\ln \tan\left(\dfrac{\pi}{4} + \dfrac{\Phi}{2}\right) = c \ln \tan\left(\dfrac{\pi}{4} + \dfrac{\phi}{2}\right)\left(\dfrac{1 - e\sin\phi}{1 + e\sin\phi}\right)^{e/2} + K \qquad (13)$$

The constant K is removed by requiring that Φ and ϕ are coincidently equal to zero. Thus, from Equation 13,

$$\tan\left(\frac{\pi}{4} + \frac{\Phi}{2}\right) = \left\{\tan\left(\frac{\pi}{4} + \frac{\phi}{2}\right)\left(\frac{1 - e\sin\phi}{1 + e\sin\phi}\right)^{e/2}\right\}^{c} \tag{14}$$

Note that this integral was encountered before, in Section IV, Chapter 3, for the loxodromic curve on the spheroid.

It remains to find the value of the constant c for a particular transformation from the spheroid to the conformal sphere.

Consider a Taylor's series expansion of the constant m^2 about the origin. By origin we mean, in this development, the latitude selected as the origin of the map. Recall that the partial derivatives of m^2 with respect to λ are zero. Then,

$$m^2 = m_o^2 + \left(\frac{\partial m^2}{\partial \phi}\right)_o \Delta\phi + \frac{1}{2}\left(\frac{\partial^2 m^2}{\partial \phi^2}\right)_o (\Delta\phi)^2 + \dots \tag{15}$$

Also from Equation 10,

$$m = \frac{cR_c\Phi}{R_p\cos\phi} \tag{16}$$

At the origin of the map, $m_o = 1$, by definition of the conformal projection. This aspect of map projections is explored in Chapter 7 on the theory of distortion. Considering Equation 10 at the origin,

$$m_o = 1 = \frac{cR_c\cos\Phi_o}{R_p\cos\phi_o} \tag{17}$$

Let

$$\frac{\partial m}{\partial \phi} = cR_c\left\{\frac{\partial}{\partial \phi}\left[\frac{\cos\Phi}{R_p\cos\phi}\right]\right\} = 0 \tag{18}$$

A second relation is now needed. Take the derivative of Equation 14 and divide out $c \cdot R_c$.

$$-\frac{\sin\Phi}{R_p\cos\phi}\frac{\partial\Phi}{\partial\phi} = \frac{\cos\Phi}{(R_p\cos\phi)^2}\left(-R_p\sin\phi + \cos\phi\frac{\partial R_p}{\partial\phi}\right) = 0 \tag{19}$$

Then, take the partial derivative of R_p and substitute this into Equation 19.

$$\sin\Phi \, \frac{\partial \Phi}{\partial \phi} = -\frac{\cos\Phi}{R_p\cos\phi}\left[-R_p\sin\phi + \frac{\cos\phi ae^2\sin\phi\cos\phi}{(1 - e^2\sin^2\phi)^{3/2}}\right]$$

$$= -\frac{\cos\Phi\sin\phi}{R_p\cos\phi}\left[-R_p + \frac{ae^2\cos^2\phi}{(1 - e^2\sin^2\phi)^{3/2}}\right]$$

$$= \frac{\cos\Phi\sin\phi}{R_p\cos\phi}\left[\frac{a(1 - e^2\sin^2\phi) - ae^2\cos^2\phi}{(1 - e^2\sin^2\phi)^{3/2}}\right]$$

$$= \frac{\cos\Phi\sin\phi}{R_p\cos\phi}\left[\frac{a(1 - e^2)}{(1 - e^2\sin^2\phi)^{3/2}}\right] \tag{20}$$

Substitute Equation 28 of Chapter 3 into Equation 20.

$$\sin\Phi \, \frac{\partial \Phi}{\partial \phi} = \cos\Phi \, \frac{R_m\sin\phi}{R_p\cos\phi} \tag{21}$$

Evaluate Equation 21 at the origin of the map. R_{po} and R_{mo} are the values of R_p and R_m, respectively, at ϕ_o.

$$\sin\Phi_o\left(\frac{\partial \Phi}{\partial \phi}\right)_o = \cos\Phi_o \, \frac{R_{mo}\sin\phi_o}{R_{po}\cos\phi_o} \tag{22}$$

Substitute Equation 10 into Equation 22.

$$\frac{\sin\Phi_o\cos\Phi_o c R_{mo}}{R_{po}\cos\phi_o} = \frac{\cos\Phi_o R_{mo}\sin\phi_o}{R_{po}\cos\phi_o}$$

$$\sin\phi_o = c\sin\Phi_o \tag{23}$$

The next step is to obtain the second partial derivative of m and equate this to zero. This is accomplished by using Equation 21 to obtain

$$\tan\phi_o = \frac{R_{po}}{R_{mo}} \, \tan\Phi_o \tag{24}$$

From Equation 23, develop the trigonometric terms

$$\sin\Phi_o = \frac{\sin\phi_o}{c}$$

$$\cos\Phi_o = \sqrt{1 - \frac{\sin^2\phi_o}{c^2}} \tag{25}$$

Substitute Equations 21 and 28 of Chapter 3 and Equation 25 into Equation 24. Then, simplify the results.

$$\frac{\sin\phi_o}{\cos\phi_o} = \sqrt{\frac{R_{po}}{R_{mo}} \frac{\sin\Phi_o}{\cos\Phi_o}}$$

$$= \sqrt{\frac{\dfrac{a}{(1 - e^2\sin^2\phi_o)^{1/2}}}{\dfrac{a(1 - e^2)}{(1 - e^2\sin^2\phi_o)^{3/2}}} \cdot \frac{\dfrac{\sin\phi_o}{c}}{\sqrt{1 - \dfrac{\sin^2\phi_o}{c^2}}}}$$

$$\frac{1}{\cos\phi_o} = \sqrt{\left(\frac{1 - e^2\sin^2\phi_o}{1 - e^2}\right)\left(\frac{1}{c^2 - \sin^2\phi_o}\right)}$$

$$\frac{1}{\cos^2\phi_o} = \left(\frac{1 - e^2\sin^2\phi_o}{1 - e^2}\right)\left(\frac{1}{c^2 - \sin^2\phi_o}\right)$$

$$\cos^2\phi_o = \frac{(1 - e^2)(c^2 - \sin^2\phi_o)}{1 - e^2\sin^2\phi_o}$$

$$\cos^2\phi_o(1 - e^2\sin^2\phi_o) = (1 - e^2)(c^2 - \sin^2\phi_o)$$

$$\cos^2\phi_o - e^2\cos^2\phi_o\sin^2\phi_o = (1 - e^2)c^2 - \sin^2\phi_o + e^2\sin^2\phi_o$$

$$\cos^2\phi_o + \sin^2\phi_o - e^2\sin^2\phi_o(1 + \cos^2\phi_o) = c^2(1 - e^2)$$

$$1 - e^2(1 - \cos^2\phi_o)(1 + \cos^2\phi_o) = c^2(1 - e^2)$$

$$1 + e^2(\cos^4\phi_o - 1) = c^2(1 - e^2)$$

$$c^2 = \frac{1 - e^2 + e^2\cos^4\phi_o}{1 - e^2}$$

$$c = \left(1 + \frac{e^2\cos^4\phi_o}{1 - e^2}\right)^{1/2} \tag{26}$$

The final step is to find the radius of the conformal sphere. From Equation 24,

$$\frac{\sin\phi_o}{\cos\phi_o} = \sqrt{\frac{R_{po}}{R_{mo}} \cdot \frac{\sin\Phi_o}{\cos\Phi_o}} \tag{27}$$

From Equation 26,

$$\frac{c\sin\Phi_o}{\cos\phi_o} = \sqrt{\frac{R_{po}}{R_{mo}}\frac{\sin\Phi_o}{\cos\Phi_o}}$$

$$c = \frac{\cos\phi_o}{\cos\Phi_o}\left(\frac{R_{po}}{R_{mo}}\right)^{1/2} \tag{28}$$

Eliminate c between Equation 17 and Equation 28.

$$1 = \frac{R_c}{\sqrt{R_{po}R_{mo}}}$$

$$R_c = \sqrt{R_{po}R_{mo}} \tag{29}$$

The final forms of the equations relating latitude and longitude on the conformal sphere to that on the reference spheroid are as follows. From the second of Equation 4,

$$\Lambda = \lambda\left(\frac{1 + e^2\cos^4\phi_o}{1 - e^2}\right)^{1/2} \tag{30}$$

From Equation 14,

$$\tan\left(\frac{\pi}{4} + \frac{\Phi}{2}\right) = \left\{\tan\left(\frac{\pi}{4} + \frac{\Phi}{2}\right)\left(\frac{1 - e\sin\phi}{1 + e\sin\phi}\right)^{e/2}\right\}\left(\frac{1 + e^2\cos^4\phi_o}{1 - e^2}\right)^{1/2} \tag{31}$$

Equations 29, 30, and 31 indicate that conformal radius, longitude, and latitude depend on the choice of the latitude of the origin of the projection. This is a vast difference from the case of the authalic sphere.

A. Example 1

Let $\phi_o = 45°$. From Equations 21 and 28 of Chapter 3, for the WGS-72 spheroid, $R_{po} = 6,388,836$ m and $R_{mo} = 6,367,380$ m. Then,

$$R_c = \sqrt{R_{po}R_{mo}} = \sqrt{(6,388,836)(6,367,380)}$$

$$= 6,478,099 \text{ meters}$$

IV. LAMBERT CONFORMAL, ONE STANDARD PARALLEL[8,9,10,12]

The Lambert conformal projection with one standard parallel makes use of a cone tangent to the model of the Earth. This projection is the conformal analogue of the equal area Albers projection with one standard parallel. This projection is derived for a spheroidal model of the Earth. Then, the eccentricity in the mapping equation is set equal to zero to obtain equations based on a spherical model of the Earth. In addition, the inverse transformation is given for the spherical model of the Earth.

The differential geometry approach is used in the derivation of the mapping coordinates. The fundamental quantities for the spheroidal model of the Earth are

$$\left. \begin{array}{l} e = R_m^2 \\[2ex] g = R_p^2 \cos^2\phi \end{array} \right\} \tag{32}$$

The fundamental quantities for the developable conical surface, in terms of polar coordinates, are

$$\left. \begin{array}{l} E' = 1 \\[2ex] G' = \rho^2 \end{array} \right\} \tag{33}$$

Three conditions are now imposed. The first is that the first polar coordinate, ρ, is a function of latitude alone. The second is that the second polar coordinate, θ, is a linear function of $\Delta\lambda$. The third is that $\theta = 0$ when $\Delta\lambda = 0$. These are stated mathematically as

$$\rho = \rho(\phi) \tag{34}$$

$$\theta = c_1\Delta\lambda$$

$$= c_1(\lambda - \lambda_o) \tag{35}$$

Next, develop the partial derivative from Equation 35.

$$\frac{\partial\theta}{\partial\lambda} = c_1 \tag{36}$$

where c_1 is the constant of the cone.

Refer next to the fundamental transformation relations of Section X, Chapter 2, to obtain

$$E = \left(\frac{\partial\rho}{\partial\phi}\right)^2 E' \left.\begin{array}{c}\\\\\\\end{array}\right\}$$

$$G = \left(\frac{\partial\theta}{\partial\lambda}\right)^2 G' \left.\begin{array}{c}\\\\\\\end{array}\right\} \tag{37}$$

Substitute Equations 33 into Equation 37.

$$E = \left(\frac{\partial\rho}{\partial\phi}\right)^2 \left.\begin{array}{c}\\\\\\\end{array}\right\}$$

$$G = \left(\frac{\partial\theta}{\partial\lambda}\right)^2 \rho^2 \left.\begin{array}{c}\\\\\\\end{array}\right\} \tag{38}$$

Substitute Equation 36 into the second of Equations 38.

$$G = c_1^2\rho^2 \tag{39}$$

Substitute Equations 32, the first of Equation 38, and Equation 39 into Equation 1, the condition of conformality

$$\frac{\left(\frac{\partial\rho}{\partial\phi}\right)^2}{R_m^2} = \frac{c_1^2\rho^2}{R_p^2\cos^2\phi} = m^2 \tag{40}$$

where m is the index of distortion.

Take the square root of Equation 40 and convert the result to an ordinary differential equation.

$$\frac{d\rho}{\rho} = -\frac{R_m c_1}{R_p\cos\phi}\,d\phi \tag{41}$$

The minus sign is chosen since ρ decreases as ϕ increases.

Equation 41 can be integrated by the method developed in Section VII for the Mercator projection to obtain

$$\ln\rho = -c_1 \ln\left\{\tan\left(\frac{\pi}{4} + \frac{\phi}{2}\right)\left(\frac{1 - e\sin\phi}{1 + e\sin\phi}\right)^{e/2}\right\} + \ln c_2$$

$$\rho = c_2\left\{\tan\left(\frac{\pi}{4} - \frac{\phi}{2}\right)\left(\frac{1 + e\sin\phi}{1 - e\sin\phi}\right)^{e/2}\right\}^{c_1} \tag{42}$$

The constants c_1 and c_2 must be evaluated now. Begin with the evaluation of constant c_2. At the origin of the Cartesian coordinate system of the map

(ϕ_o, λ_o), the cone is tangent to the spheroid. Thus, similar to the development of Section XIV, Chapter 2,

$$\rho_o = R_{po}\cot\phi_o \qquad (43)$$

Evaluate Equation 42 at ϕ_o and equate to Equation 43.

$$R_{po}\cot\phi_o = c_2\left\{\tan\left(\frac{\pi}{4} - \frac{\phi_o}{2}\right)\left(\frac{1 + e\sin\phi_o}{1 - e\sin\phi_o}\right)^{e/2}\right\}^{c_1}$$

$$c_2 = \frac{R_{po}\cot\phi_o}{\left\{\tan\left(\frac{\pi}{4} - \frac{\phi_o}{2}\right)\left(\frac{1 + e\sin\phi_o}{1 - e\sin\phi_o}\right)^{e/2}\right\}^{c_1}} \qquad (44)$$

Then, from Equation 40,

$$m = \frac{c_1\rho}{R_p\cos\phi} \qquad (45)$$

At the origin (ϕ_o, λ_o), as is treated in detail in Chapter 7, $m_o = 1$. This implies that $(\partial m/\partial\phi)_o = 0$. Differentiate Equation 45 and evaluate this at the origin.

$$\frac{\partial m}{\partial\phi} = c_1\left(\frac{\partial\rho}{\partial\phi}\right)\frac{1}{R_p\cos\phi} + c_1\rho\frac{R_m\sin\phi}{R_p^2\cos^2\phi} = 0$$

$$c_1\left(\frac{\partial\rho}{\partial\phi}\right)_o + \frac{c_1\rho_oR_{mo}}{R_p\cos\phi_o} = 0 \qquad (46)$$

From Equation 41 we obtain

$$\left(\frac{\partial\rho}{\partial\phi}\right)_o = -\frac{c_1\rho_oR_{mo}\sin\phi_o}{R_{po}\cos\phi_o} \qquad (47)$$

Substitute Equation 47 into Equation 46.

$$-\frac{c_1^2\rho_oR_{mo}}{R_{po}\cos\phi_o} + \frac{c_1\rho_oR_{mo}\sin\phi_o}{R_{po}\cos\phi_o} = 0$$

$$c_1 = \sin\phi_o \qquad (48)$$

Substitute Equations 43 and 48 into Equation 46.

$$c_2 = \frac{\rho_o}{\left\{ \tan\left(\frac{\pi}{4} - \frac{\phi_o}{2}\right)\left(\frac{1 - e\sin\phi_o}{1 - e\sin\phi_o}\right)^{3/2} \right\}^{\sin\phi_o}} \tag{49}$$

Substitute Equation 49 into Equation 42 to obtain the first polar coordinate.

$$\rho = \rho_o \left[\frac{\tan\left(\frac{\pi}{4} - \frac{\phi}{2}\right)\left(\frac{1 + e\sin\phi}{1 - e\sin\phi}\right)^{e/2}}{\tan\left(\frac{\pi}{4} - \frac{\phi_o}{2}\right)\left(\frac{1 + e\sin\phi_o}{1 - e\sin\phi_o}\right)^{e/2}} \right]^{\sin\phi_o} \tag{50}$$

Substitute Equation 48 into Equation 35 to obtain the second

$$\theta = \Delta\lambda\sin\phi_o \tag{51}$$

Since the polar coordinates are now in our possession, we can go to Equation 7 of Chapter 1 to obtain the plotting equations. First, it is necessary to substitute the value for R_{po} into Equation 43. Then, with the aid of Equations 49 and 50, we have

$$\left. \begin{aligned} x &= \frac{a \cdot S\cot\phi_o}{(1 - e^2\sin^2\phi_o)^{1/2}} \\ &\quad \times \left[\frac{\tan(\pi/4 - \phi/2)}{\tan(\pi/4 - \phi_o/2)} \frac{\left(\frac{1 + e\sin\phi}{1 - e\sin\phi}\right)^{e/2}}{\left(\frac{1 + e\sin\phi_o}{1 - e\sin\phi_o}\right)^{e/2}} \right]^{\sin\phi_o} \sin(\Delta\lambda\sin\phi_o) \\ y &= \frac{a \cdot S\cot\phi_o}{(1 - e^2\sin^2\phi_o)^{1/2}} \\ &\quad \times \left\{ 1 - \left[\frac{\tan(\pi/4 - \phi/2)}{\tan(\pi/4 - \phi_o/2)} \frac{\left(\frac{1 + e\sin\phi}{1 - e\sin\phi}\right)^{e/2}}{\left(\frac{1 + e\sin\phi_o}{1 - e\sin\phi_o}\right)^{e/2}} \right]^{\sin\phi_o} \cos(\Delta\lambda\sin\phi_o) \right\} \end{aligned} \right\} \tag{52}$$

Equation 52 gives the grid in Figure 1. All meridians are straight lines, converging at the apex of the cone. The parallels are a set of concentric circles. Only the standard parallel at ϕ_o is true scale. The y axis is along the central meridian. The x axis is perpendicular to the y axis at ϕ_o. Due to the relatively rapid increase of distortion north and south of the standard parallel, this projection is not recommended for practical use.

To obtain the plotting equations based on a spherical model of the Earth,

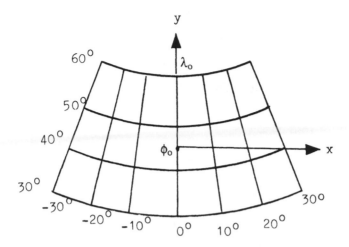

FIGURE 1. Lambert conformal projection, one standard parallel.

it is only necessary to substitute e = o and R = a into Equations 52. These reduce to

$$x = R \cdot Scot\phi_o \left[\frac{\tan(\pi/4 - \phi/2)}{\tan(\pi/4 - \phi_o/2)} \right]^{\sin\phi_o} \sin(\Delta\lambda\sin\phi_o)$$

$$y = R \cdot Scot\phi_o \left\{ 1 - \left[\frac{\tan(\pi/4 - \phi/2)}{\tan(\pi/4 - \phi_o/2)} \right]^{\sin\phi_o} \cos(\Delta\lambda\sin\phi_o) \right\}$$

(53)

The inverse transformation for the sphere may be obtained from Equations 53. First define f_1 as

$$f_1 = \left[\frac{\tan(\pi/4 - \phi/2)}{\tan(\pi/4 - \phi_o/2)} \right]^{\sin\phi_o}$$

(54)

to simplify the derivation. Substitute Equation 54 into Equation 53 to obtain

$$x = R \cdot Scot\phi_o \cdot f_1 \cdot \sin(\Delta\lambda\sin\phi_o)$$

$$y = R \cdot Scot\phi_o [1 - f_1 \cdot \cos(\Delta\lambda\sin\phi_o)]$$

(55)

The object is to eliminate f_1, which contains unknown terms, and solve for $\Delta\lambda$. To this end, Equations 55 are manipulated as follows.

$$\frac{x}{R \cdot Scot\phi_o} = f_1 \cdot \sin(\Delta\lambda\sin\phi_o)$$

(56)

$$1 - \frac{y}{R \cdot Sco t\phi_o} = f_1 \cdot \cos(\Delta\lambda \sin\phi_o) \qquad (57)$$

Divide Equation 56 by Equation 57 and finally solve for $\Delta\lambda$.

$$\tan(\Delta\lambda \sin\phi_o) = \frac{\dfrac{x}{R \cdot Sco t\phi_o}}{1 - \dfrac{y}{R \cdot Sco t\phi_o}}$$

$$\Delta\lambda \sin\phi_o = \tan^{-1}\left(\frac{x}{R \cdot Sco t\phi_o - y}\right)$$

$$\Delta\lambda = \frac{1}{\sin\phi_o} \cdot \tan^{-1}\left(\frac{x}{R \cdot Sco t\phi_o - y}\right) \qquad (58)$$

Equation 58 in conjunction with $\lambda = \Delta\lambda + \lambda_o$ uniquely defines λ.

Once λ is known, return to Equation 56 to find a relation for ϕ. First, rearrange Equation 56 in terms of f_1, and then equate to Equation 54.

$$f_1 = \frac{x}{R \cdot Sco t\phi_o \sin(\Delta\lambda \sin\phi_o)} \qquad (59)$$

Finally, solve for ϕ.

$$\tan(\pi/4 - \phi/2) = \tan(\pi/4 - \phi_o/2)\left[\frac{x}{R \cdot Sco t\phi_o \sin(\Delta\lambda \sin\phi_o)}\right]^{1/\sin\phi_o}$$

$$\frac{\pi}{4} - \frac{\phi}{2} = \tan^{-1}\left\{\tan(\pi/4 - \phi_o/2)\left[\frac{x}{R \cdot Sco t\phi_o \sin(\Delta\lambda \sin\phi_o)}\right]^{1/\sin\phi_o}\right\} \qquad (60)$$

$$\phi = \frac{\pi}{2} - 2\tan^{-1}\left\{\tan(\pi/4 - \phi_o/2)\left[\frac{x}{R \cdot Sco t\phi_o \sin(\Delta\lambda \sin\phi_o)}\right]^{1/\sin\phi_o}\right\} \qquad (61)$$

The geographic coordinate ϕ is in radians.

A. Example 2

Given: Lambert conformal, one standard parallel; $\phi_o = 45°$, $\lambda_o = 10°$, $\lambda = 30°$, $\phi = 50°$, WGS-72 spheroid, S = 1:10,000,000.

Find: the Cartesian plotting coordinates.

$a = 6,378,135$ m

$e = 0.081819$

$\rho_o = R_{po}\cot\phi_o$

$\quad = (6,388,836)(1) = 6,388,836$ m

$$\rho = \rho_o\left\{\frac{\tan\left(\dfrac{\pi}{4} - \dfrac{\phi}{2}\right)\left(\dfrac{1 + e\sin\phi}{1 - e\sin\phi}\right)^{e/2}}{\tan\left(\dfrac{\pi}{4} - \dfrac{\phi_o}{2}\right)\left(\dfrac{1 + e\sin\phi_o}{1 - e\sin\phi_o}\right)^{e/2}}\right\}^{\sin\phi_o}$$

$$= (6,388,836)\left\{\frac{\tan\left(45° - \dfrac{50}{2}\right)\left(\dfrac{1 + 0.081819\sin50}{1 - 0.081819\sin50}\right)^{0.04091}}{\tan\left(45° - \dfrac{45}{2}\right)\left(\dfrac{1 + 0.081819\sin45}{1 - 0.081819\sin45}\right)^{0.04091}}\right\}^{\sin45°}$$

$$= (6,388,836)\left\{\frac{(0.36397)\left(\dfrac{1.06268}{0.93732}\right)^{0.04091}}{(0.41421)\left(\dfrac{1.05785}{0.94214}\right)^{0.04091}}\right\}^{0.70711}$$

$$= (6,388,836)\left\{\frac{(0.36397)\,(1.0051)}{(0.41421)\,(1.0048)}\right\}^{0.70711}$$

$$= (6,388,836)(0.91282) = 5,831,841 \text{ m}$$

$\theta = \Delta\lambda\sin\phi_o = (30 - 10)\sin45°$

$\quad = 14.14°$

$x = \rho S\sin\theta = \left(\dfrac{5,831,841}{10,000,000}\right)(\sin14.14)$

$\quad = 0.1420$ m

$y = \rho S(\rho_o - \cos\theta) = \dfrac{6,388,836 - (5,831,841)(\cos14.14)}{10,000,000}$

$\quad = 0.0735$ m

V. LAMBERT CONFORMAL, TWO STANDARD PARALLELS[8,9,10,12]

The Lambert conformal projection with two standard parallels has the mapping cone secant to the model of the Earth at two parallel circles. This projection is derived below for the spheroidal model of the Earth. Then, the spherical case is obtained by considering a ellipse of zero eccentricity. The inverse transformation is derived for the spherical model of the Earth. The development of this projection emphasizes the conceptual quality of secancy for conics.

For the spheroidal model of the Earth, the plotting equations are obtained by a mathematical manipulation of the equations for the Lambert conformal projection for one standard parallel. The derivation begins by finding a relation for the constant of the cone.

Let the two standard parallels be chosen as ϕ_1 and ϕ_2, where $\phi_2 > \phi_1$. Then, from Equations 47 and 50,

$$\left.\begin{aligned} m &= \rho_1 \frac{\sin\phi_o}{R_{p1}\cos\phi_1} \\[2mm] &= \rho_2 \frac{\sin\phi_o}{R_{p2}\cos\phi_2} \end{aligned}\right\} \tag{62}$$

$$\frac{\rho_1}{\rho_2} = \frac{R_{p1}\cos\phi_1}{R_{p2}\cos\phi_2} \tag{63}$$

From Equation 52,

$$\frac{\rho_1}{\rho_2} = \left\{ \frac{\tan\left(\dfrac{\pi}{4} - \dfrac{\phi_1}{2}\right)\left(\dfrac{1 + e\sin\phi_1}{1 - e\sin\phi_1}\right)^{e/2}}{\tan\left(\dfrac{\pi}{4} - \dfrac{\phi_2}{2}\right)\left(\dfrac{1 + e\sin\phi_2}{1 - e\sin\phi_1}\right)^{e/2}} \right\}^{\sin\phi_o} \tag{64}$$

From Equations 63 and 64,

$$\frac{R_{p1}\cos\phi_1}{R_{p2}\cos\phi_2} = \left\{ \frac{\tan\left(\dfrac{\pi}{4} - \dfrac{\phi_1}{2}\right)\left(\dfrac{1 + e\sin\phi_1}{1 - e\sin\phi_1}\right)^{e/2}}{\tan\left(\dfrac{\pi}{4} - \dfrac{\phi_2}{2}\right)\left(\dfrac{1 + e\sin\phi_2}{1 - e\sin\phi_2}\right)^{e/2}} \right\}^{\sin\phi_o}$$

$$\ln\left(\frac{R_{p1}\cos\phi_1}{R_{p2}\cos\phi_2}\right) = \sin\phi_o \ln\left\{ \frac{\tan\left(\dfrac{\pi}{4} - \dfrac{\phi_1}{2}\right)\left(\dfrac{1 + e\sin\phi_1}{1 - e\sin\phi_1}\right)^{e/2}}{\tan\left(\dfrac{\pi}{4} - \dfrac{\phi_2}{2}\right)\left(\dfrac{1 + e\sin\phi_2}{1 - e\sin\phi_2}\right)^{e/2}} \right\}$$

$$\sin\phi_o = \frac{\ln\left(\dfrac{R_{p1}\cos\phi_1}{R_{p2}\cos\phi_2}\right)}{\ln\left\{\dfrac{\tan(\pi/4 - \phi_1/2)[(1 + e\sin\phi_1)/(1 - e\sin\phi_1)]^{e/2}}{\tan(\pi/4 - \phi_2/2)[(1 + e\sin\phi_2)/(1 - e\sin\phi_2)]^{e/2}}\right\}} \tag{65}$$

Next, substitute Equation 21 of Chapter 3 into Equation 65 to obtain a relation for the constant of the cone.

$$\sin\phi_o = \frac{\ln\left\{\left(\dfrac{\cos\phi_1}{\cos\phi_2}\right)\left(\dfrac{1 + e^2\sin^2\phi_2}{1 - e^2\sin^2\phi_1}\right)^{1/2}\right\}}{\ln\left\{\dfrac{\tan(\pi/4 - \phi_1/2)[(1 + e\sin\phi_1)/(1 - e\sin\phi_1)]^{e/2}}{\tan(\pi/4 - \phi_2/2)[(1 + e\sin\phi_2)/(1 - e\sin\phi_2)]^{e/2}}\right\}} \quad (66)$$

So far, the conical surface has not been constrained to be secant at ϕ_1 and ϕ_2. This is done by requiring $m = 1$ at ϕ_1 and ϕ_2. This is substituted into Equation 62 to obtain

$$\frac{\rho_1\sin\phi_o}{R_{p1}\cos\phi_1} = \frac{\rho_2\sin\phi_o}{R_{p2}\cos\phi_2} = 1$$

$$\left.\begin{array}{l} \rho_1\sin\phi_o = R_{p1}\cos\phi_1 \\[2mm] \rho_2\sin\phi_o = R_{p2}\cos\phi_2 \end{array}\right\} \quad (67)$$

Since the constant of the cone is dependent only on ϕ_1 and ϕ_2, Equation 66 applies for Equation 67.

The next step is to define an auxiliary variable, ψ, to make the derivation less unwieldy. To this end, let $\phi = \phi_1$ in Equation 52. We then obtain

$$\rho_1 = \rho_o\left\{\frac{\tan\left(\dfrac{\pi}{4} - \dfrac{\phi_1}{2}\right)\left(\dfrac{1 + e\sin\phi_1}{1 - e\sin\phi_1}\right)^{e/2}}{\tan\left(\dfrac{\pi}{4} - \dfrac{\phi_o}{2}\right)\left(\dfrac{1 + e\sin\phi_o}{1 - e\sin\phi_o}\right)^{e/2}}\right\}^{\sin\phi_o} \quad (68)$$

Substitute Equation 61 into Equation 60.

$$\sin\phi_o \cdot \rho_o\left\{\frac{\tan\left(\dfrac{\pi}{4} - \dfrac{\phi_1}{2}\right)\left(\dfrac{1 + e\sin\phi_1}{1 - e\sin\phi_1}\right)^{e/2}}{\tan\left(\dfrac{\pi}{4} - \dfrac{\phi_o}{2}\right)\left(\dfrac{1 + e\sin\phi_o}{1 - e\sin\phi_o}\right)^{e/2}}\right\}^{\sin\phi_o} = R_{p1}\cos\phi_1 \quad (69)$$

Let the auxiliary variable be defined such that

$$\psi = \frac{\rho_o}{\left[\tan\left(\dfrac{\pi}{4} - \dfrac{\phi_o}{2}\right)\left(\dfrac{1 + e\sin\phi_o}{1 - e\sin\phi_o}\right)^{e/2}\right]^{\sin\phi_o}} \quad (70)$$

Substitute Equation 70 into Equation 69 and solve for ψ.

$$\psi = \frac{R_{p1}\cos\phi_1}{\sin\phi_o\left[\tan\left(\dfrac{\pi}{4} - \dfrac{\phi_1}{2}\right)\left(\dfrac{1 + e\sin\phi_1}{1 - e\sin\phi_1}\right)^{e/2}\right]^{\sin\phi_o}} \tag{71}$$

Then substitute Equation 21 of Chapter 3 into Equation 71 to eliminate the curvature term.

$$\psi = \frac{\dfrac{a\cos\phi_1}{(1 - e^2\sin^2\phi_1)^{1/2}}}{\sin\phi_o\left[\tan\left(\dfrac{\pi}{4} - \dfrac{\phi_1}{2}\right)\left(\dfrac{1 + e\sin\phi_1}{1 - e\sin\phi_1}\right)^{e/2}\right]^{\sin\phi_o}} \tag{72}$$

By a similar manipulation, we can also obtain

$$\psi = \frac{\dfrac{a\cos\phi_2}{(1 - e^2\sin^2\phi_2)^{1/2}}}{\sin\phi_o\left[\tan\left(\dfrac{\pi}{4} - \dfrac{\phi_2}{2}\right)\left(\dfrac{1 + e\sin\phi_2}{1 - e\sin\phi_2}\right)^{e/2}\right]^{\sin\phi_o}} \tag{73}$$

We now have the means of defining the polar coordinates, θ and ρ, for the projection. These are:

$$\theta = \Delta\lambda\sin\phi_o$$

$$\rho = \psi\left[\tan\left(\frac{\pi}{4} - \frac{\phi}{2}\right)\left(\frac{1 + e\sin\phi}{1 - e\sin\phi}\right)^{e/2}\right]^{\sin\phi_o} \tag{74}$$

Equations 74 are then used in conjunction with Equations 7 of Chapter 1 to obtain the Cartesian plotting equations. The derived equations are repeated below to demonstrate the algorithm required to obtain the plotting coordinates. This algorithm is based on defining the y axis as along the central meridian, and the x axis perpendicular to the y axis at ϕ_1.

1. Constant of the cone

$$\sin\phi_o = \frac{\ln\left[\left(\dfrac{\cos\phi_1}{\cos\phi_2}\right)\left(\dfrac{1 - e^2\sin^2\phi_2}{1 + e^2\sin^2\phi_1}\right)\right]^{1/2}}{\ln\left[\dfrac{\tan(\pi/4 - \phi_1/2)[(1 + e\sin\phi_1)/(1 - e\sin\phi_1)]^{e/2}}{\tan(\pi/4 - \phi_2/2)[(1 + e\sin\phi_2)/(1 - e\sin\phi_2)]}\right]}$$

2. Polar angle

$$\theta = \Delta\lambda\sin\phi_o$$

3. Auxiliary function

$$\psi = \frac{\dfrac{a\cos\phi_1}{(1 - e^2\sin^2\phi_1)^{1/2}}}{\sin\phi_o\left[\tan\left(\dfrac{\pi}{4} - \dfrac{\phi_1}{2}\right)\left(\dfrac{1 + e\sin\phi_1}{1 - e\sin\phi_1}\right)^{e/2}\right]^{\sin\phi_o}}$$

4. Polar radius to origin

$$\rho_1 = \psi\left[\tan\left(\frac{\pi}{4} - \frac{\phi_1}{2}\right)\left(\frac{1 + e\sin\phi_1}{1 - e\sin\phi_1}\right)^{e/2}\right]^{\sin\phi_o}$$

5. Polar radius to latitude ϕ

$$\rho = \psi\left[\tan\left(\frac{\pi}{4} - \frac{\phi}{2}\right)\left(\frac{1 + e\sin\phi}{1 - e\sin\phi}\right)^{e/2}\right]^{\sin\phi_o}$$

6. Cartesian plotting coordinates

$$x = S \cdot \rho\sin\theta$$

$$y = S \cdot (\rho_1 - \rho\cos\theta)$$

The grid of Figure 2 results from the evaluation of the algorithms above. The meridians are straight lines radiating from the apex of the cone. The parallels of latitude are concentric circles. The scale between the standard parallels is less than true scale. The scale above and below the standard parallels is more than true scale. The use of the two-standard parallel case results in a better total distribution of distortion. In the figure, note the location of the axes for the mapping equations.

Table 1 gives representative plotting coordinates for the projection. Each choice of a set of standard parallels results in a separate plotting table and grid.

The algorithm for the plotting equations based on a spherical model of the Earth results from substituting $e = 0$ and $R = a$ into the equations for the spheroidal model. The resulting algorithm is as follows.

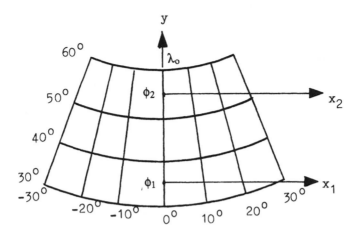

FIGURE 2.   Lambert conformal projection, two standard parallels.

TABLE 1
Normalized Plotting Coordinates for the Lambert Conformal Projection,
Two Standard Parallels
($\phi_1 = 35°$, $\phi_2 = 55°$, $\lambda_o = 0°$, $R \cdot S = 1$)

Tabulated Values: x/y

Longitude	Latitude			
	30°	**40°**	**50°**	**60°**
0°	0.000	0.000	0.000	0.000
	−0.126	0.120	0.347	0.559
10°	0.158	0.128	0.100	0.073
	−0.116	0.128	0.353	0.564
20°	0.314	0.254	0.198	0.146
	−0.087	0.152	0.371	0.577
30°	0.465	0.375	0.293	0.216
	−0.038	0.191	0.402	0.600
40°	0.609	0.492	0.384	0.282
	0.028	0.245	0.444	0.631
50°	0.743	0.600	0.468	0.345
	0.112	0.312	0.497	0.670
60°	0.866	0.699	0.546	0.402
	0.212	0.393	0.560	0.716

1. Constant of the cone

$$\sin\phi_o = \frac{\ln\left(\dfrac{\cos\phi_1}{\cos\phi_2}\right)}{\ln\left[\dfrac{\tan(\pi/4 - \phi_1/2)}{\tan(\pi/4 + \phi_2/2)}\right]}$$

2. Polar angle

$$\theta = \Delta\lambda\sin\phi_o$$

3. Auxiliary function

$$\psi = \frac{R\cos\phi_1}{\sin\phi_o\left[\tan\left(\dfrac{\pi}{4} - \dfrac{\phi_1}{2}\right)\right]^{\sin\phi_o}}$$

4. Polar radius to origin

$$\rho_1 = \psi[\tan(\pi/4 - \phi_1/2)]^{\sin\phi_o}$$

5. Polar radius to latitude ϕ

$$\rho_1 = \psi[\tan(\pi/4 - \phi/2)]^{\sin\phi_o}$$

6. Cartesian plotting coordinates

$$\left.\begin{array}{l} x = S \cdot \rho\sin\theta \\[2mm] y = S \cdot (\rho_1 - \rho\cos\theta) \end{array}\right\}$$

The inverse transformation for the spherical model of the Earth is obtained from the equations for the algorithm above. First, $\sin\phi_o$ is developed as in step 1. Then, ρ_1 is obtained in an alternate fashion as

$$\rho_1 = \frac{R\cos\phi_1}{\sin\phi_o} \tag{75}$$

Then, the equations for the Cartesian plotting coordinates are rearranged to give

$$\left. \begin{array}{l} \sin\theta = \dfrac{x}{S \cdot \rho} \\[3mm] \cos\theta = \dfrac{\rho_1 - y \cdot S}{\rho} \end{array} \right\} \qquad (76)$$

The tangent of θ is obtained from Equations 76 as a means of eliminating ρ. This leads to a solution for λ, since Equation 75 applies.

$$\tan\theta = \frac{x/S}{\rho_1 - y/S}$$

$$\theta = \tan^{-1}\left(\frac{x/S}{\rho_1 - y/S}\right)$$

$$\lambda = \lambda_o + \frac{1}{\sin\phi_o} \cdot \tan^{-1}\left(\frac{x/S}{\rho_1 - y/S}\right) \qquad (77)$$

The latitude is found by the equation of Step 5 and the relation from the first of Equations 76.

$$\rho = \frac{x}{S \cdot \sin\theta} = \psi[\tan(\pi/4 - \phi/2)]^{\sin\phi_o}$$

$$\tan(\pi/4 - \phi/2) = \left(\frac{x}{S \cdot \psi\sin\theta}\right)^{1/\sin\phi_o}$$

$$\frac{\pi}{4} - \frac{\phi}{2} = \tan^{-1}\left[\left(\frac{x}{S\psi\sin\theta}\right)^{1/\sin\phi_o}\right]$$

$$\phi = \frac{\pi}{2} - 2\tan^{-1}\left[\left(\frac{x}{S\psi\sin\theta}\right)^{1/\sin\phi_o}\right] \qquad (78)$$

Equation 78 is evaluated with the aid of ψ obtained as Step 3.

A similar procedure may be used to obtain the inverse transformation for the spheroidal model of the Earth. However, an alternative approach is given in Section VII. In that section, an approximate inverse relation is developed.

The Lambert conformal projection with two standard parallels is the conformal projection of choice for mid-latitudes. It has been used extensively for aircraft navigation charts in these regions, and it has great use as the basis of some state plane coordinate systems, as described in Section VII.

VI. STEREOGRAPHIC[3,8,9,10,11,12]

The stereographic projection entails the transformation directly from the model of the Earth onto a plane. It is the important conformal azimuthal projection. The stereographic projection is a perspective projection and permits a purely geometrical construction. This is illustrated for the polar stereographic projection in Figure 1, Chapter 6. The mapping plane is tangent to the sphere at the equator. The principle of the stereographic projection requires that the projection point be diametrically across from the point of tangency. A typical ray from S to a point P on the Earth is transformed to the position Ps on the plane. Thus, the entire projection can be derived by elementary trigonometry for the spherical Earth model. However, for the spheroidal case, the projection cannot be both perspective and conformal. The approach used in this section is to obtain the equations for the spheroidal case by mathematical manipulation.

Three variations of the direct stereographic projections are derived. These are the polar, the oblique, and the equatorial. The polar case is considered for both a spheroidal and a spherical model of the Earth. In addition, the inverse transformation is given for the polar stereographic projection.

We begin the derivations with the polar stereographic projection for a spheroidal model of the Earth. We consider the plane as one of the limiting forms of the cone, and then apply the equations already derived for the Lambert conformal projection with one standard parallel. This is done by letting the parallel of tangency for the spheroidal case shrink to a polar point of tangency in order to derive the polar stereographic projection. To this end, let $\phi_o = 90°$ in Equations 45, 50, 52, and 53. We then obtain the following:

$$\sin\phi_o = 1 \tag{79}$$

$$\theta = \Delta\lambda \tag{80}$$

$$\rho_o = R_{po}\cot\phi_o$$

$$= \frac{a}{\sqrt{1 - e^2}} \cdot \cot\phi_o \tag{81}$$

and

$$\rho = \frac{a\cot\phi_o}{\sqrt{1 - e^2}} \left\{ \frac{\tan\left(\frac{\pi}{4} - \frac{\phi}{2}\right)\left(\frac{1 + e\sin\phi}{1 - e\sin\phi}\right)^{e/2}}{\tan\left(\frac{\pi}{4} - \frac{\phi_o}{2}\right)\left(\frac{1 + e}{1 - e}\right)^{e/2}} \right\}$$

$$= \tan\left(\frac{\pi}{4} - \frac{\phi}{2}\right)\left(\frac{1 + e\sin\phi}{1 - e\sin\phi}\right)^{e/2} \cdot \frac{a}{\sqrt{1 - e^2}}\left(\frac{1 + e}{1 - e}\right)^{e/2}$$

$$\cdot \frac{\tan\left(\frac{\pi}{4} - \frac{\phi_o}{2}\right)}{\tan\phi_o} \tag{82}$$

In Equation 82, take the limit of

$$\frac{\tan\left(\frac{\pi}{4} + \frac{\phi_o}{2}\right)}{\tan\phi_o}$$

as ϕ_o approaches 90°.

$$\underset{\phi_o \to 90°}{\text{Lim}} \frac{\tan\left(\frac{\pi}{4} - \frac{\phi_o}{2}\right)}{\tan\phi_o} = \underset{\phi_o \to 90°}{\text{Lim}} \left(\frac{1 + \tan\phi_o/2}{1 - \tan\phi_o/2}\right)\left(\frac{1 - \tan^2\phi_o/2}{2\tan\phi_o/2}\right)$$

$$= \underset{\phi_o \to 90°}{\text{Lim}} \frac{(1 + \tan\phi_o/2)^2}{2\tan\phi_o/2} = 2 \qquad (83)$$

Substitute Equation 83 into Equation 82.

$$\rho = \frac{2a}{\sqrt{1 - e^2}} \left(\frac{1 - e}{1 + e}\right)^{e/2} \tan\left(\frac{\pi}{4} - \frac{\phi}{2}\right)\left(\frac{1 + e\sin\phi}{1 - e\sin\phi}\right)^{e/2} \qquad (84)$$

The Cartesian plotting equations follow from Equations 6 of Chapter 1. Substituting Equations 80 and 84, we obtain

$$\left.\begin{array}{l} x = \dfrac{2a \cdot S}{\sqrt{1 - e}} \left(\dfrac{1 - e}{1 + e}\right)^{e/2} \tan\left(\dfrac{\pi}{4} - \dfrac{\phi}{2}\right)\left(\dfrac{1 + e\sin\phi}{1 - e\sin\phi}\right)^{e/2} \sin\Delta\lambda \\[4mm] y = -\dfrac{2a \cdot S}{\sqrt{1 - e}} \left(\dfrac{1 - e}{1 + e}\right)^{e/2} \tan\left(\dfrac{\pi}{4} - \dfrac{\phi}{2}\right)\left(\dfrac{1 + e\sin\phi}{1 - e\sin\phi}\right)^{e/2} \cos\Delta\lambda \end{array}\right\} \quad (85)$$

where S is the scale factor.

The plotting equations for a spherical model of the Earth follow directly from Equations 85 by substituting e = 0 and R = a.

$$\left.\begin{array}{l} x = 2R \cdot S \cdot \tan\left(\dfrac{\pi}{4} - \dfrac{\phi}{2}\right)\sin\Delta\lambda \\[4mm] y = -2R \cdot S \cdot \tan\left(\dfrac{\pi}{4} - \dfrac{\phi}{2}\right)\cos\Delta\lambda \end{array}\right\} \quad (86)$$

Table 2 contains the plotting coordinates for the polar stereographic projection. A single table is all that is necessary for the polar case.

Figure 3 gives an example of the polar stereographic projection. Note that the meridians are straight lines converging on the pole, and the parallels are concentric circles centered on the pole. The spacing between the parallels

TABLE 2
Normalized Plotting Coordinates for the Polar Stereographic Projection ($\phi_o = 90°$, $\lambda_o = 0°$, $R \cdot S = 1$)

Tabulated Values: x/y

Longitude	Latitude			
	60°	70°	80°	90°
0°	0.000	0.000	0.000	0.000
	−0.536	−0.353	−0.175	0.000
10°	0.093	0.061	0.030	0.000
	−0.528	−0.347	−0.172	0.000
20°	0.183	0.121	0.060	0.000
	−0.504	−0.331	−0.164	0.000
30°	0.268	0.176	0.087	0.000
	−0.464	−0.305	−0.152	0.000
40°	0.344	0.277	0.112	0.000
	−0.411	−0.270	−0.134	0.000
50°	0.411	0.270	0.134	0.000
	−0.344	−0.227	−0.112	0.000
60°	0.464	0.305	0.152	0.000
	−0.268	0.176	−0.087	0.000
70°	0.504	0.331	0.164	0.000
	−0.183	−0.121	−0.060	0.000
80°	0.528	0.347	0.172	0.000
	−0.093	−0.061	−0.030	0.000
90°	0.536	0.353	0.175	0.000
	0.000	0.000	0.000	0.000

increases as one goes towards the equator. The x axis is along the 90° meridian. The y axis is along the 180° meridian.

For the polar stereographic projection, the equations for the inverse transformation from Cartesian coordinates to geographical coordinates follow from Equation 86 for the spherical case. Dividing the first of Equation 86 by the second, we obtain

$$\Delta\lambda = \tan^{-1}\left(\frac{x}{-y}\right) \qquad (87)$$

Summing the squares of Equation 86,

$$\left(\frac{x}{2R \cdot S}\right)^2 + \left(\frac{y}{2R \cdot S}\right)^2 = \tan^2\left(\frac{\pi}{4} - \frac{\phi}{2}\right) \qquad (88)$$

After some manipulation of Equation 87, the result is

$$\phi = \frac{\pi}{2} - 2\tan^{-1}\sqrt{\frac{x^2 + y^2}{4R^2 \cdot S^2}} \qquad (89)$$

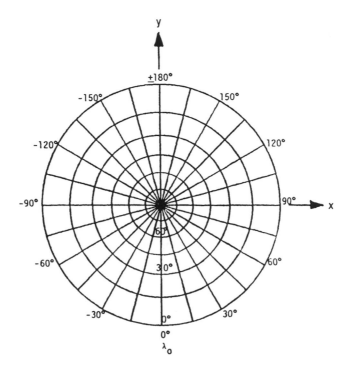

FIGURE 3.   Stereographic projection, polar case.

Next, the oblique stereographic projection based on a spherical model of the Earth is obtained by applying the transformation formulas of Section XII, Chapter 2. Let the latitude and longitude of the pole of the auxiliary coordinate system be ϕ_p and λ_p, respectively. Write Equation 86, including the scale factor, with auxiliary variables α and h, as

$$\left. \begin{array}{l} x = 2R \cdot S \cdot \tan\left(\dfrac{\pi}{4} - \dfrac{h}{2}\right)\sin\alpha \\[4mm] y = 2R \cdot S \cdot \tan\left(\dfrac{\pi}{4} - \dfrac{h}{2}\right)\cos\alpha \end{array} \right\} \tag{90}$$

Then, from Section XII, Chapter 2,

$$\left. \begin{array}{l} \alpha = \tan^{-1} \dfrac{\sin(\lambda - \lambda_p)}{\cos\phi_p\tan\phi - \sin\phi_p\cos(\lambda - \lambda_p)} \\[4mm] h = \sin^{-1}\{\sin\phi\sin\phi_p + \cos\phi\cos\phi_p\cos(\lambda - \lambda_p)\} \end{array} \right\} \tag{91}$$

Equations 90 and 91 produce a grid such as the one in Figure 4. The only straight line in this grid is the central meridian. The y axis is along the central meridian. The x axis intersects the y axis at ϕ_p.

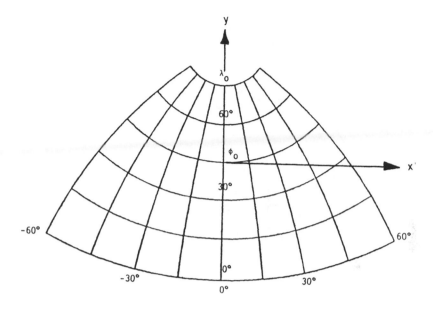

FIGURE 4.   Stereographic projection, oblique case.

The equatorial case follows from the oblique case when $\phi_p = 0°$. When this is substituted into Equation 91, the equations simplify to give

$$\left.\begin{array}{l} \alpha = \tan^{-1}\left[\dfrac{\sin(\lambda - \lambda_p)}{\tan\phi}\right] \\[3mm] h = \sin^{-1}[\cos\phi\cos(\lambda - \lambda_p)] \end{array}\right\} \qquad (92)$$

Figure 5 shows an equatorial stereographic projection. The equator and central meridian are the only straight lines on the grid. All the other lines are arcs. The y axis is along the central meridian. The x axis is along the equator.

Of the three variations of the stereographic projection, the polar case has been the most useful. It is the conformal projection used to display polar areas. The polar stereographic projection can also be developed by placing the mapping plane secant to the Earth. The parallel of secancy then becomes true scale. Equations 86 are then multiplied by a factor which accomplishes this. The use of a secant mapping plane results in a better distribution of distortion.

A. Example 3

Given: $\lambda_o = 0°, \lambda = 25°, \phi = 80°, R = 6,371,007\text{ m}, S = 1:5,000,000.$

Find: the Cartesian coordinates for the polar stereographic projection of a spherical earth model.

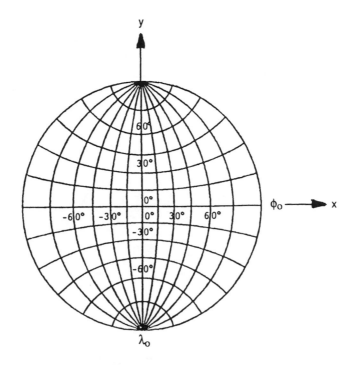

FIGURE 5.   Stereographic projection equatorial case.

$$\Delta\lambda = \lambda - \lambda_o$$

$$= 25 - 0 = 25°$$

$$= 2R \cdot S \tan\left(45° - \frac{\phi}{2}\right)$$

$$= \frac{(2)(6,371,007)}{5,000,000} \tan\left(45° - \frac{80}{2}\right)$$

$$= (2.54840)(0.087489)$$

$$= 0.22296 \text{ m}$$

$$x = \rho\sin\Delta\lambda$$

$$= (0.22296)(\sin 25°)$$

$$= (0.22296)(0.422618)$$

$$= 0.0942 \text{ m}$$

$$y = -\rho\cos\Delta\lambda$$

$$= -(0.22296)(\cos 25°)$$

$$= -(0.22296)(0.906308)$$

$$= -0.2021 \text{ m}$$

B. Example 4

Given: $\lambda_o = 0°$, $\lambda = 25°$, $\phi = 80°$, $S = 1:5,000,000$, $a = 6,378,135$ m, $= 0.081819$.

Find: the Cartesian coordinates for the polar stereographic projection of a spherical earth model.

$$\Delta\lambda = \lambda - \lambda_o$$

$$= 25 - 0 = 25°$$

$$\rho = \frac{2aS}{1 - e^2} \left(\frac{1 - e}{1 + e}\right)^{e/2} \tan\left(\frac{\pi}{4} - \frac{\phi}{2}\right)\left(\frac{1 + e\sin\phi}{1 - e\sin\phi}\right)^{e/2}$$

$$= (2)\left(\frac{6,378,135}{5,000,000}\right)\left(\frac{1 + 0.081819}{1 + 0.081819}\right)^{0.081819/2}$$

$$\times \tan\left(45° - \frac{80}{2}\right)\left[\frac{1 + (0.081819)\sin80°}{1 - (0.081819)\sin80°}\right]^{0.081819/2}$$

$$= \left(\frac{2.55125}{0.993306}\right)\left(\frac{0.918181}{1.081819}\right)^{0.04091} \times (0.087489)$$

$$\times \left(\frac{1.080576}{0.919424}\right)^{0.04091}$$

$$= (2.56844)(0.993313)(0.087489)(1.00343)$$

$$= 0.22397 \text{ m}$$

$$x = \rho\sin\Delta\lambda$$

$$= (0.22397)(\sin25°)$$

$$= 0.0946 \text{ m}$$

$$y = -\rho\cos\Delta\lambda$$

$$= -(0.22397)(\cos25°)$$

$$= -0.2030 \text{ m}$$

VII. MERCATOR[3,6,8,9,10,11,12]

The Mercator projection, devised in 1569 by Gerardus Mercator, marked the beginning of modern map projections. It has served as an aid to navigation from the age of ocean exploration to the age of space exploration.

In this section, all three of the Mercator variations, the regular, the oblique, and the transverse, are considered. The general procedure for a transformation from the spherical model of the Earth is to derive the regular Mercator projection, and then to rotate the mapping surface to obtain the oblique and transverse cases. The regular Mercator projection is also derived for the transformation from a reference spheroid. The inverse transformation is also given for the spherical case for the regular Mercator projection.

The Mercator projection is a cylindrical projection. In the regular version to be derived, the cylinder is tangent to the model of the Earth at the equator. The Mercator projection can be considered as a semigraphical projection. As indicated in Figure 10, Chapter 1, the projection can be obtained graphically in terms of a variably located projection point. This projection point, as a function of latitude, must be obtained mathematically. However, the most practical method of developing the projection is by a strictly mathematical approach.

The projection equations are now derived for the spherical Earth model. Let the Cartesian mapping equations be defined by the following relations:

$$\left.\begin{array}{l} x = RS(\lambda - \lambda_o) \\[2mm] y = y(\phi) \end{array}\right\} \tag{93}$$

With the x coordinate arbitrarily defined at the beginning of the derivation, the y coordinate is determined by the condition of conformality.

The differential forms of Equation 93 are

$$\left.\begin{array}{l} dx = RSd\lambda \\[2mm] dy = \dfrac{dy}{d\phi}\, d\phi \end{array}\right\} \tag{94}$$

Consider the first fundamental form for the cylinder.

$$(ds)^2 = (dy)^2 + (dx)^2 \tag{95}$$

Substitute Equations 94 into Equation 95 to obtain

$$(ds)^2 = \left(\dfrac{dy}{d\phi}\right)^2 (d\phi)^2 + R^2 S^2 (d\lambda)^2 \tag{96}$$

From Equation 96, the first fundamental quantities are given for the plotting surface as

$$\left.\begin{array}{l} E = \left(\dfrac{dy}{d\phi}\right)^2 \\[3mm] G = R^2 S^2 \end{array}\right\} \tag{97}$$

TABLE 3
Normalized Plotting Coordinates for the Equatorial Mercator Projection ($\phi_o = 0°$, $\lambda_o = 0°$, $R \cdot S = 1$)

Latitude or longitude	0°	30°	60°	90°	120°	150°	180°
x	0.000	0.524	1.047	1.571	2.094	2.618	3.142
y	0.000	0.549	1.317	—	—	—	—

The first fundamental quantities for the spherical model of the Earth are

$$\left. \begin{array}{l} e = R^2 \\ \\ g = R^2\cos^2\phi \end{array} \right\} \qquad (98)$$

Equations 97 and 98 are then substituted into Equation 1 and simplified

$$\frac{R^2S^2}{R^2\cos^2\phi} = \frac{\left(\dfrac{dy}{d\phi}\right)^2}{R^2}$$

$$dy = \frac{RSd\phi}{\cos\phi} \qquad (99)$$

Equation 99 is easily integrated.

$$y = R \cdot S \ln \tan\left(\frac{\pi}{4} - \frac{\phi}{2}\right) + c \qquad (100)$$

Impose the condition that $y = 0$ when $\phi = 0$. Then, $c = 0$ in Equation 100, and the second plotting equation is obtained.

$$y = R \cdot S \ln \tan\left(\frac{\pi}{4} + \frac{\phi}{2}\right) \qquad (101)$$

The first of Equations 93 and Equation 101 are the plotting equations for this transformation. A plotting table for this projection is in Table 3. This is a universal table based on an arbitrary choice of $\lambda_o = 0°$.

Figure 6 contains the grid for the regular Mercator projection. The x axis is along the equator ($\phi_o = 0°$). The y axis is along the central meridian ($\lambda_o = 0°$). The meridians are equally spaced straight lines. The parallels are unequally spaced straight lines, orthogonal to the meridians. The spacing

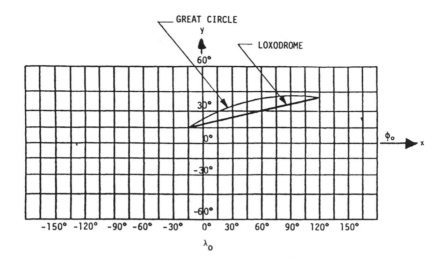

FIGURE 6.   Equatorial Mercator projection.

between the parallels increases as one proceeds toward either pole. Convergency of the meridians does not occur, and distortion becomes excessive in either poleward direction. The pole can never be reached.

The distortion at high latitudes appears especially in these areas. However, on both sides of the equator, distortion is minimal, and this projection is especially suitable for maps of the equatorial regions.

Two lines of importance on a regular Mercator projection are the loxodrome and the great circle. The loxodrome, or rhumbline, between two positions on the Mercator is a straight line which intersects all meridians at a constant angle. Note on Figure 6 that the loxodrome is portrayed as being shorter than the great circle distance between the same two points. On the spehrical model of the Earth, the great circle distance is shorter than the loxodrome. The great circle is a plane curve, and the loxodrome is a curve with torsion not equal to zero. However, the loxodrome is a more convenient concept for practical navigation.

Looking ahead to Chapter 6, on the gnomonic projection, the great circle is a straight line and the loxodrome is a curved line. Often the regular Mercator and the gnomonic projections are used together for practical navigation. In Chapter 8, there is an extended example showing distances on the loxodrome and the great circle on the Earth, the Mercator projection, and the gnomonic projection.

Turn next to inverse transformation equations for the regular Mercator projection based on a spherical Earth. The inverse relations follow simply from the plotting equations. From the first of Equations 93,

$$\lambda = \lambda_o + \frac{x}{RS} \tag{102}$$

From Equation 101,

$$\phi = 2\left[\tan^{-1}(\epsilon^{y/RS}) - \frac{\pi}{4}\right] \tag{103}$$

where $\epsilon = 2.7183$.

Consider next a derivation of plotting coordinates based on a spheroidal model of the Earth. The object of the derivation is to equate scale along the meridian and the parallel at an arbitrary point to obtain an equation which gives the y coordinate on the map.

The first step is to define an element of distance along a parallel of the spheroid. This is

$$dp_s = \frac{a\cos\phi d\lambda}{(1 - e^2\sin^2\phi)^{1/2}} \tag{104}$$

The corresponding elemental distance along the parallel of the map is

$$dp_m = ad\lambda \tag{105}$$

The scale along the parallel is given by

$$m_p = \frac{dp_m}{dp_s} \tag{106}$$

Substitute Equations 104 and 105 into Equation 106 to obtain

$$m_p = \frac{(1 - e^2\sin^2\phi)^{1/2}}{\cos\phi} \tag{107}$$

Next consider distances along the meridian. On the spheroid, the elemental distance along the meridian is given by

$$dm_s = \frac{a(1 - e^2)d\phi}{(1 - e^2\sin^2\phi)^{3/2}} \tag{108}$$

On the map, let the elemental distance along the spheroid be given by

$$dm_m = dy \tag{109}$$

The scale along the meridian is given by

$$m_m = \frac{dm_m}{dm_s} \tag{110}$$

Substitute Equations 108 and 109 into Equation 110.

$$m_m = \frac{(1 - e^2 \sin^2 \phi)^{3/2}}{a(1 - e^2) d\phi} \, dy \tag{111}$$

From the definition of conformality, the scale along the meridian and along the parallel must be equal at any point. Thus, equate Equations 107 and 111 and simplify.

$$\frac{(1 - e^2 \sin^2 \phi)^{3/2} dy}{a(1 - e^2) d\phi} = \frac{(1 - e^2 \sin^2 \phi)^{1/2}}{\cos\phi}$$

$$dy = \frac{a(1 - e^2) d\phi}{(1 - e^2 \sin^2 \phi) \cos\phi} \tag{112}$$

The y coordinate is found by integrating Equation 112 between zero and an arbitrary latitude. This is accomplished by first expanding Equation 112 by partial fractions.

$$y = \int_0^\phi \frac{a(1 - e^2) d\phi}{(1 - e^2 \sin^2 \phi) \cos\phi}$$

$$= a \left\{ \int_0^\phi \frac{d\phi}{\cos\phi} + \frac{e}{2} \int_0^\phi \frac{-e \cos\phi d\phi}{1 - e\sin\phi} - \frac{e}{2} \int_0^\phi \frac{e \cos\phi d\phi}{1 + e\sin\phi} \right\}$$

$$= a \left\{ \int_0^\phi \frac{d\phi}{\sin(\pi/2 + \phi)} + \frac{e}{2} \int_0^\phi \frac{-e \cos\phi d\phi}{1 - e\sin\phi} - \frac{e}{2} \int_0^\phi \frac{e \cos\phi d\phi}{1 + e\sin\phi} \right\}$$

$$= a \left\{ \int_0^\phi \frac{\cos\left(\dfrac{\pi}{4} + \dfrac{\phi}{2}\right)}{\sin\left(\dfrac{\pi}{4} + \dfrac{\phi}{2}\right)} \frac{d\phi}{2} - \int_0^\phi \frac{\sin\left(\dfrac{\pi}{4} + \dfrac{\phi}{2}\right)}{\cos\left(\dfrac{\pi}{4} + \dfrac{\phi}{2}\right)} \frac{d\phi}{2} \right.$$

$$\left. + \frac{e}{2} \int_0^\phi \frac{-e \cos\phi d\phi}{1 - e\sin\phi} - \frac{e}{2} \int_0^\phi \frac{e \cos\phi d\phi}{1 - e\sin\phi} \right\}$$

$$= a \left\{ \ln[\sin(\pi/4 + \phi/2)] - \ln[\cos(\pi/4 + \phi/2)] \right.$$

$$\left. + \frac{e}{2} \ln(1 - e\sin\phi) - \frac{e}{2} \ln(1 + e\sin\phi) \right\}$$

$$= a \left\{ \ln[\tan(\pi/4 + \phi/2)] + \frac{e}{2} \ln\left(\frac{1 - e\sin\phi}{1 + e\sin\phi}\right) \right\}$$

$$= a \ln \left[\tan\left(\frac{\pi}{4} + \frac{\phi}{2}\right) \left(\frac{1 - e\sin\phi}{1 + e\sin\phi}\right)^{e/2} \right] \tag{113}$$

TABLE 4
Normalized Plotting Coordinates for the Oblique Mercator Projection ($\phi_o = 45°$, $\lambda_o = 0°$, $R \cdot S = 1$)

Tabulated Values: x/y

Longitude	Latitude				
	0°	10°	20°	30°	40°
0°	0.000	0.124	0.252	0.388	0.536
	0.000	0.123	0.247	0.369	0.490
10°	0.124	0.246	0.368	0.494	0.629
	−0.123	0.002	0.127	0.252	0.377
20°	0.252	0.372	0.488	0.605	0.726
	−0.247	−0.116	0.015	0.145	0.276
30°	0.388	0.504	0.614	0.721	0.830
	−0.369	−0.229	−0.091	0.047	0.186
40°	0.536	0.647	0.749	0.845	0.940
	−0.400	−0.337	−0.187	−0.040	0.107

The x coordinate can be found from the integral

$$x = a \int_{\lambda_o}^{\lambda} d\lambda$$

$$= a(\lambda - \lambda_o) \tag{114}$$

Equation 113 reduces to Equation 101 if e = o and the value for S is included. With the inclusion of the scale factor, S, Equations 113 and 114 are the plotting equations for the spheroidal case. At the scale of Figure 6, the resulting grid would be imperceptibly different form the given figure.

The oblique and transverse Mercator projections, using the spherical model, are obtained by a mathematical manipulation of the plotting equations for the equatorial Mercator projection. The plotting equations for the oblique Mercator projection are given as

$$x = R \cdot S \cdot \tan^{-1}\left[\frac{\tan\phi\cos\phi_p + \sin\phi_p\sin\Delta\lambda}{\cos\Delta\lambda}\right]$$

$$y = \frac{R \cdot S}{2} \ln\left[\frac{1 + \sin\phi_p\sin\phi - \cos\phi_p\cos\phi\sin\Delta\lambda}{1 - \sin\phi_p\sin\phi + \cos\phi_p\cos\phi\sin\Delta\lambda}\right] \tag{115}$$

where ϕ_p is the latitude of the pole, and λ_o is displaced 90° from λ_p, the longitude of the pole.

A plotting table for a representative oblique Mercator grid is given in Table 4. Each plotting table is unique to a user's choice of the inclination of

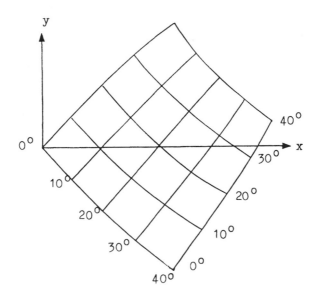

FIGURE 7. Oblique Mercator projection.

the projection pole. A sample oblique Mercator grid is given in Figure 7. For the oblique Mercator projection, there are no straight parallels and only two straight meridians. In spite of this, orthogonality is maintained with respect to the intersection of meridians and parallels. Note the location of the coordinate axes in Figure 7.

For the transverse Mercator projection, the plotting equations are as follows.

$$
\left.
\begin{aligned}
x &= \frac{R \cdot S}{2} \ln\left(\frac{1 + \cos\phi\sin\Delta\lambda}{1 - \cos\phi\sin\Delta\lambda}\right) \\[2mm]
y &= R \cdot S \tan^{-1}\left(\frac{\tan\phi}{\cos\Delta\lambda}\right)
\end{aligned}
\right\} \tag{116}
$$

Table 5 gives the plotting coordinates for a small-scale transverse Mercator projection. In this case, a universally applicable table may be calculated.

The transverse Mercator grid resulting from Table 5 is presented as Figure 8. The central meridian and the equator are the only straight lines. However, the curved parallels and meridians intersect orthogonally. The x axis is along the equator. The y axis is along the central meridian.

For the spheroidal case, the plotting equations are very complex for the oblique and transverse projections.

A. Example 5

Given: $R = 6,371,007$ m, $S = 1:20,000,000$, $\lambda_o = 0°$, $\lambda = 15°$, $\phi = 30°$.

TABLE 5
Normalized Plotting Coordinates for the Transverse Mercator Projection ($\phi_o = 0°$, $\lambda_o = 0°$, $R \cdot S = 1$)

Tabulated Values: x/y

Longitude	Latitude				
	0°	15°	30°	45°	60°
0°	0.000	0.000	0.000	0.000	0.000
	0.000	0.262	0.534	0.785	1.047
15°	0.265	0.255	0.228	0.185	0.130
	0.000	0.270	0.539	0.803	1.062
30°	0.549	0.527	0.464	0.369	0.255
	0.000	0.300	0.588	0.857	1.107
45°	0.881	0.835	0.712	0.549	0.369
	0.000	0.362	0.685	0.955	1.183
60°	1.317	1.209	0.973	0.713	0.464
	0.000	0.492	0.957	1.107	1.290

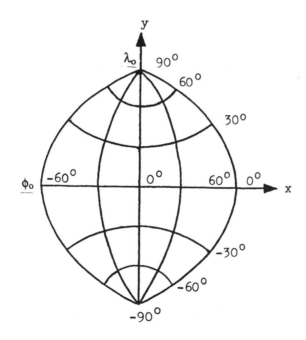

FIGURE 8. Transverse Mercator projection.

Find: x and y for an equatorial Mercator projection of the spherical Earth.

$$\Delta\lambda = \lambda - \lambda_o$$

$$= 15 - 0 = 15°$$

$$x = RS\Delta\lambda$$

$$= \left(\frac{6,371,007}{20,000,000}\right) \cdot \left(\frac{15}{57.295}\right)$$

$$= 0.0834 \text{ m}$$

$$y = RS\ln\tan\left(\frac{\pi}{4} + \frac{\phi}{2}\right)$$

$$= \left(\frac{6,371,007}{20,000,000}\right)\ln\tan\left(45° + \frac{30}{2}\right)$$

$$= (0.318550)(0.549306)$$

$$= 0.1750 \text{ m}$$

B. Example 6

Given: $a = 6,378,135$, $S = 1:20,000,000$, $\lambda_o = 0°$, $\phi = 30°$, $\lambda = 15°$, $e = 0.081819$.

Find: the Cartesian coordinates for an equatorial Mercator projection of a spheroidal Earth model.

$$\Delta\lambda = \lambda - \lambda_o$$

$$= 15 - 0 = 15°$$

$$x = RS\Delta\lambda$$

$$= \left(\frac{6,378,135}{20,000,000}\right) \cdot \left(\frac{15}{57.295}\right)$$

$$= 0.0835 \text{ m}$$

$$y = a \cdot S \cdot \ln\left[\tan\left(\frac{\pi}{4} + \frac{\phi}{2}\right)\left(\frac{1 - e\sin\phi}{1 + e\sin\phi}\right)^{e/2}\right]$$

$$= \left(\frac{6,378,135}{20,000,000}\right)\ln\left[\tan\left(45° + \frac{30}{2}\right)\right.$$

$$\left. \times \left(\frac{1 - (0.081819)(\sin30°)}{1 + (0.081819)(\sin30°)}\right)^{0.081819/2}\right]$$

$$= (0.318907)\ln\left[(1.73205) \times \left(\frac{1 - 0.04091}{1 + 0.04091}\right)^{0.04091} \right]$$

$$= (0.318907)\ln\left[(1.73205)\left(\frac{0.95909}{1.04091}\right)^{0.04091} \right]$$

$$= (0.318907)\ln[(1.73205)(0.996656)]$$

$$= (0.318907) \cdot (0.545956)$$

$$= 0.1741 \text{ m}$$

C. Example 7

Given: an equatorial Mercator projection based on a spherical Earth model; λ_o = 0°, R = 6,371,007 m, S = 1:20,000,000, x = 0.2 m, y = 0.25 m.

Find: the geographic coordinates.

$$\Delta\lambda = \frac{x}{RS}$$

$$= \frac{(0.2)(20,000,000)(57.295)}{6,371,007}$$

$$= 35°.972$$

$$\lambda = \Delta\lambda + \lambda_o$$

$$= 35.972 + 0 = 35°.972$$

$$\phi = 2\left[\tan^{-1}\left(2.7183^{y/RS}\right) - \frac{\pi}{4} \right]$$

$$= 2\left[\tan^{-1}\left(2.7183^{(0.25)(20,000,000)/6,371,007}\right) - 45° \right]$$

$$= 2[\tan^{-1}(2.7183)^{(0.78480)} - 45°]$$

$$= 2[\tan^{-1}(2.19198) - 45°]$$

$$= 2(65.477 - 45)$$

$$= (2)(20.477)$$

$$= 40°.954$$

D. Example 8

Given: λ_o = 0°, λ = 30°, ϕ = 30°, R = 6,371,007 m, S = 1:20,000,000.

Find: the Cartesian coordinates for a transverse Mercator projection.

$$\Delta\lambda = \lambda - \lambda_o$$

$$= 30 - 0 = 30°$$

$$x = \frac{RS}{2}\ln\left(\frac{1 + \cos\phi\sin\Delta\lambda}{1 - \cos\phi\sin\Delta\lambda}\right)$$

$$= \frac{(6,371,007)}{(2)(20,000,000)}\ln\left(\frac{1 + \cos30°\sin30°}{1 - \cos30°\sin30°}\right)$$

$$= (0.159275)\ln\left[\frac{1 + (0.866025)(0.500000)}{1 - (0.866025)(0.500000)}\right]$$

$$= (0.159275)\ln\left(\frac{1.433013}{0.566987}\right)$$

$$= (0.159275)(0.927198)$$

$$= 0.1477 \text{ m}$$

$$y = RS\tan^{-1}\left(\frac{\tan\phi}{\cos\Delta\lambda}\right)$$

$$= \left(\frac{6,371,007}{20,000,000}\right) \cdot \left(\frac{1}{57.295}\right)\tan^{-1}\left(\frac{\tan30°}{\cos30°}\right)$$

$$= (0.00555939)\tan^{-1}\left(\frac{0.577350}{0.866025}\right)$$

$$= (0.00555939)(33.6900)$$

$$= 0.1873 \text{ m}$$

E. Example 9

Given: $\phi_p = 45°$, $\phi = 30°$, $\Delta\lambda = 30°$, $R = 6,371,007$ m, $S = 1:20,000,000$.

Find: the Cartesian coordinates for an oblique Mercator projection.

$$x = RS\tan^{-1}\left[\frac{\tan\phi\cos\phi_p + \sin\phi_p\sin\Delta\lambda}{\cos\Delta\lambda}\right]$$

$$= \frac{6,371,007}{20,000,000}\tan^{-1}\left[\frac{\tan30°\cos45° + \sin45°\sin30°}{\cos30°}\right]$$

$$= (0.318550)\tan^{-1}\left\{\frac{[(0.577350)(0.707107) + (0.707107)(0.5)]}{0.866025}\right\}$$

$$= (0.318550)\tan^{-1}\left\{\frac{0.408248 + 0.353554}{0.866025}\right\}$$

$$= (0.318550)\tan^{-1}\left(\frac{0.761802}{0.866025}\right)$$

$$= (0.318550)(0.721595)$$

$$= 0.22986 \text{ m}$$

$$y = \frac{RS}{2} \ln\left[\frac{1 + (\sin\phi_p\sin\phi - \cos\phi_p\cos\phi\sin\Delta\lambda)}{1 - (\sin\phi_p\sin\phi - \cos\phi_p\cos\phi\sin\Delta\lambda)}\right]$$

$$\sin45°\sin30° - \cos45°\cos30°\sin30°$$

$$= (0.707107)(0.5) - (0.707107)(0.866025)(0.5)$$

$$= 0.353554 - 0.306186 = 0.047368$$

$$y = \left(\frac{0.318550}{2}\right)\ln\left[\frac{1 + 0.047368}{1 - 0.047368}\right]$$

$$= (0.159275)\ln\left[\frac{1.047368}{0.952632}\right]$$

$$= (0.159275)(0.094807)$$

$$= 0.01514 \text{ m}$$

VIII. STATE PLANE COORDINATES[2,4,5]

A state plane coordinate system provides a convenient means of locating mapping positions on the planar mapping surface. The state plane systems are based on a rectangular grid defined for each state of the U.S. These systems permit the methods of plane surveying to be extended over great distances. At the same time, a precision approaching that of geodetic surveying is maintained.

In the continental U.S. the two map projections used for the state plane coordinate systems are the transverse Mercator secant to the model of the Earth and the Lambert conformal with two standard parallels. Since both are conformal projections, angle and size are maintained at a point.

The transverse Mercator projection defines a grid zone roughly 158 mi east and west. This bounds the distortion, which increases at distances from the central meridian. Distortion is not a function of latitude, so north-south extension is unlimited. Thus, the transverse Mercator projection is well suited for states of major north-to-south extent, such as Illinois. There are two true scale lines running roughly north and south in each transverse Mercator zone. Between these lines, which are the lines of secancy, the distances are less than true scale. Outside of the true scale lines, the distances are greater than true scale.

The Lambert conformal projection with two standard parallels defines a

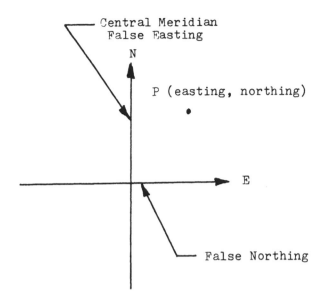

FIGURE 9.   State plane coordinate system geometry.

grid zone roughly 158 mi north and south. This bounds the distortion which is a function of latitude. Since distortion is not a function of longitude, east-west extension is unlimited. Thus, the Lambert conformal projection is suited for states of major east-to-west extent, such as Tennessee. The standard parallels, running east and west, of this conical projection are true scale. Between the true scale lines, the distances are less than true scale. North and south of the true-scale lines, the distances are more than true scale. The mid-latitude location of the U.S. dictated the use of the Lambert conformal for states of major east-to-west extent.

For states nearly square in shape, either projection could be used. However, the Lambert conformal projection permits somewhat less complex equations and is thus the projection of choice.

The transverse Mercator and Lambert conformal zones are bounded by state and county boundaries. The number of zones per state ranges from two to eight, depending on the dimensions of the state, except for the very small states. The zones are designed to maintain scale distortion within 1 part in 10,000.

In order to obtain the precision required, the mapping equations have been developed for a spheroidal model of the Earth. The constants for the Clarke 1866 ellipsoid have been incorporated in the equations for the transformations.

Figure 9 contains the coordinate system applicable to both the Lambert conformal and transverse Mercator grids. The central meridian is given a false easting, or x coordinate. This forces all eastings of arbitrary points, such as P, to be positive numbers. The northings are the y coordinates. In a similar manner, they are all adjusted to be positive numbers.

The grid resulting from the direct transformation equations for either have several characteristics which need to be noted. First, only the central or reference meridian on the grid is parallel to the corresponding meridian on the Earth. Other meridians are inclined or are curved lines. Second, azimuths on the grid are not exactly true. For precision work, corrections are necessary; and grids from state to state are not overlapping. One cannot create a matching mosaic.

Equations for both the direct and inverse transformations have been developed for each of the projections. The U.S. Coast and Geodetic Survey has developed a method for accomplishing these transformations in a manner well suited for the computer. The direct and inverse transformations for the transferse Mercator and Lambert conformal projections are given below. For each projection, a set of constants is defined. These constants are unique for each state plane coordinate system, since they contain the local geodetic information. The user should consult Reference 2 for the locality of interest.

Beginning with the transverse Mercator projection, the transformation constants are defined as follows:

- T_1 is the false easting, or x coordinate, of the central meridian, in feet.
- T_2 is the central meridian expressed in seconds.
- T_3 is the degree and minutes portion, in minutes, of the rectifying latitude, ω_0, for ϕ_0, the latitude of the origin. (See the explanation of ω'' under L_7 of the Lambert projection below.)
- T_4 is the remainder of ω_0, i.e., the seconds.
- T_5 is the scale along the central meridian.
- $T_6 = (1/6\, R_m R_p T_5^2) \times 10^{15}$. ($R_m$ and R_p are computed for the mean latitude of the area in the zone, by Equations 28 and 21 of Chapter 3, respectively.)

For the direct transverse Mercator transformation, from geographic position to state plane coordinates, the algorithm is as follows:

$$S_1 = \frac{30.92241\,724\cos\phi}{(1 - 0.00676\,86580\sin^2\phi)^{1/2}}\left[T_2 - \lambda - 3.9174\left(\frac{T^2 - \lambda}{10^4}\right)^3\right]$$

$$\text{(in seconds)}$$

$$S_m = S_1 + 4.0831\left(\frac{S_1}{10^5}\right)^3$$

$$x = T_1 + 3.28083\,333\,S_m T_5 + \left(\frac{3.28083\,333\,S_m T_5}{10^5}\right)^3 T_6$$

(T_1 and x will be in the millions in New Jersey; x may be in millions in certain other states.)

$$\phi_1 = \phi + \frac{25.52381}{10^{10}} S_m^2 (1 - 0.00676\,86580\sin^2\phi)^2\tan\phi \quad (\phi \text{ in seconds})$$

$$\phi_2 = \phi + \frac{25.52381}{10^{10}} S_m^2 (1 - 0.00676\,86580\sin^2\phi_1)^2\tan\phi_1$$

$$y = 101.27940\,65\,T_5\{60(\phi' - T_3) + \phi'' - T_4 - [1,052.89388\,2$$
$$- (4.48334\,4 - 0.02352\,0\cos^2\phi_2)\cos^2\phi_2]\sin\phi_2\cos\phi_2\}$$

(ϕ' is degrees and minutes of ϕ_2 in whole minutes and

ϕ'' is the remainder of ϕ_2 in seconds)

For the inverse transverse Mercator transformation, from state plane coordinates to geographic position, the algorithm is as follows:

$$S_{gl} = x - T_1 - T_6\left(\frac{x - T_1}{10^5}\right)^3$$

$$S_m = \frac{0.30480\,06099}{T_5}\left[x - T_1 - T_6\left(\frac{S_{gl}}{10^5}\right)^3\right]$$

$\omega' = T_3$ (degrees and minutes of ω in whole minutes)

$$\omega'' = T_4 + \frac{0.00987\,36755\,53}{T_5} y \quad (\text{remainder of } \omega \text{ in seconds})$$

$$\omega = \omega' + \omega''$$

$(\phi')' = T_3$ (degrees and minutes of ϕ and ϕ' in whole minutes)

$(\phi')'' = \omega'' + [1,047.54671\,0 + (6.19276\,0$

$\qquad + 0.05091\,2\cos^2\omega)\cos^2\omega]\sin\omega\cos\omega$

(remainder of ϕ' in seconds)

$$\phi' = (\phi')' + (\phi')''$$

$$(\phi)'' = (\phi')'' - 25.52381(1 - 0.00676\,86580\sin^2\phi')^2\left(\frac{S_m}{10^5}\right)\tan\phi'$$

$$\phi = (\phi')' + (\phi)''$$

$$S_a = S_m - 4.0831\left(\frac{S_m}{10^5}\right)^3$$

$$S_1 = S_m - 4.0831\left(\frac{S_a}{10^5}\right)^3$$

$$\Delta\lambda_1 = \frac{S_1(1 - 0.00676\,86580\sin^2\phi)^{1/2}}{30.9224\,724\cos\phi}$$

$$\Delta\lambda_a = \Delta\lambda_1 + 3.9174\left(\frac{\Delta\lambda_1}{10^4}\right)^3$$

$$\lambda'' = T_2 - \Delta\lambda_1 - 3.9174\left(\frac{\Delta\lambda_a}{10^4}\right)^3$$

A similar set of transformation constants for the Lambert conformal projection are defined as follows:

- L_1 is the false easting, or x coordinates, of the central meridian.
- L_2 is the central meridian expressed in seconds.
- L_3 is the map radius of the central parallel (ϕ_o).
- L_4 is the map radius of the lowest parallel of the projection table plus the y value on the central meridian at this parallel. This y value is zero in most, but all cases.
- L_5 is the scale (m) of the projection along the central parallel (ϕ_o).
- L_6 is the c computed from the basic equations for the Lambert projection with two standard parallels.
- L_7 is the degrees and minutes portion, in minutes, of the rectifying latitude for ϕ_o, where c = $\sin\phi_o$, Equation 66. Rectifying latitude is defined as $\omega'' = \phi'' - 525.32978\sin2\phi + 0.557478\sin4\phi - 0.000735\sin6\phi$
- L_8 is the remainder of ω_o, i.e., the seconds.
- $L_9 = (1/6R_mR_p) \times 10^{16}$.
- $L_{10} = \tan\phi_o/24(R_mR_p)^{3/2} \times 10^{24}$.
- $L_{11} = (5 + 3\tan^2\phi_o)/120R_mR_p^3 \times 10^{32}$.

where R_m and R_p (of L_9, L_{10}, and L_{11}) are given by Equations 28 and 21 of Chapter 3, respectively, evaluated at ϕ_o.

The algorithm for the direct Lambert conformal transformation, from geographic position to state plane coordinates, is given below:

$$S = 101.27940\,65\{60(L_7 - \phi') + L_8 - \phi'' + 1,052.89388\,2$$
$$- (4.48334\,4 - 0.02352\,0\cos^2\phi)\cos^2\phi]\sin\phi\cos\phi\}$$

(ϕ' is the degrees and minutes of ϕ expressed in whole minutes and ϕ'' is the remainder of ϕ in seconds)

$$R = L_3 + sL_5\left\{1 + \left(\frac{S}{10^8}\right)^2\left[L_9 - \left(\frac{S}{10^8}\right)L_{10} + \left(\frac{S}{10^8}\right)^2L_{11}\right]\right\}$$

$$\theta = L_6(L_2 - \lambda) \quad (\theta \text{ and } \lambda \text{ are in seconds})$$

$$x = L_4 + R\sin\theta$$

$$y = L_4 - R + 2R\sin^2\frac{\theta}{2}$$

For the inverse Lambert conformal transformation, from state plane coordinates to geographic position, the algorithm is as follows:

$$\theta = \tan^{-1}\left(\frac{x - L_1}{L_4 - y}\right)$$

$$\lambda = L_2 - \frac{\theta}{L_6} \quad (\theta \text{ and are in seconds})$$

$$R = \frac{L_4 - y}{\cos\theta}$$

$$S_1 = \frac{L_4 - L_3 - y + 2R\sin^2\dfrac{\theta}{2}}{L_5}$$

$$S_2 = \frac{S_1}{1 + \left(\dfrac{S_1}{10^8}\right)^2 L_9 - \left(\dfrac{S_1}{10^8}\right)^3 L_{10} + \left(\dfrac{S_2}{10^8}\right)^4 L_{11}}$$

$$S_3 = \frac{S_1}{1 + \left(\frac{S_2}{10^8}\right)^2 L_9 - \left(\frac{S_2}{10^8}\right)^2 L_{10} + \left(\frac{S_2}{10^8}\right)^4 L_{11}}$$

$$S = \frac{S_1}{1 + \left(\frac{S_3}{10^8}\right)^2 L_9 - \left(\frac{S_3}{10^8}\right)^3 L_{10} + \left(\frac{S_3}{10^8}\right)^4 L_{11}}$$

$\omega' = L_7 - 600$ (degrees and minutes of ω in whole minutes)

$\omega'' = 36{,}000 + L_8 - 0.00987\,36755\,53\,S$ (remainder of ω in seconds)

$\omega = \omega' + \omega''$

$\phi' = L_7 - 600$ (degrees and minutes of ϕ in whole minutes)

$\phi'' = \omega'' + [1{,}047.54671\,0 + (6.19276\,0$

$+\ 0.05091\,2\cos^2\omega)\cos^2\omega]\sin\omega\cos\omega$

(remainder of ϕ in seconds)

$\phi = \phi' + \phi''$

A FORTRAN IV program incorporating the previous four algorithms is given in Section V, Chapter 9. Numerical examples of the evaluation of these algorithms are given in Reference 2.

IX. MILITARY GRID SYSTEMS[4,7,13]

Large-scale military maps used by the U.S. use two conformal projections, the transverse Mercator and the polar stereographic. Specific points on the surface of the Earth can be located by a reference to latitude and longitude on these projections. However, since the parallels and meridians on the map may be curved and unevenly spaced lines, the precise location of a point will require the use of a grid overlay with curved lines which applies at only specific zones of latitude and longitude.

A rectangular grid is superimposed on military maps to assist in the location of points. This concept is similar to that of the state plane coordinate system for the U.S. introduced in the previous section. For the military maps, the worldwide grid system adopted by the Department of the Army makes use of the universal transverse Mercator (UTM) grid and the universal polar stereograph (UPS) grid superimposed on the respective projections.

A grid system has certain advantages over the use of geographic coordinates. First, every grid square is of the same size and shape. Second, linear values can be used rather than angular values.

For the UTM grid, the world is divided into 60 north to south zones. Each zone covers a 6°-wide strip of longitude. The maximum extent of the zone was chosen to minimize distortion. The zones are numbered consecutively, beginning at zone 1 between 180 and 174° west to zone 60 between 174 and 180° east longitude.

For the U.S., the zones range from zone 10 on the West Coast to zone 19 in New England.

Each zone is then divided into 19 segments with an 8° difference in latitude, plus an additional segment at the extreme north with a 12° difference in latitude. The rows of these segments are lettered from south to north by the letters C through X (with omission of the letters I and O to avoid confusion). By specifying a letter and a number, each element in the UTM system is uniquely identified.

Within each zone, the location of a point is made by specifying its x coordinate, the easting, and its y coordinate, the northing. The unit of the coordinates is meters. The central meridian of each zone is assigned the false easting value of 500,000 m. This is done to avoid negative numbers at the western edge of the zone. The easting numbers increase from west to east. For north-south values in the Northern Hemisphere, the equator has a false northing value of 0 m, and the northings increase towards the North Pole. In the Southern Hemisphere, the equator has a false northing value of 10,000,000 m, and the northings decrease towards the South Pole. The coordinate system for the UTM grid is illustrated in Figure 10a. Of course, the scale of the grid must be consistent with the scale of the base transverse Mercator projection. An adaptation of the plotting equations from Section VII are used to calculate the location of a point in this grid.

For the UTM grid, the mapping cylinder is secant to the Earth. The local scale factor for the central meridian, halfway between the zone boundary meridians, is 0.9996. This limits maximum variation to 1 part in 1000.

The UPS grid is applicable in the polar regions of the Earth, that is, north of 84° N latitude and south of 80° S latitude. The coordinate systems for the UPS system are given in Figure 10b. In the northern case, the origin is the North Pole, and in the southern case, the origin is the South Pole. In both cases, the 0-to-180° meridian line has a false easting of 2,000,000 m, and the 90-to-270° meridian line has a false northing of 2,000,000 m.

Each 0-to-180° meridian divides its respective polar grid into two zones. From Figure 10b it is seen that the two zones for the southern polar cap are labeled A and B. For the northern polar cap, the zones are Y and Z. Thus, for the UPS, a single letter uniquely identifies a zone.

For the UPS grid, a point is located by an adaptation of the plotting equations of Section VI. In this projection, the mapping plane is secant to

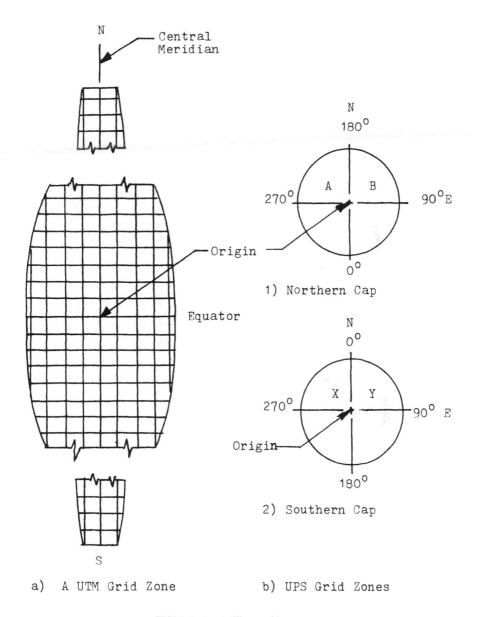

a)   A UTM Grid Zone   b) UPS Grid Zones

FIGURE 10.   Military grid system.

the Earth. The local central scale factor is 0.994. This gives standard parallels of ± 81°06′52.3″.

An abridged form of the plotting equations for the UTM and UPS is given below. For the UTM, with the mapping cylinder secant to the Earth, the northing (N) and easting (E) equations are

$$N = 0.9996Y + FN \left.\vphantom{\begin{matrix}a\\b\end{matrix}}\right\} \tag{117}$$
$$E = 0.9996X + FE$$

where X and Y are the basic coordinates of a point, as calculated below, and FN and FE are the false northing and false easting, respectively.

The equations for the coordinates to be used in Equation 118 are given by the following for a spheroidal Earth model.

$$x = R_p \left[\Delta\lambda\cos\phi + \frac{\Delta\lambda^3\cos^3\phi}{6} (1 - \tan^2\phi + \eta^2) \right]$$

$$y = d_m + R_p \left[\frac{\Delta\lambda^2}{2} \sin\phi\cos\phi + \frac{\Delta\lambda^4}{24} \sin\phi\cos^3\phi(5 - \tan^2\phi + 9\eta^2) \right] \tag{118}$$

The terms in Equation 118 is defined as follows. $\Delta\lambda = \lambda - \lambda_o$ is the longitude difference from the central meridian of the zone containing the point of interest; d_m is the distance along the meridian from the equator to the latitude ϕ as defined by Equation 37 of Chapter 3; from Equation 21, Chapter 3,

$$R_p = \frac{a}{\sqrt{1 - e^2\sin^2\phi}} \tag{119}$$

is the radius of curvature in the prime meridian; and

$$\eta^2 = \frac{e^2}{1 - e^2} \cos^2\phi \tag{120}$$

is an auxiliary variable.

The transformation equations for northing and easting for the UPS are

$$N = Y + 2,000,000 \left.\vphantom{\begin{matrix}a\\b\end{matrix}}\right\} \tag{121}$$
$$E = X + 2,000,000$$

The X and Y coordinates to be used in Equation 121 are given as

$$X = \rho\sin\Delta\lambda$$
$$Y = \rho\cos\Delta\lambda \qquad \text{(North UPS zone)} \tag{122}$$
$$Y = -\rho\cos\Delta\lambda \qquad \text{(South UPS zone)}$$

In these equations, ρ is the polar coordinate for a secant polar stereographic projection based on a spheroidal Earth model.

A. Example 10

Given: $a = 6,378,135$ m, $e = 0.081819$, $\phi = 2°$, $\Delta\lambda = 2°$.

Find: the UTM coordinates.

$$e^2 = 0.0066943 \qquad \cos = 0.999391$$

$$e^4 = 0.0000448 \qquad \sin\phi = 0.034900$$

$$FE = 500,000 \text{ m} \qquad \tan\phi = 0.034921$$

$$FN = 0 \text{ m}$$

$$R_p = \frac{a}{\sqrt{1 - e^2\sin^2\phi}} = \frac{6,378,135}{\sqrt{1 - (0.0069943)(0.034900)^2}}$$

$$= \frac{6,378,135}{\sqrt{1 - 0.0000085}} = \frac{6,378,135}{0.999996}$$

$$= 6,378,162 \text{ m}$$

$$\eta^2 = \frac{e^2\cos^2\phi}{1 - e^2} = \frac{(0.0069943)(0.999391)^2}{1 - 0.0069943}$$

$$= \frac{0.0069858}{0.993006} = 0.007035$$

$$f_1 = (1 - 0.25e^2 - 0.046875e^4)\phi$$

$$= [1 - (0.25)(0.0069943) - (0.046875)(0.0000448)]\left(\frac{2}{57.29578}\right)$$

$$= (1 - 0.001746 - 0.000002)\left(\frac{2}{57.29578}\right)$$

$$= (0.998252)\left(\frac{2}{57.29578}\right)$$

$$= 0.034846$$

$$f_2 = (0.375e^2 + 0.093758e^4)\sin4°$$

$$= [(0.375)(0.0069943) + (0.093758)(0.0000448)](0.069756)$$

$$= (0.0026229 + 0.000042)(0.069756)$$

$$= (0.0026271)(0.069756) = 0.000183$$

$$= [(0.375)(0.0069943) + (0.093758)(0.0000448)](0.069756)$$

$$= (0.0026229 + 0.000042)(0.069756)$$

$$= (0.0026271)(0.069756) = 0.000183$$

$$f_3 = 0.058594e^4\sin 8°$$

$$= (0.058594)(0.0000448)(0.139173)$$

$$= 0$$

$$d_m = a(f_1 - f_2 + f_3)$$

$$= (6,378,135)(0.034846 - 0.000183 + 0)$$

$$= 221,085 \text{ m}$$

$$f_x = 1 - \tan^2\phi + \eta^2 = 1 - (0.034921)^2 + 0.007035$$

$$= 1 - 0.001220 + 0.007035 = 1.005815$$

$$f_y = 5 - \tan^2\phi + 9\eta^2 = 5 - 0.001220 + (9)(0.007035)$$

$$= 4.998780 + 0.063315 = 5.062095$$

$$X = R_p\left[\Delta\lambda\cos\phi + \frac{\Delta\lambda^3\cos^3\phi f_x}{6}\right]$$

$$= (6,378,162)\left[\frac{(2)(0.999391)}{57.29578} + \left(\frac{(2)(0.999391)}{57.29578}\right)^3\left(\frac{1.005815}{6}\right)\right]$$

$$= (6,378,162)[0.0348853 + 0.0000071]$$

$$= (6,378,162)(0.0348924) = 222,549 \text{ m}$$

$$Y = d_m + R_p\left[\frac{\Delta\lambda^2}{2}\sin\phi\cos\phi + \frac{\Delta\lambda^4}{24}\sin\phi\cos^3\phi \cdot fy\right]$$

$$= 221,085 + (6,378,162)\left[\left(\frac{2}{57.29578}\right)^2\frac{(0.034900)(0.999391)}{2}\right.$$

$$\left. + \left(\frac{2}{57.29578}\right)^4\frac{(0.034900)(0.999391)^3(5.062095)}{24}\right]$$

$$= 221,085 + (6,378,162)(0.0000212 + 0)$$

$$= 221,085 + 135 = 221,220 \text{ m}$$

$$E = 0.9996X + FE$$

$$= (0.9996)(222,549) + 500,000$$

$$= 222,460 + 500,000 = \underline{722,460 \text{ m}}$$

$$N = 0.9996Y + FN$$
$$= (0.9996)(221,220) + 0$$
$$= \underline{221,132 \text{ m}}$$

At this point, a scale factor would be applied to E and N to reduce to map scale.

REFERENCES

1. **Bowditch, N.,** American Practical Navigator, Defense Mapping Agency Hydrographic Center, U.S. Gov't. Printing Office, Washington, D.C., 1977.
2. **Claire, C. N.,** State Plane Coordinates by Automatic Data Processing, Publ. 62-4, U.S. Coast and Geodetic Survey, U.S. Gov't. Printing Office, Washington, D.C., 1973.
3. **Deetz, C. H. and Adams, O. S.,** Elements of Map Projection, Spec. Publ. 68, U.S. Coast and Geodetic Survey, U.S. Gov't. Printing Office, Washington, D.C., 1944.
4. **Malling, D. H.,** *Coordinate Systems and Map Projections,* George Philip & Son, 1973.
5. **Moffit, F. H. and Bouchard, H.,** *Surveying,* Intext, Wiley Interscience, New York, 1975.
6. **Nelson, H.,** Five Easy Projections, GTE-Sylvania, Mountain View, Ca., 1979.
7. **O'Brien, L. P.,** The Universal Transverse Mercator Grid Formulas for Cartographic Applications Programmers, Defense Mapping Agency IAGS Chile Project, U.S. Gov't. Printing Office, Washington, D.C., 1986.
8. **Richardus, P. and Adler, R. K.,** *Map Projections for Geodesists, Cartographers, and Geographers,* North-Holland, Amsterdam, 1972.
9. **Snyder, J. P.,** Map projections used by the U.S. Geological Survey, *U.S. Geol. Surv. Bull.,* 1532, 1983.
10. **Snyder, J. P.,** Map Projections — A Working Manual, *U.S. Geol. Surv. Prof. Pap.,* 1395, 1987.
11. **Steers, J. A.,** *An Introduction to the Study of Map Projections,* University of London, London, 1962.
12. **Thomas, P. D.,** Conformal Projections in Geodesy and Cartography, Spec. Publ. 251, U.S. Coast and Geodetic Survey, U.S. Gov't. Printing Office, Washington, D.C., 1952.
13. *Map Reading,* Department of the Army Field Manual FM 21-26, Department of the Army, Washington, D.C., 1969.

Chapter 6

CONVENTIONAL PROJECTIONS

I. INTRODUCTION

Conventional projections form the third major type of map projection. These are projections which are neither equal area or conformal. Some of the conventional projections were devised in order to preserve some special quality which is more important to a particular application than the qualities of equality of area or conformality. Others were produced in order to obtain a projection which is either mathematically or graphically simple. Since the category of convention is a catchall, there is a wide variety of projections in this class.

The most useful of the conventional projections are the gnomonic, the azimuthal equidistant, the Miller, the polyconic, the Robinson, and the orthographic. Of lesser interest are the conical projections such as the perspective, and the simple conics of one- and two-standard parallels, the cylindrical projections such as the perspective, the plate Carrée, and the Carte parallelogrammatique, and a selection of specialty projections, such as the Globular, the Van der Grinten, the Cassini, and the aerial perspective. This class of projections contains both world maps and maps limited to local areas of coverage. The world maps are represented by both cylindrical and pseudocylindrical projections.

II. SUMMARY OF PROCEDURES

Unlike the equal area and conformal projections, there is no simple underlying principle that applies to the various conventional projections. Thus, there is no specific general procedure to be applied in all cases. Rather, the procedure must be developed on a projection-by-projection basis. The best that can be done is to point out the quality to be preserved in the most important of the projections, or at least the rationale behind the selection of some projection point.

The outstanding feature of the gnomonic projection is that all great circles on the Earth appear as straight lines on the projection. To bring this about, the projection point is taken at the center of the spherical Earth. The gnomonic is an azimuthal projection. A similar location for the projection point is encountered in the perspective conical and perspective cylindrical projections.

For the orthographic projection, the projection point is taken at infinity. The resulting projection is that of engineering drawing.

In the azimuthal equidistant projection, azimuths and distances from the center of the projection are to be preserved. This is also true for the Cassini projection. For the Cassini projection, however, the plotting surface is a cylinder rather than a plane.

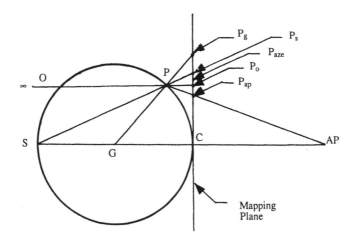

FIGURE 1.   Summary of projection points for conventional projections.

The Miller projection results from a mathematical manipulation of the plotting equations for the regular Mercator projection. The polyconic projection results from applying the procedure for a simple tangent conic at an infinity of latitudes. The Robinson projection was devised to realistically represent land masses.

It should be obvious that for the projections above, and the rest of the projections in this section, an individual procedure is needed in each case.

One can consider the variation of projection point for several conventional projections. This is done in Figure 1. Consider an arbitrary point P on the model of the Earth. The projection points for the gnomonic (G), orthographic (O), and aerial perspective (AP) projections are given on the figure. These result in the projected points P_g, P_o, and P_{ap} for the respective projections. In addition, the projection point for the stereographic projection (S) in Chapter 5 leads to the projected point P_s. Finally the point P_{aze} is the result of applying the criterion for the azimuthal equidistant project, that is, $CP = CP_{aze}$.

III. GNOMONIC

The gnomonic projection is an azimuthal projection. It may be produced by a graphical construction in which the projection point is the center of the spherical model of the Earth. A ray from the projection point intersects the surface of the Earth, and then the mapping plane, as indicated in Figure 1. This choice of a projection point gives the prime characteristic of the gnomonic projection: all great circles project as straight lines.

There are three cases of the gnomonic projection, depending on the location of the mapping plane. These are the polar, with the plane tangent at a pole, the equatorial, with the plane tangent at any point on the equator, and the oblique, with the plane tangent at any point on the surface of the Earth,

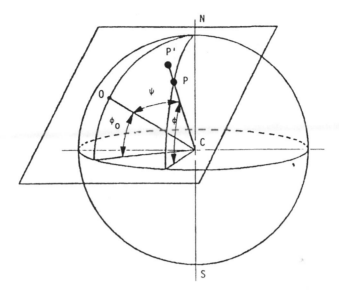

FIGURE 2. Geometry for the oblique gnomonic projection.

except the poles or the equator. The direct transformations are derived for all three cases. In addition, the inverse transformations are derived for the polar and equatorial cases.

The mathematical derivation of the gnomonic projection begins with the oblique case. Then, the polar and equatorial cases are obtained by particularizing the equations for the oblique case.

Figure 2 portrays the geometry required for deriving the oblique gnomonic projection. Let a mapping plane be tangent to the sphere at point 0, the coordinates of which are (ϕ_o, λ_o). The Cartesian axes on the plane are such that x is east, and y is north. Let an arbitrary point P, with coordinates (ϕ, λ) be projected onto the plane to become P', with mapping coordinates (x, y).

It is now necessary to define two auxiliary angles to aid in the derivation. Define the first auxiliary angle. ψ, as the angle between the two radii CO and CP. From the figure, with R as the radius of the sphere,

$$OP' = R \tan\psi \tag{1}$$

The second auxiliary angle, θ, is defined on the mapping plane. This angle orients line OP' with respect to the x axis of the map. This relation is

$$x = OP' \cos\theta \tag{2}$$

Substitute Equation 1 into Equation 2.

$$x = R\tan\psi\cos\theta$$

$$= \frac{R\sin\psi}{\cos\psi}\cos\theta \tag{3}$$

Also, we have

$$y = OP'\sin\theta \tag{4}$$

Substitute Equation 1 into Equation 4 to obtain

$$y = R\tan\psi\sin\theta$$

$$= \frac{R\sin\psi}{\cos\psi}\sin\theta \tag{5}$$

The next step is to find ψ and θ in terms of ϕ, ϕ_o, λ, and λ_o. To this end, spherical trigonometrical formulas are applied. First, from Figure 2, by the use of the law of sines, with $\Delta\lambda = \lambda - \lambda_o$,

$$\frac{\sin\Delta\lambda}{\sin\psi} = \frac{\sin(90° - \theta)}{\sin(90° - \phi)}$$

$$= \frac{\cos\theta}{\cos\phi}\sin\psi\cos\theta$$

$$\sin\psi\cos\theta = \sin\Delta\lambda\cos\phi \tag{6}$$

Apply the law of cosines to obtain the following:

$$\cos\psi = \cos(90° - \phi_o)\cos(90° - \phi)$$

$$+ \sin(90° - \phi_o)\sin(90° - \phi)\cos\Delta\lambda$$

$$= \sin\phi_o\sin\phi + \cos\phi_o\cos\phi\cos\Delta\lambda \tag{7}$$

Apply Equation 6 to obtain

$$\sin\psi\cos(90° - \theta) = \sin(90° - \phi_o)\cos(90° - \phi)$$

$$- \cos(90° - \phi_o)\sin(90° - \phi)\cos\Delta\lambda$$

$$\sin\psi\sin\theta = \cos\theta_o\sin\phi - \sin\phi_o\cos\phi\cos\Delta\lambda \tag{8}$$

These relations yield the Cartesian plotting coordinates for the oblique gnomonic projection.

TABLE 1
Normalized Plotting Coordinates for the Gnomonic Projection, Oblique Case (ϕ_o = 45°, λ_o = 0°, R · S = 1)

Tabulated Values: x/y

	Latitude					
Longitude	**0°**	**15°**	**30°**	**45°**	**60°**	**75°**
0°	0.000	0.000	0.000	0.000	0.000	0.000
	−1.000	−0.577	−0.268	0.000	0.268	0.577
15°	0.379	0.297	0.237	0.186	0.136	0.078
	−1.000	−0.566	−0.252	0.017	0.284	0.589
30°	0.816	0.624	0.490	0.379	0.272	0.154
	−1.000	−0.527	−0.200	0.072	0.333	0.623
45°	1.414	1.026	0.779	0.586	0.410	0.225
	−1.000	−0.450	−0.101	0.172	0.420	0.681
60°	2.449	1.595	1.137	0.816	0.549	0.289
	−1.000	−0.302	0.072	0.333	0.552	0.764
75°	5.728	2.593	1.634	1.085	0.686	0.342
	−1.000	0.017	0.381	0.589	0.740	0.870

Substitute Equations 6 and 7 into Equation 3, and Equations 7 and 8 into Equation 5. Also, introduce the scale factor S. The results are

$$\left.\begin{aligned} x &= \frac{R \cdot S\cos\phi\sin\Delta\lambda}{\sin\phi_o\sin\phi + \cos\phi_o\cos\phi\cos\Delta\lambda} \\ y &= \frac{R \cdot S(\cos\phi_o\sin\phi - \sin\phi_o\cos\phi\cos\Delta\lambda)}{\sin\phi_o\sin\phi + \cos\phi_o\cos\phi\cos\Delta\lambda} \end{aligned}\right\} \qquad (9)$$

Table 1 has been generated using Equations 9 for the selection of ϕ_o = 45°. Each selection of ϕ_o by the user requires a separate plotting table. There is no universal plotting table. The grid resulting from Table 1 is given in Figure 3. In this grid, the y axis is along the central meridian. The x axis is perpendicular to the y axis at ϕ_o. All the meridians and the equator are straight lines, since they are great circles. All parallels are characteristic curves.

The direct transformation for the gnomonic polar projection may be obtained from the plotting equations for the oblique gnomonic projection by letting ϕ_o = 90° in Equations 9. The resulting equations are

$$\begin{aligned} x &= \frac{R \cdot S\cos\phi\sin\Delta\lambda}{\sin\phi} \\ &= R \cdot S\cot\phi\sin\Delta\lambda \end{aligned} \qquad (10)$$

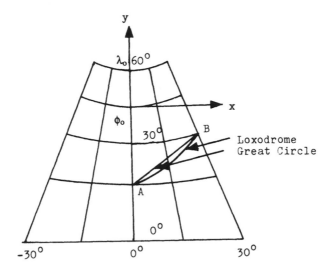

FIGURE 3.   Gnomonic projection, oblique case.

$$y = -\frac{R \cdot S\cos\phi\cos\Delta\lambda}{\sin\phi}$$

$$= -R \cdot S\cot\phi\cos\Delta\lambda \qquad (11)$$

A polar gnomonic grid is given in Figure 4 and is based on Equations 10 and 11. In this case, all meridians again are straight lines. The parallels are concentric circles, the spacing of which increases as the latitude decreases. Thus, the distortion becomes extreme as the equator is approached. The equator itself can never be portrayed on the gnomonic polar projection, since a ray from the center of the Earth to any point on the equator will be parallel to the projection plane. Table 2 is the plotting table for the polar projection. A universal plotting table can be generated for the polar case. In Figure 4, the positive x axis is along the 90° meridian. The positive y axis is along the 180° meridian.

The inverse relations for the polar case, for the transformation from Cartesian to geographic coordinates, follow from the inversion of Equations 10 and 11.

$$\left.\begin{array}{l} \Delta\lambda = \tan^{-1}\left(-\dfrac{x}{y}\right) \\[2em] \phi = \tan^{-1}\sqrt{\dfrac{(R \cdot S)^2}{x^2 + y^2}} \end{array}\right\} \qquad (12)$$

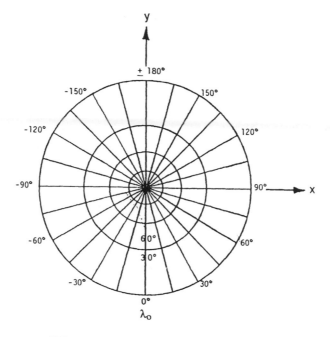

FIGURE 4.   Gnomonic projection, polar case.

The equatorial gnomonic projection is also obtained from the oblique gnomonic projection plotting equations. This time, let $\phi_o = 0°$ in Equations 9. The resulting equations are

$$x = \frac{R \cdot S\cos\phi\sin\Delta\lambda}{\cos\phi\cos\Delta\lambda}$$

$$= R \cdot S\tan\Delta\lambda \tag{13}$$

$$y = \frac{R \cdot S\sin\phi}{\cos\phi\cos\Delta\lambda} \tag{14}$$

Figure 5 gives the grid for the equatorial gnomonic projection based on Equations 13 and 14. In this grid, the y axis is along the central meridian, and the x axis is along the equator. The meridians and the equator of the grid are straight lines. The meridians are perpendicular to the equator. The spacing of the meridians increases with distance from the central meridian. The parallels are characteristic curved lines. The spacing of the parallels increases with distance from the equator. Table 3 gives the plotting coordinates for the projection. In this case, a universal table is possible.

TABLE 2
Normalized Plotting Coordinates for the Gnomonic Projection, Polar Case ($\phi_o = 90°$, $\lambda_o = 0°$, $R \cdot S = 1$)

Tabulated Values: x/y

Longitude	Latitude			
	60°	70°	80°	90°
0°	0.000	0.000	0.000	0.000
	−0.577	−0.364	−0.176	0.000
10°	0.100	0.063	0.031	0.000
	−0.569	−0.358	−0.174	0.000
20°	0.197	0.124	0.060	0.000
	−0.543	−0.342	−0.166	0.000
30°	0.289	0.182	0.888	0.000
	−0.500	−0.315	−0.153	0.000
40°	0.371	0.234	0.113	0.000
	−0.442	−0.279	−0.135	0.000
50°	0.442	0.279	0.135	0.000
	−0.371	−0.234	−0.113	0.000
60°	0.500	0.315	0.153	0.000
	−0.289	−0.182	−0.088	0.000
70°	0.543	0.342	0.166	0.000
	−0.197	−0.124	−0.060	0.000
80°	0.569	0.538	0.174	0.000
	−0.100	−0.063	−0.031	0.000
90°	0.577	0.364	0.176	0.000
	0.000	0.000	0.000	0.000

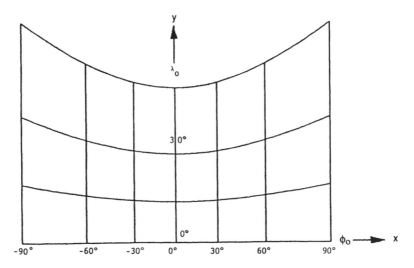

FIGURE 5.   Gnomonic projection, equatorial case.

TABLE 3
Normalized Plotting Coordinates for the Gnomonic Projection, Equatorial Case ($\phi_o = 0°$, $\lambda_o = 90°$, $R \cdot S = 1$)

Tabulated Values: x/y

Longitude	Latitude					
	0°	15°	30°	45°	60°	75°
0°	0.000	0.000	0.000	0.000	0.000	0.000
	0.000	0.268	0.577	1.000	1.732	3.732
15°	0.268	0.268	0.268	0.268	0.268	0.268
	0.000	0.277	0.598	1.035	1.793	3.864
30°	0.577	0.577	0.577	0.577	0.577	0.577
	0.000	0.309	0.666	1.155	2.000	4.309
45°	1.000	1.000	1.000	1.000	1.000	1.000
	0.000	0.379	0.816	1.414	2.449	5.278
60°	1.732	1.732	1.732	1.732	1.732	1.732
	0.000	0.536	1.115	2.000	3.464	7.464
75°	3.732	3.732	3.732	3.732	3.732	3.732
	0.000	1.035	2.231	3.864	6.692	14.420

The inverse transformation for the equatorial gnomonic projection is obtained by inverting Equations 13 and 14. The resulting equations are

$$
\left.
\begin{aligned}
&= \tan^{-1}\left(\frac{x}{R \cdot S}\right) \\
\phi &= \tan^{-1}\left(\frac{y\cos\Delta\lambda}{R \cdot S}\right)
\end{aligned}
\right\}
\qquad (15)
$$

The characteristic that all great circles appear as straight lines in gnomonic projection has led to the use of this projection in navigation. Refer again to Figure 3. On this figure, the great circle distance between points P_1 and P_2 is given. The loxodrome is also shown. The great circle is a straight line, but not to true scale. The loxodrome appears as a curved line. As on the sphere itself, the loxodrome is longer than the great circle distance. Compare Figure 3 to Figure 6, Chapter 5, for the equatorial Mercator projection. Recall that the loxodrome on the equatorial Mercator projection appears straight and shorter and the great circle distance appears curved and longer.

Another characteristic of the gnomonic projection is that the azimuth from the point of tangency to any other point on the projection is true. This has led to the use of the oblique gnomonic projection with CRT displays of position of other points relative to the point of tangency. The drawback here is that range along the great circle is not true and linear distortion significantly increases with actual distance from the point of tangency. The effects of distortion are obvious from Figure 1. Points 90° from the point of tangency of the mapping plane cannot be portrayed. Even at 60°, from the point of tangency of the mapping plane, distortion is severe.

A. Example 1

Given: $\phi_o = 45°$, $\lambda_o = 0°$, $\lambda = 10°$, $\phi = 50°$, $R = 6,371,007$ m, $S = 1:5,000,000$.

Find: the Cartesian coordinates in the oblique gnomonic projection.

$$\Delta\lambda = \lambda - \lambda_o = 10 - 0 = 10°$$

$$
\begin{aligned}
x &= \frac{RS\cos\phi\sin\Delta\lambda}{\sin\phi_o\sin\phi + \cos\phi_o\cos\phi\cos\Delta\lambda} \\[2mm]
&= \frac{\left(\dfrac{6,371,007}{5,000,000}\right)\cos50°\sin10°}{\sin45°\sin50° + \cos45°\cos50°\cos10°} \\[2mm]
&= \frac{(1.27420)(0.642788)(0.173648)}{(0.707107)(0.766044) + (0.707107)(0.642788)(0.984808)} \\[2mm]
&= \frac{0.142225}{0.989290} \\[2mm]
&= 0.1438 \text{ m}
\end{aligned}
$$

$$
\begin{aligned}
y &= \frac{RS(\cos\phi_o\sin\phi - \sin\phi_o\cos\phi\cos\Delta\lambda}{\sin\phi_o\sin\phi + \cos\phi_o\cos\phi\cos\Delta\lambda} \\[2mm]
&= \frac{(1.27420)[(0.707107)(0.766044) - (0.707107)(0.642788)(0.984808)]}{0.989290} \\[2mm]
&= \frac{(1.27420)(0.098605)}{0.989290} \\[2mm]
&= 0.1270 \text{ m}
\end{aligned}
$$

B. Example 2

Given: $\lambda_o = 0°$, $\lambda = 15°$, $\phi = 80°$, $R = 6,371,007$ m, $S = 1:5,000,000$.

Find: x and y for a polar gnomonic projection.

$$\Delta\lambda = \lambda - \lambda_o$$

$$= 15 - 0 = 15°$$

$$
\begin{aligned}
x &= RS\cot\phi\sin\Delta\lambda \\[2mm]
&= \left(\frac{6,371,007}{5,000,000}\right)\cot80°\sin15°
\end{aligned}
$$

$$= (1.27420)(0.176327)(0.258819)$$

$$= 0.0588 \text{ m}$$

$$y = - R S \cot\phi \cos\Delta\lambda$$

$$= -(1.27420)(0.176327)\cos 15°$$

$$= -(1.27420)(0.176327)(0.965925)$$

$$= -0.2170 \text{ m}$$

C. Example 3

Given: $\lambda_o = 0°, \lambda = 15°, \phi = 30°, R = 6,371,007 \text{ m}, S = 1:5,000,000.$
Find: x and y for an equatorial gnomonic projection.

$$\Delta\lambda = \lambda - \lambda_o$$

$$= 15 - 0 = 15°$$

$$x = R S \tan\Delta\lambda$$

$$= \left(\frac{6,371,007}{5,000,000}\right)\tan 15°$$

$$= (1.27420)(0.267949)$$

$$= 0.3414 \text{ m}$$

$$y = \frac{R S \tan\phi}{\cos\Delta\lambda}$$

$$= \frac{(1.27420)\tan 30°}{\cos 15°}$$

$$= \frac{(1.27420)(0.577350)}{0.965926}$$

$$= 0.7616 \text{ m}$$

IV. AZIMUTHAL EQUIDISTANT[2,4,6]

The azimuthal equidistant projection is an extremely useful azimuthal projection. The projection is defined by the requirement that all distances and azimuths from the origin of the projection to any other points be true. Thus, all great circles through the origin are true-scale lines. The projection equations are derived for a spherical model of the Earth by stating the basic requirement mathematically. Three cases of the azimuthal equidistant projection are considered: the oblique, the polar, and the equatorial. By far, the oblique case is the most important for practical use. The procedure for the derivations is

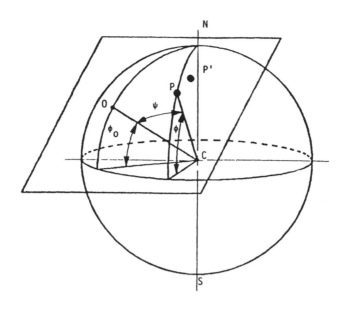

FIGURE 6. Geometry for the oblique azimuthal equidistant projection.

to obtain the oblique case, and then simplify the plotting equations for the limiting cases of the polar and the equatorial.

The geometry required for the derivation of the oblique azimuthal equidistant projection is given in Figure 6. The point of tangency of the mapping plane to the spherical earth is at 0. This is the origin of the projection. The geographical coordinates of the origin are ϕ_o and λ_o. The x axis is positive eastward, and the y axis is positive northward. Let an arbitrary point P on the sphere have geographic coordinates ϕ and λ. This point is transformed to P' on the mapping plane.

Again, it is necessary to define two auxiliary angles, ψ and θ. Let the angle between the two radii CO and CP be ψ. The curved length of the line on the sphere is

$$OP = R\psi \tag{16}$$

From the basic requirement for the azimuthal equidistant projection, the corresponding straight line distance on the map, OP', is defined by

$$OP' = OP \tag{17}$$

Substitute Equation 17 into Equation 16 to obtain

$$OP' = R\psi \tag{18}$$

The second auxiliary angle, θ, is used to orient OP′ with respect to the x axis. This is applied, along with Equation 18, to obtain

$$x = OP'\cos\theta$$

$$y = OP'\sin\theta$$

$$\left.\begin{array}{l} x = R\psi\cos\theta \\[4pt] y = R\psi\sin\theta \end{array}\right\} \tag{19}$$

The next step is to find relations for ψ and θ in terms of the geographic coordinates. The relation for ψ is found from Equation 7

$$\psi = \cos^{-1}(\sin\phi_o\sin\phi + \cos\phi_o\cos\phi\cos\Delta\lambda) \tag{20}$$

where $\Delta\lambda = \lambda - \lambda_o$. Since ψ is restricted to the range from 0 to 180°, ψ is uniquely defined. Then, $\sin\psi$ is available immediately. Equations 6 and 7 can then be used to obtain θ.

$$\cos\theta = \frac{\sin\Delta\lambda\cos\phi}{\sin\psi} \tag{21}$$

$$\sin\theta = \frac{\cos\phi_o\sin\phi - \sin\phi_o\cos\phi\cos\Delta\lambda}{\sin\psi} \tag{22}$$

The evaluations of the plotting equations are easiest if they are handled as an algorithm based on Equations 19 to 22. First, a simplification should be introduced to remove the θ term, and the scale factor, S, must be included.

$$\left.\begin{array}{l} \psi = \cos^{-1}\sin\phi_o\sin\phi + \cos\phi_o\cos\phi\cos\Delta\lambda) \\[8pt] x = R \cdot S \cdot \left(\dfrac{\psi}{\sin\psi}\right)\sin\Delta\lambda\cos\phi \\[12pt] y = R \cdot S \cdot \left(\dfrac{\psi}{\sin\psi}\right)(\cos\phi_o\sin\phi - \sin\phi_o\cos\phi\cos\Delta\lambda) \end{array}\right\} \tag{23}$$

Equations 23 have been evaluated for an arbitrary choice of ϕ_o to obtain Table 4. Every choice of ϕ_o produces a different set of plotting values. No universal grid is possible.

A representative grid based on the plotting table is shown in Figure 7. The y axis is along the central meridian. The x axis is perpendicular to the y axis and intersects it at ϕ_o. Only the central meridian is a straight line. All other meridians, all parallels, and the equator are curved lines. However, any straight line drawn on the map from the origin to any other point is true scale

TABLE 4
Normalized Plotting Coordinates for the Azimuthal Equidistant Projection, Oblique Case ($\phi_o = 45°$, $\lambda_o = 0°$, $R \cdot S = 1$)

Tabulated Values: x/y

Longitude	Latitude				
	15°	**30°**	**45°**	**60°**	**75°**
0°	0.000	0.000	0.000	0.000	0.000
	−0.524	−0.262	0.000	0.262	0.524
15°	0.264	0.228	0.184	0.131	0.070
	−0.503	−0.242	0.017	0.275	0.531
30°	0.523	0.451	0.362	0.257	0.137
	−0.442	−0.184	0.069	0.315	0.554
45°	0.771	0.660	0.526	0.371	0.195
	−0.339	−0.086	0.154	0.380	0.591
60°	1.001	0.849	0.669	0.466	0.243
	−0.190	0.054	0.273	0.469	0.640
75°	1.203	1.006	0.782	0.537	0.275
	0.008	0.235	0.424	0.579	0.700

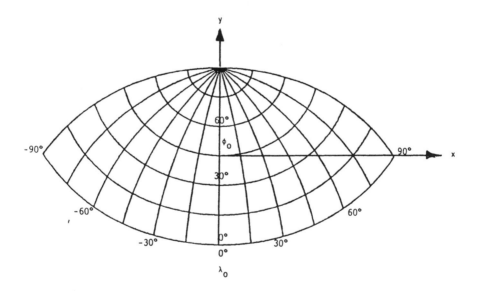

FIGURE 7.   Azimuthal equidistant projection, oblique case.

and true azimuth. Distance between any two points, neither at the origin, is not true scale, nor is the aximuth true.

The oblique azimuthal equidistant projection has seen much use as rocket and missile firing charts and air route planning charts. It is also frequently used in CRT displays in which some moving object, such as a sensor platform, is taken as the origin of the mapping coordinate system. An extended example of such a use is explored in Chapter 8.

The polar azimuthal equidistant projection is obtained from the oblique case by setting $\phi_o = 90°$ in the equations for ψ and θ. We then have from Equation 20

$$\psi = \cos^{-1}(\sin\phi)$$
$$\cos\psi = \sin\phi$$
$$= \cos(\pi/2 - \phi)$$
$$\psi = \frac{\pi}{2} - \phi \qquad (24)$$

Similarly, from Equations 21 and 22, we have

$$\cos\theta = \frac{\sin\Delta\lambda\cos\phi}{\sin(\pi/2 - \phi)}$$
$$= \sin\Delta\lambda \qquad (25)$$

$$\sin\theta = -\frac{\cos\phi\cos\Delta\lambda}{\sin(\pi/2 - \phi)}$$
$$= -\cos\Delta\lambda \qquad (26)$$

Substituting Equations 24 to 26 into Equation 19, and including the scale factor, S, the plotting equations for the polar case are

$$\left.\begin{array}{l} x = R \cdot S \cdot (\pi/2 - \phi)\sin\Delta\lambda \\ y = -R \cdot S \cdot (\pi/2 - \phi)\cos\Delta\lambda \end{array}\right\} \qquad (27)$$

where ϕ is in radians.

A plotting table for the polar case is given as Table 5. A single table applies for the polar projection. Figure 8 is the grid. All meridians are straight lines. The parallels are concentric circles, equally spaced. The positive x axis coincides with the 90° meridian. The positive y axis coincides with the 180° meridian. The polar azimuthal equidistant projection has the same spacing as found on polar graph paper.

The polar azimuthal equidistant projection is of only limited use. One of its best uses is as a standard of comparison of distortion when it is compared to other azimuthal projections. This is done in Chapter 7.

TABLE 5
Normalized Plotting Coordinates for the Azimuthal
Equidistant Projection, Polar Case
$$(\phi_o = 90°, \quad \lambda = 0°, \quad R \cdot S = 1)$$

Tabulated Values: x/y

Longitude	Latitude			
	60°	70°	80°	90°
0°	0.000	0.000	0.000	0.000
	−0.524	−0.349	−0.175	0.000
10°	0.091	0.061	0.030	0.000
	−0.516	−0.344	−0.172	0.000
20°	0.179	0.119	0.060	0.000
	−0.492	−0.328	−0.164	0.000
30°	0.262	0.175	0.087	0.000
	−0.453	−0.302	−0.151	0.000
40°	0.337	0.224	0.112	0.000
	−0.401	−0.267	−0.134	0.000
50°	0.401	0.267	0.134	0.000
	−0.377	−0.224	−0.112	0.000
60°	0.453	0.302	0.151	0.000
	−0.262	−0.175	−0.087	0.000
70°	0.492	0.328	0.164	0.000
	−0.179	−0.119	−0.060	0.000
80°	0.516	0.344	0.172	0.000
	−0.091	−0.061	−0.030	0.000
90°	0.524	0.349	0.175	0.000
	0.000	0.000	0.000	0.000

The equatorial azimuthal equidistant projection is also obtained from the oblique case. This time, $\phi_o = 0°$ is substituted into the relations for ψ and θ. From Equation 20,

$$\psi = \cos^{-1}(\cos\phi\cos\Delta\lambda) \tag{28}$$

From Equation 22,

$$\sin\theta = \frac{\sin\phi}{\sin\psi} \tag{29}$$

Equations 19, 20, 28, and 29 are best evaluated by means of an algorithm. Again, some simplification is called for to eliminate the θ term, and the scale factor, S, must be included.

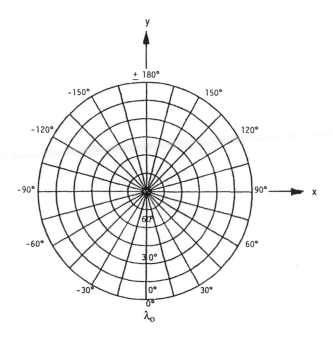

FIGURE 8.   Azimuthal equidistant projection, polar case.

$$\psi = \cos^{-1}(\cos\phi\cos\Delta\lambda)$$

$$x = R \cdot S \cdot \left(\frac{\psi}{\sin\psi}\right)\sin\Delta\lambda\cos\phi$$

$$y = R \cdot S \cdot \left(\frac{\psi}{\sin\psi}\right)\sin\phi \tag{30}$$

A. Example 4

Given: $\phi_o = 45°$, $\lambda_o = 0°$, $\lambda = 10°$, $\phi = 50°$, $R = 6,371,007$ m, $S = 1:5,000,000$.

Find: the Cartesian coordinates on the oblique azimuthal equidistant projection.

$$\Delta\lambda = \lambda - \lambda_o$$

$$= 10 - 0 = 10°$$

$$\psi = \cos^{-1}(\sin\phi_o\sin\phi + \cos\phi_o\cos\phi\cos\Delta\lambda)$$

$$= \cos^{-1}(\sin45°\sin50° + \cos45°\cos50°\cos10°)$$

$$= \cos^{-1}[(0.707107)(0.766044) + (0.707107)(0.642788)(0.984808)]$$

$$= \cos^{-1}(0.541675 + 0.447615)$$

$$= \cos^{-1}(0.989290)$$

$$= 8°.3930$$

$$\frac{\psi}{\sin\psi} = \frac{8.3930}{(57.295)(0.145962)} = 1.00360$$

$$x = R \cdot S \cdot \left(\frac{\psi}{\sin\psi}\right)\sin\Delta\lambda\cos\phi$$

$$= \left(\frac{6,371,007}{5,000,000}\right)(1.00360)\sin10°\cos50°$$

$$= (1.27878)(0.173648)(0.642788)$$

$$= 0.1427 \text{ m}$$

$$y = R \cdot S \cdot \left(\frac{\psi}{\sin\psi}\right)(\cos\phi_o\sin\phi - \sin\phi_o\cos\phi\cos\Delta\lambda)$$

$$= (1.27878)[(0.707107)(0.766044) - (0.707107)(0.642788)(0.984808)]$$

$$= (1.27878)(0.541675 - 0.447615)$$

$$= (1.27878)(0.094060)$$

$$= 0.1203 \text{ m}$$

B. Example 5

Given: $\lambda_o = 0$, $\phi = 80°$, $\lambda = 30°$, $R = 6,371,007$ m, $S = 1{:}5,000,000$.

Find: the Cartesian coordinates in the polar azimuthal equidistant projection.

$$x = R \cdot S \cdot (\pi/2 - \phi)\sin\Delta\lambda$$

$$= \left(\frac{6,371,007}{5,000,000}\right)\left(\frac{\pi}{2} - \frac{80}{57.295}\right)\sin30°$$

$$= \left(\frac{6,371,007}{5,000,000}\right)(1.57080 - 1.39628)(0.500000)$$

$$= \left(\frac{6,371,007}{5,000,000}\right)(0.17452)(0.500000)$$

$$= 0.1106 \text{ m}$$

$$y = -R \cdot S \cdot (\pi/2 - \phi)\cos\Delta\lambda$$

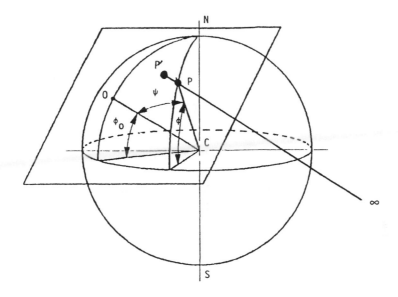

FIGURE 9. Geometry for the orthographic projection.

$$= -\left(\frac{6,371,007}{5,000,000}\right)(0.17452)\cos 30°$$

$$= -\left(\frac{6,371,007}{5,000,000}\right)(0.17452)(0.866025)$$

$$= 0.1916 \text{ m}$$

V. ORTHOGRAPHIC[4-6,8-10]

The orthographic projection is yet another azimuthal projection. This projection is derived with a sphere as the model for the Earth. This projection may be obtained graphically, as indicated in Figure 1. The projection point is taken at infinity. Thus, all rays passing through the sphere and ending on the projection plane are parallel. The geometry of the projection limits the area portrayed to a hemisphere or less.

Again, three cases of this projection are of interest. These are the oblique, the polar, and the equatorial. The direct transformation is first derived for the oblique case. Then, the polar and equatorial projections are obtained by particular substitutions into the plotting equations for the oblique case.

The geometry required for the derivation of the oblique orthographic projection is given in Figure 9. Again, it is necessary to use two auxiliary angles, ψ and θ. The first auxiliary angle, ψ, is the angle between the radii CO and CP. From the figure,

$$O'P = R \cdot S \sin\psi \tag{31}$$

From the configuration of parallel lines, $O'P = OP'$. This is substituted in Equation 31 to obtain

$$OP' = R \cdot S \sin\psi \tag{32}$$

It is then necessary to introduce the second auxiliary angle, θ, to orient the line OP' with respect to the x axis in the figure. Note that the x axis is eastward, and the y axis is northward. The orientation of OP' yields the Cartesian coordinates

$$\left.\begin{aligned}
x &= OP'\cos\theta \\
y &= OP'\sin\theta
\end{aligned}\right\} \tag{33}$$

Substitute Equation 32 into Equations 33.

$$\left.\begin{aligned}
x &= R \cdot S \sin\psi\cos\theta \\
y &= R \cdot S \sin\psi\sin\theta
\end{aligned}\right\} \tag{34}$$

It is now necessary to introduce the relation between the auxiliary angles and the geographic coordinates. Substitute Equations 6 and 8 into Equation 34 and simplify.

$$\left.\begin{aligned}
x &= R \cdot S \cdot \cos\phi\sin\Delta\lambda \\
y &= R \cdot S \cdot (\cos\phi_o\sin\phi - \sin\phi_o\cos\phi\cos\Delta\lambda)
\end{aligned}\right\} \tag{35}$$

These are Cartesian plotting equations for the oblique orthographic projection.

A plotting table resulting from the evaluation of Equations 35 is given in Table 6. Each choice of a different point of tangency for the mapping plane results in a different plotting table. There is no universal table.

A grid resulting from the plotting table is shown in Figure 10. The central meridian is the only straight line. The y axis is along the central meridian. The x axis is perpendicular to the y axis at ϕ_o.

The oblique orthographic projection has recently been used in CRT displays in which the origin is the location of a moving tracking sensor. This use is questionable since neither distance nor angle is preserved in this pro-

TABLE 6
Normalized Plotting Coordinates for the Orthographic
Projection, Oblique Case ($\phi_o = 45°$, $\lambda_o = 0°$, $R \cdot S = 1$)

Tabulated Values: x/y

Longitude	Latitude				
	15°	**30°**	**45°**	**60°**	**75°**
0°	0.000	0.000	0.000	0.000	0.000
	−0.500	−0.259	0.000	0.259	0.500
15°	0.250	0.224	0.183	0.129	0.067
	−0.477	−0.238	0.017	0.271	0.506
30°	0.483	0.433	0.353	0.250	0.129
	−0.408	−0.177	0.067	0.306	0.524
45°	0.683	0.612	0.500	0.354	0.183
	−0.300	−0.079	0.146	0.362	0.554
60°	0.836	0.750	0.612	0.433	0.224
	−0.158	0.047	0.250	0.436	0.592
75°	0.933	0.836	0.683	0.483	0.250
	0.006	0.195	0.370	0.521	0.636

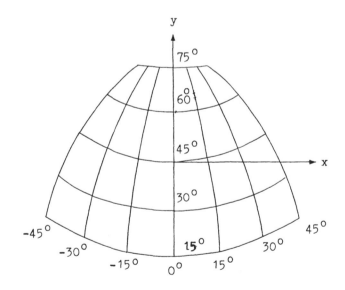

FIGURE 10.   Orthographic projection, oblique case.

TABLE 7
Normalized Plotting Coordinates for the Orthographic
Projection, Polar Case ($\phi_o = 90°$, $\lambda_o = 0°$, $R \cdot S = 1$)

Tabulated Values: x/y

Longitude	Latitude			
	60°	70°	80°	90°
0°	0.000	0.000	0.000	0.000
	−0.500	−0.342	−0.174	0.000
10°	0.082	0.059	0.030	0.000
	−0.492	−0.337	−0.171	0.000
20°	0.171	0.117	0.059	0.000
	−0.470	−0.321	−0.163	0.000
30°	0.250	0.171	0.087	0.000
	−0.433	−0.296	−0.150	0.000
40°	0.321	0.220	0.117	0.000
	−0.383	−0.262	−0.133	0.000
50°	0.383	0.262	0.133	0.000
	−0.321	−0.220	−0.117	0.000
60°	0.433	0.296	0.150	0.000
	−0.250	−0.171	−0.087	0.000
70°	0.470	0.321	0.163	0.000
	−0.171	−0.117	−0.059	0.000
80°	0.492	0.336	0.171	0.000
	−0.087	−0.059	0.030	0.000
90°	0.500	0.342	0.174	0.000
	0.000	0.000	0.000	0.000

jection. The projection cannot be used in small-scale displays without introducing deceptive distortion. The most practical use of the oblique orthographic is in a restricted area about the point of tangency.

The polar orthographic projection can be obtained from the oblique case by letting $\phi_o = 90°$ in Equations 35. Then, the Cartesian plotting equations are

$$\left. \begin{aligned} x &= R \cdot S \cdot \cos\phi \sin\Delta\lambda \\ y &= -R \cdot S \cdot \cos\phi \cos\Delta\lambda \end{aligned} \right\} \tag{36}$$

Equations 36 were used to produce Table 7. This table is unique for the projection. The resulting grid is given in Figure 11. The meridians are straight lines radiating from the center. The parallels are concentric circles. As the equator is approached, the parallel circles are greatly compressed, and distortion becomes extreme. The x axis is along the 90° meridian, and the y axis is along the 180° meridian.

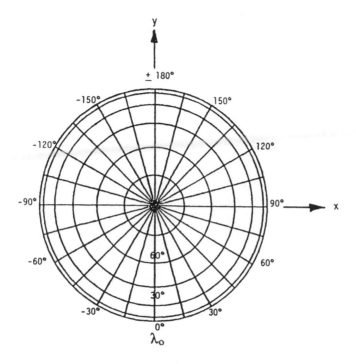

FIGURE 11.   Orthographic projection, polar case.

The polar orthographic projection has very limited practical use. Its greatest asset is as a striking visual portrayal of the effect of distortion in the projection per se.

The equatorial orthographic projection is another limiting case of the oblique projection. The Cartesian plotting equations follow rather simply from Equations 35 by substituting $\phi_o = 0°$.
This gives

$$x = R \cdot S \cdot \cos\phi\sin\Delta\lambda \left.\vphantom{\begin{array}{c}a\\b\end{array}}\right\}$$
$$y = R \cdot S \cdot \sin\phi \qquad\qquad \tag{37}$$

Equations 37 are the basis for the grid of Figure 12. In this figure, only the equator and central meridian are straight lines. The parallels and all other meridians are characteristic curves. The y axis is along the central meridian, and the x axis is along the equator. The plotting table for the equatorial orthographic projection is given as Table 8. This also is a unique table.

In Figure 12, notice that the distortion becomes extreme at the margins of the grid. If the Earth is portrayed on a map, the result is a view similar to that obtained by a viewer in a deep-space orbit. One of the most practical

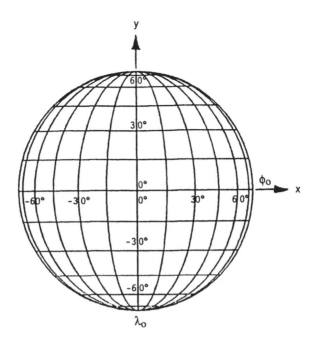

FIGURE 12.   Orthographic projection, equatorial case.

applications of this projection was the lunar mosaic produced by the U.S. Air Force. The map was similar to a telescopic view of the moon.

A. Example 6

Given: $\lambda_o = 0°$, $\lambda = 30°$, $\phi = 80°$, $S = 1:5,000,000$, $R = 6,371,007$ m.

Find: the Cartesian coordinates for the polar orthographic projection.

$$\Delta\lambda = \lambda - \lambda_o$$

$$= 30 - 0 = 30°$$

$$x = R \cdot S \cdot \cos\phi\sin\Delta\lambda$$

$$= \left(\frac{6,371,007}{5,000,000}\right)\cos80°\sin30°$$

$$= \left(\frac{6,371,007}{5,000,000}\right)(0.173648)(0.500000)$$

$$= 0.1106 \text{ m}$$

$$y = -R \cdot S \cdot \cos\phi\cos\Delta\lambda$$

TABLE 8
Normalized Plotting Coordinates for the Orthographic
Projection, Equatorial Case ($\phi_o = 0°$, $\lambda_o = 0°$, $R \cdot S = 1$)

Tabulated Values: x/y

Longitude	0°	10°	20°	30°
0°	0.000	0.000	0.000	0.000
	0.000	0.174	0.342	0.500
10°	0.174	0.171	0.163	0.150
	0.000	0.174	0.342	0.500
20°	0.342	0.337	0.321	0.296
	0.000	0.174	0.342	0.500
30°	0.500	0.492	0.470	0.433
	0.000	0.174	0.342	0.500
40°	0.643	0.633	0.604	0.557
	0.000	0.174	0.342	0.500
50°	0.766	0.755	0.720	0.663
	0.000	0.174	0.342	0.500
60°	0.866	0.853	0.814	0.750
	0.000	0.174	0.342	0.500
70°	0.940	0.925	0.883	0.814
	0.000	0.174	0.342	0.500
80°	0.985	0.970	0.925	0.853
	0.000	0.174	0.342	0.500
90°	1.000	0.985	0.940	0.866
	0.000	0.174	0.342	0.500

Latitude

$$= -\left(\frac{6,371,007}{5,000,000}\right)\cos80°\cos30°$$

$$= -\left(\frac{6,371,007}{5,000,000}\right)(0.173648)(0.866025)$$

$$= -0.1916 \text{ m}$$

B. Example 7

Given: $\lambda_o = 0°$, $\lambda = 30°$, $\phi = 30°$, $S = 1{:}5,000,000$, $R = 6,371,007$ m.

Find: the Cartesian coordinates for the equatorial orthographic projection.

$$\Delta\lambda = \lambda - \lambda_o$$

$$= 30 - 0 = 30°$$

$$x = R \cdot S \cdot \cos\phi\sin\Delta\lambda$$

$$= \left(\frac{6,371,007}{5,000,000}\right)\cos30°\sin30°$$

$$= \left(\frac{6,371,007}{5,000,000}\right)(0.866025)(0.500000)$$

$$= 0.5517 \text{ m}$$

$$y = R \cdot S \cdot \sin\phi$$

$$= \left(\frac{6,371,007}{5,000,000}\right)\sin30°$$

$$= 0.6371 \text{ m}$$

C. Example 8

Given: $\phi_o = 45°$, $\lambda_o = 0°$, $\lambda = 30°$, $\phi = 30°$, $S = 1:10,000,000$, $R = 6,371,007$ m.

Find: the Cartesian coordinates for the oblique orthographic projection.

$$\Delta\lambda = \lambda - \lambda_o$$

$$= 30 - 0 = 30°$$

$$x = R \cdot S \cdot \cos\phi\sin\Delta\lambda$$

$$= \left(\frac{6,371,007}{10,000,000}\right)\cos30°\sin30°$$

$$= \left(\frac{6,371,007}{10,000,000}\right)(0.866025)(0.500000)$$

$$= 0.2758 \text{ m}$$

$$y = R \cdot S \cdot (\cos\phi_o\sin\phi - \sin\phi_o\cos\phi\cos\Delta\lambda)$$

$$y = \left(\frac{6,371,007}{10,000,000}\right)(\cos45°\sin30° - \sin45°\cos30°\cos30°)$$

$$= \left(\frac{6,371,007}{10,000,000}\right)[(0.707107)(0.500000)$$

$$- (0.707107)(0.866025)(0.866025)]$$

$$= \left(\frac{6,371,007}{10,000,000}\right)(0.353553 - 0.530330)$$

$$= -\left(\frac{6,371,007}{10,000,000}\right)(0.176777)$$

$$= -0.0949 \text{ m}$$

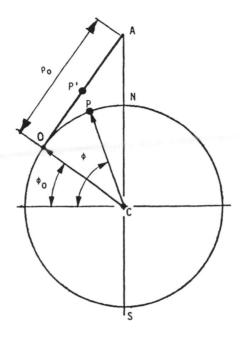

FIGURE 13. Geometry for the simple conical
projection with one standard parallel.

VI. SIMPLE CONIC, ONE STANDARD PARALLEL[2,10]

The defining criterion for the simple conic projection with one standard
parallel is that the distances along every meridian are true scale.

The geometry for the simple conical projection, with one standard parallel
is displayed in Figure 13. The cone is tangent to the spherical model of the
Earth at latitude ϕ. The location of the circle of tangency is represented by
point 0.

From Section XIV, Chapter 2, the constant of the cone is

$$c = \sin\phi_o$$

Thus, the first polar coordinate is

$$\theta = \Delta\lambda\sin\phi_o \tag{38}$$

Also, the radius from the apex of the cone to the point of tangency, line
AO, is given by

$$\rho_o = R\cot\phi_o \tag{39}$$

where R is the radius of the Earth.

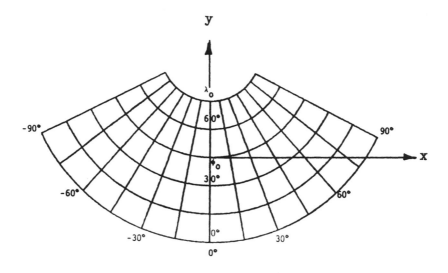

FIGURE 14.   Simple conical projection with one standard parallel.

To obtain the spacing of the parallels, let the central meridian be truly divided. This implies that length OP on the sphere equals length OP′ on the map. This criterion is used with Equation 39 to obtain the second polar coordinate.

$$\rho = \rho_o - R \cdot (\phi - \phi_o) \tag{40}$$

where ϕ and ϕ_o are in radians.

Equations 38 and 40 are used in conjunction with the equations of Section XI, Chapter 1, to obtain the plotting equations.

$$x = R \cdot S \, [\cot\phi_o - (\phi - \phi_o)]\sin(\Delta\lambda\sin\phi_o)$$

$$y = R \cdot S\{\cot\phi_o - [\cot\phi_o - (\phi - \phi_o)] \times \cos(\Delta\lambda\sin\phi_o)\} \tag{41}$$

The grid produced by Equation 41 is given in Figure 14. All of the meridians are straight lines. The parallels are concentric circles, equally spaced. The parallel at ϕ_o is the only true-scale parallel. The y axis is along the central meridian. The x axis is perpendicular to the y axis at latitude ϕ_o. This projection has occasionally been used in atlases.

VII. SIMPLE CONICAL, TWO STANDARD PARALLELS[2,10]

The simple conical projection with two standard parallels also maintains equal spacing along every meridian. However, in this case, the cone is conceptually secant to the spherical model of the Earth.

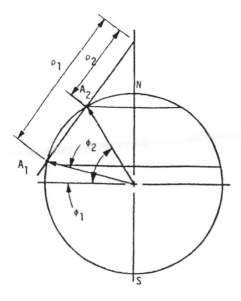

FIGURE 15.   Geometry for the simple conical projection with two standard parallels.

The derivation of this projection follows from Figure 15. The secant cone
defines two true-scale standard parallels of ϕ_1 and ϕ_2, with $\phi_2 > \phi$. From
the equal spacing along the central meridian criterion, we obtain the following.

$$\rho_1 - \rho_2 = R(\phi_2 - \phi_1) \tag{42}$$

From the similar triangles in Figure 15,

$$\frac{\rho_1}{\rho_2} = \frac{R\cos\phi_1}{R\cos\phi_2}$$

$$= \frac{\cos\phi_1}{\cos\phi_2}$$

$$\rho_2 = \rho_1 \frac{\cos\phi_2}{\cos\phi_1} \tag{43}$$

Substitute Equation 43 into Equation 42 to obtain

$$\rho_1 \cdot \left(1 - \frac{\cos\phi_2}{\cos\phi_1}\right) = R(\phi_2 - \phi_1)$$

$$\rho_1 = \frac{R(\phi_2 - \phi_1)}{1 - \dfrac{\cos\phi_2}{\cos\phi_1}} \tag{44}$$

The next step is to find the first polar coordinate, ρ, as the function of the latitude of an arbitrary point. The criterion of equal spacing along the central meridian is again invoked.

$$\rho = \rho_1 - R(\phi - \phi_1) \tag{45}$$

Substitute Equation 44 into Equation 45.

$$\rho = R\left[\frac{\phi_2 - \phi_1}{1 - \dfrac{\cos\phi_2}{\cos\phi_1}} - (\phi - \phi_1)\right] \tag{46}$$

Then, it is necessary to find the constant of the cone for the configuration. From the requirement of the circle of parallel to be true at ϕ_1,

$$2\pi \cdot R \cdot \cos\phi_1 = 2\pi \cdot c_1 \rho_1$$

$$c_1 = \frac{R\cos\phi_1}{\rho_1} \tag{47}$$

Substitute Equation 44 into Equation 47.

$$c_1 = \frac{R\cos\phi_1}{R(\phi_2 - \phi_1)} \cdot \left(1 - \frac{\cos\phi_2}{\cos\phi_1}\right)$$

$$= \cos\phi_1 \frac{\left(1 - \dfrac{\cos\phi_2}{\cos\phi_1}\right)}{\phi_2 - \phi_1}$$

$$= \frac{\cos\phi_1 - \cos\phi_2}{\phi_2 - \phi_1} \tag{48}$$

We now have the relations for a polar representation of the map point. The final step is to obtain the Cartesian plotting equations. These are, from Equations 7, Chapter 1,

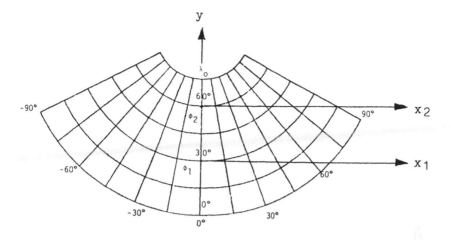

FIGURE 16. Simple conical projection with two standard parallels.

$$x = R \cdot S \left\{ \frac{\phi_2 - \phi_1}{1 - \dfrac{\cos\phi_2}{\cos\phi_1}} - \phi + \phi_1 \right\} \sin\left[\frac{\Delta\lambda(\cos\phi_1 - \cos\phi_2)}{\phi_2 - \phi_1} \right]$$

$$y = R \cdot S \left\{ \frac{\phi_2 - \phi_1}{1 - \dfrac{\cos\phi_2}{\cos\phi_1}} \right.$$

$$\left. - \left[\frac{\phi_2 - \phi_1}{1 - \dfrac{\cos\phi_2}{\cos\phi_1}} - \phi + \phi_1 \right] \cos\left[\frac{\Delta\lambda(\cos\phi_1 - \cos\phi_2)}{\phi_2 - \phi_1} \right] \right\} \quad (49)$$

Figure 16 is a sample grid. The y axis is along the central meridian. The x axis is perpendicular to the y axis at latitude ϕ_1. Again, the meridians are straight lines, and the parallels are equally spaced concentric circles. As always, the use of two standard parallels permits a better distribution of distortion.

This projection has often been used for atlas maps where it was not necessary to have either conformality or equality of area and where the distortion could be tolerated.

VIII. CONICAL PERSPECTIVE[10]

The conical perspective projection can be obtained by graphical means. For this projection, the projection point is the center of the spherical Earth.

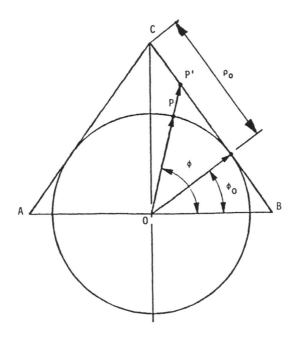

FIGURE 17.   Geometry for the conical perspective projection.

In this respect, the conical perspective projection is similar to the gnomonic and cylindrical perspective projections. While other points can be chosen as the center of projection, only the projection point at the center of the Earth is considered.

The geometry for the conical perspective projection is shown in Figure 17. The cone is tangent at latitude ϕ_o, and the central meridian has longitude λ_o. The geometry given in the figure is used to develop the distance, ρ, from the apex of the cone to an arbitrary point P. The distance to an arbitrary latitude is

$$\rho = \rho_o - R \cdot \tan(\phi - \phi_o) \qquad (50)$$

The constant of the cone and the radius to the parallel circle of tangency are the same as for the simple conic with one standard parallel.

$$\left. \begin{array}{l} c = \sin\phi_o \\[2ex] \rho_o = R \cdot \cot\phi_o \end{array} \right\} \qquad (51)$$

Equations 50 and 51 are used in conjunction with Equation 7 of Chapter 1 to produce the Cartesian plotting equations.

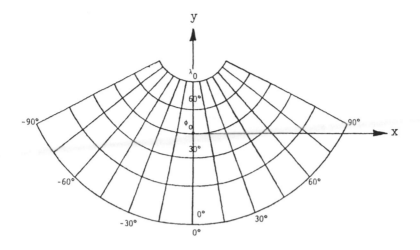

FIGURE 18.   Conical perspective projection.

$$x = R \cdot S \cdot [\cot\phi_o - \tan(\phi - \phi_0)]\sin(\Delta\lambda\sin\phi_o)$$

$$y = R \cdot S \cdot \{\cot\phi_o - [\cot\phi_o - \tan(\phi - \phi_0)]\cos(\Delta\lambda\sin\phi_o)\}$$

$$(52)$$

where S is the scale factor.

Equations 52 give the grid of Figure 18. The parallels are concentric circles, and the meridians are straight lines. The spacing of the parallels increases in either direction from the standard parallel. Thus, distortion increases significantly as one moves north or south of the circle of tangency. This projection has been used for purely illustrative purposes. Note the location of the y axis along the central meridian. The x axis is perpendicular to the y axis at latitude ϕ_0.

IX. POLYCONIC[1,8,9,11]

The polyconic projection is based on a variation of the simple conic projection. It has been noted for every conical projection with one standard parallel that the circle of tangency is true scale; but distortion rapidly increases with distance above and below the chosen circle of tangency. One approach is to use multiple-tangent cones, as illustrated in Figure 19. Each cone is of such a limited north-south extent that distortion is not allowed beyond certain bounds. The strips of map defined by each cone, however, would not be continuous. The polyconic projection goes a step further in that an infinity of tangent cones is used. Thus, each parallel is a standard parallel. And the cones, defined as being infinitesimally small in width, permit a continuous map.

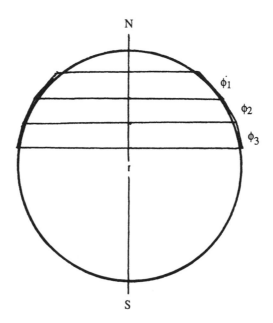

FIGURE 19.   Geometry for the polyconic projection.

The second criterion imposed in the polyconic projection is that the parallels are spaced on the central meridian at the same distance as on the Earth. In this respect, the polyconic projection is similar to the simple conic projection. The direct transformation is derived below for both the spherical and the spheroidal models of the Earth.

For the spherical model of the Earth with radius, R, we begin by deriving the relations for the polar coordinates. The central meridian is true length and has longitude λ_o. Choose some latitude ϕ_o as the origin of the coordinate system. Then, the distance along the meridian for a spherical Earth is

$$d = R \cdot (\phi - \phi_o) \tag{53}$$

From Equation 152 of Chapter 2, the radius to the point of tangency for an arbitrary latitude ϕ, or the first polar coordinate, is

$$\rho = R \cdot \cot\phi \tag{54}$$

and the constant of the cone is, from Equation 155 of Chapter 2,

$$c = \sin\phi \tag{55}$$

The second polar coordinate is, from Equation 55,

$$\theta = \Delta\lambda\sin\phi \qquad (56)$$

The basic mapping equations are

$$\left.\begin{array}{l} x = \rho\sin\theta \\[2mm] y = d + \rho(1 - \cos\theta) \end{array}\right\} \qquad (57)$$

Substituting Equations 53, 54, and 56 into Equation 57, we obtain the Cartesian plotting for the spherical model of the Earth

$$\left.\begin{array}{l} x = R \cdot S \cdot \cot\phi\sin(\Delta\lambda\sin\phi) \\[2mm] y = R \cdot S \cdot \{\phi - \phi_o + \cot\phi[1 - \cos(\Delta\lambda\sin\phi)]\} \end{array}\right\} \qquad (58)$$

where S is the scale factor. These equations fail if $\phi = 0°$. In this case, the plotting equations are

$$\left.\begin{array}{l} x = R\Delta\lambda \\[2mm] y = - R\phi_o \end{array}\right\} \qquad (59)$$

Figure 20 is the polyconic projection. The central meridian and the equator are the only straight lines. All other meridians are curves. The central meridian, the equator, and all parallels are true scale. Thus, the distortion occurs in angles, areas, and length for all meridians except for the central meridian. The y axis on this grid is along the central meridian. The x axis intersects the y axis at latitude ϕ_o. A projection table for the grid is given in Table 9. Each choice of ϕ_o requires a separately generated projection table.

Note in Equation 58 that the so-called constant of the cone is not a constant in this projection. It varies with latitude. In this way, the polyconic projection is similar to the Bonne projection.

The polyconic is a very popular projection. Its distortion is not a function of latitude. Thus, it lends itself to portraying strips of limited width in longitude, but a latitude variation from pole to pole. Because of this, it was used in place of the transverse Mercator projection, such as for the 1:24,000 quadrangles. It is beneficial to now give the plotting equations based on a spheroid as the model of the Earth.

Consider a spheroid defined by semimajor axis, a, and eccentricity, e. The relation for the constant of the cone remains the same. However, for the spheroid, Equation 54 becomes

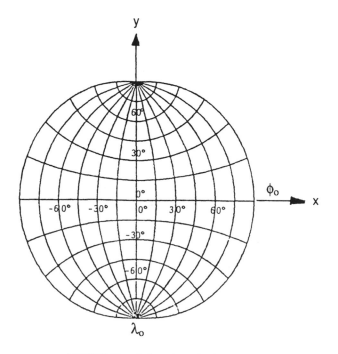

FIGURE 20.   Polyconic projection.

$$\rho = \frac{a\cot\phi}{(1 - e^2\sin^2\phi)^{1/2}} \tag{60}$$

The next changes occur in calculation of distance along the central meridian. From Equation 34 of Chapter 3,

$$d = a(1 - e^2) \int_{\phi_0}^{\phi} \frac{d\phi}{(1 - e^2\sin^2\phi)^{3/2}} \tag{61}$$

The series expansion for Equation 34 of Chapter 3 is given as Equation 37 in Chapter 3 and is not repeated here.

The Cartesian plotting equations for the spheroidal case are given by substituting Equations 56, 60, and 61 into Equation 57.

$$\left.\begin{array}{l} x = \dfrac{a\cot\phi\sin(\Delta\lambda\sin\phi)}{(1 - e^2\sin^2\phi)^{1/2}} \\[2ex] y = a(1 - e^2) \displaystyle\int_{\phi_0}^{\phi} \dfrac{d\phi}{(1 - e^2\sin^2\phi)^{3/2}} \\[2ex] \quad + \dfrac{a\cot\phi}{(1 - e^2\sin^2\phi)^{1/2}} [1 - \cos(\Delta\lambda\sin\phi)] \end{array}\right\} \tag{62}$$

TABLE 9
Normalized Plotting Coordinates for the
Polyconic Projection ($\phi_o = 0°$, $\lambda_o = 0°$, $R \cdot S = 1$)

Tabulated Values: x/y

Longitude	Latitude			
	0°	30°	60°	90°
0°	0.000	0.000	0.000	0.000
	0.000	0.524	1.047	1.571
30°	0.524	0.448	0.253	0.000
	0.000	0.583	1.106	1.571
60°	1.047	0.866	0.455	0.000
	0.000	0.756	1.269	1.571
90°	1.571	1.225	0.565	0.000
	0.000	1.031	1.504	1.571

Again, these equations fail at $\phi = 0°$. Then the plotting equations are

$$x = a\Delta\lambda$$
$$y = -a(1 - e^2) \int_{\phi_0}^{0} \frac{d\phi}{(1 - e^2\sin^2\phi)^{3/2}} \Bigg\} \qquad (63)$$

A. Example 9

Given: $\phi_o = 15°$, $\lambda_o = 0°$, $\phi = 35°$, $\lambda = 25°$, $R = 6,371,007$ m, $S = 1:20,000,000$.

Find: the Cartesian coordinates in the polyconic projection for a spherical earth model.

$$\Delta\lambda = \lambda - \lambda_o$$
$$= 25 - 0 = 25°$$
$$\theta = \Delta\lambda \cdot \sin\phi$$
$$= (25)\sin35$$
$$= (25)(0.573576)$$
$$= 14°.339$$

$$x = R \cdot S \cdot \cot\phi\sin\theta$$
$$= \left(\frac{6,371,007}{20,000,000}\right)\cot35°\sin14°.339$$

$$= \left(\frac{6,371,007}{20,000,000}\right)(1.42815)(0.247658)$$

$$= 0.1127 \text{ m}$$

$$y = R \cdot S \cdot [\phi - \phi_o + \cot\phi(1 - \cos\theta)]$$

$$= \left(\frac{6,371,007}{20,000,000}\right)\left[\frac{35 - 15}{57.295} + (1.42815)(1 - \cos14.339)\right]$$

$$= (0.318550)[0.349071 + (1.42815)(0.031153)]$$

$$= (0.318550)(0.349071 + 0.044491)$$

$$= (0.318550)(0.393562)$$

$$= 0.1254 \text{ m}$$

X. PERSPECTIVE CYLINDRICAL[10]

The first of the conventional cylindrical projections to be considered is the perspective cylindrical. This projection can be implemented graphically. The mapping cylinder is tangent to the sphere. The projection point is the center of the Earth, and the mapping surface is the cylinder. While other locations of the projection point are possible, we will consider only the case where the center of the Earth is chosen. The geometry for the projection is given in Figure 21.

The plotting equations are

$$\left. \begin{array}{l} x = R \cdot S \cdot \Delta\lambda \\[2ex] y = R \cdot S \tan\phi \end{array} \right\} \tag{64}$$

where $\Delta\lambda = \lambda - \lambda_o$ is in radians.

Figure 22 shows the grid for this projection. Distortion becomes very great as higher latitudes are reached. This projection is related to the gnomonic projection and suffers distortion in a similar manner as the tangent function increases. Added to this is the fact that the convergency of the meridians is ignored. The equator is the only true-length line. As shown on the figure, the y axis is along the central meridian. The x axis intersects the y axis at $\phi_o = 0°$, the equator.

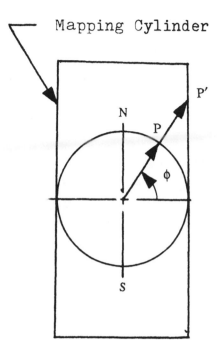

FIGURE 21. Geometry for the perspective cylindrical projection.

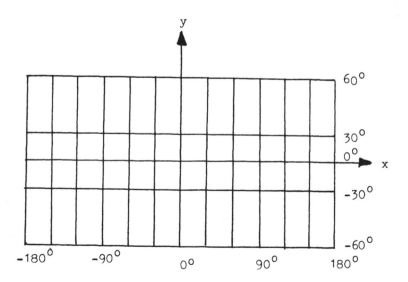

FIGURE 22. Perspective cylindrical projection.

The inverse transformation relations are simply

$$
\left.
\begin{aligned}
\Delta\lambda &= \frac{x}{R \cdot S} \\[2mm]
\phi &= \tan^{-1}\left(\frac{y}{R \cdot S}\right)
\end{aligned}
\right\} \tag{65}
$$

This projection has had limited use in altas maps.

A. Example 10

Let $R = 6{,}378{,}004$ m, $S' = 1{:}20{,}000{,}000$ m, $\Delta\lambda = 30°$, $\phi = 30°$.

$$
\begin{aligned}
x &= \left(\frac{6{,}378{,}004}{20{,}000{,}000}\right)\left(-\frac{30}{57.2958}\right) \\[2mm]
&= 0.16700 \text{ m} \\[2mm]
y &= \left(\frac{6{,}378{,}004}{20{,}000{,}000}\right)\tan 30° \\[2mm]
&= 0.18412 \text{ m}
\end{aligned}
$$

XI. PLATE CARRÉE[10]

The plate Carrée is a conventional cylindrical projection defined by a mathematical rule. The meridians are equally divided, as they are on the sphere. The equator is also equally divided. This leads to the plotting equations

$$
\left.
\begin{aligned}
x &= RS\Delta\lambda \\
y &= RS\phi
\end{aligned}
\right\} \tag{66}
$$

where $\Delta\lambda$ and ϕ are in radians.

The grid for this projection is given in Figure 23. The meridians are true-length straight lines, parallel to each other. The parallels and the equator are also straight lines, perpendicular to the meridians. The equator is also a true-length line. The mapping cylinder is assumed tangent to the sphere. The y axis is along the central meridian. The x axis is along the equator.

The distortion in length along the parallels is extreme as one goes poleward. The convergency of the meridians is ignored, and the poles are stretched into straight lines. In spite of its shortcomings, this projection is often used for a diagrammetric display of data. It is very useful for the display of enumerative data and as the index map for a geographic data base.

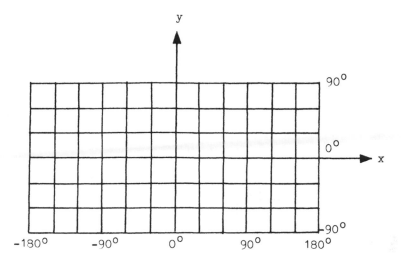

FIGURE 23. Plate Carrée projection.

The inverse relations are equally simple. They are:

$$\Delta\lambda = \frac{x}{RS} \left.\vphantom{\frac{x}{RS}}\right\}$$

$$\phi = \frac{y}{RS} \left.\vphantom{\frac{y}{RS}}\right\} \tag{67}$$

A. Example 11

Given: $R = 6,378,004$ m, $S = 1:20,000,000$, $\Delta\lambda = 30°$, $\phi = 30°$ on a plate Carrée projection.

Find: the Cartesian coordinates.

$$x = \left(\frac{6,378,000}{20,000,000}\right)\left(\frac{30}{57.2958}\right)$$

$$= 0.16700 \text{ m}$$

$$y = \left(\frac{6,378,000}{20,000,000}\right)\left(\frac{30}{57.2958}\right)$$

$$= 0.16700 \text{ m}$$

XII. CARTE PARALLELOGRAMMATIQUE[10]

The carte parallelogrammatique (or die reckteckige plattkarte) has a cylinder secant to the sphere at latitudes $\pm\phi_o$. This projection is an attempt to improve on the plate Carrée. The general configuration of the meridians and

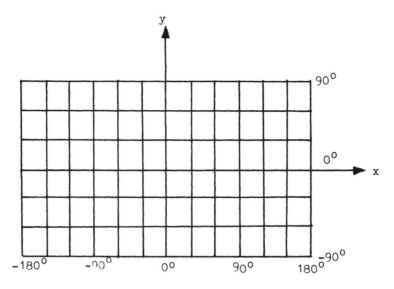

FIGURE 24.   Carte parallelogrammatique projection.

TABLE 10
Normalized Plotting Coordinates for the Plate Carrée Projections
$(\phi_o = 0°, \quad \lambda_o = 0°, \quad R \cdot S = 1)$

Latitude or longitude	0°	30°	60°	90°	120°	150°	180°
x	0.000	0.524	1.047	1.571	2.094	2.618	3.142
y	0.000	0.524	1.047	1.571	—	—	—

parallels is the same as for the Plate Carrée, but the spacing is different. The plotting equations are

$$\left. \begin{array}{l} x = RS\Delta\lambda\cos\phi_o \\ \\ y = RS\phi \end{array} \right\} \tag{68}$$

where ϕ and $\Delta\lambda$ are in radians.

The grid for this projection is given in Figure 24. The grid is composed of rectangles. The meridians and the parallels at $\pm\phi_o$ are the true-length lines. The y axis is along the central meridian. The x axis is along the equator, as indicated in the figure. Table 10 contains the universal plotting table for this projection.

The area between the standard parallels is smaller, and that poleward from each standard parallel is larger than on the model of the Earth. Again, the lack of convergency of the meridians contributes to extreme distortion near the poles.

The inverse transformation relations are simply

$$\Delta\lambda = \frac{x}{RS\cos\phi_o}$$

$$\phi = \frac{y}{RS} \qquad (69)$$

This projection has had limited use in atlases.

A. Example 12

Given: $R = 6,378,004$ m, $S = 1:20,000,000$, $\Delta\lambda = 30°$, $\phi = 30°$, $\phi_o = 30°$.

Find: the Cartesian coordinates.

$$x = \left(\frac{6,378,004}{20,000,000}\right)\left(\frac{30}{57.2958}\right)\cos30°$$

$$= 0.14453 \text{ m}$$

$$y = \left(\frac{6,378,004}{20,000,000}\right)\left(\frac{30}{57.2958}\right)$$

$$= 0.16700 \text{ m}$$

XIII. MILLER[8,9]

The plotting equations for the Miller projection are obtained by a mathematical adjustment of the plotting equations for the regular Mercator projection, based on the sphere as the model of the Earth. The equation for the x coordinate is exactly the same as for the Mercator projection.

$$x = RS(\Delta\lambda) \qquad (70)$$

The y coordinate is obtained by multiplying the latitude by 0.8 in Equation 101 of Chapter 5 and dividing the resulting quantity by 0.8. The plotting equation is

$$y = RS[\ln\tan(\pi/4 + 0.4\phi)]/0.8 \qquad (71)$$

Table 11 gives the plotting coordinates from Equations 70 and 71 for a world map. The resulting grid is given in Figure 25. From the figure, the y axis is along the central meridian. The x axis is along the equator.

The Miller projection is neither equal area or conformal. The meridians and parallels are straight lines, intersecting at right angles. Meridians are

TABLE 11

Normalized Plotting Coordinates for the Miller Projection

$(\phi_o = 0°, \quad \lambda_o = 0°, \quad R \cdot S = 1)$

Latitude or longitude	0°	30°	60°	90°	120°	150°	180°
x	0.000	0.524	1.047	1.571	2.094	2.618	3.142
y	0.000	0.540	1.197	—	—	—	—

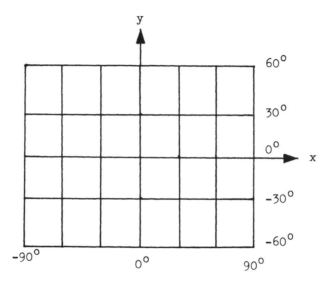

FIGURE 25.   Miller projection.

spaced equidistant. Parallels are spaced farther apart as the poles are approached. However, the adjustment permits the poles to be represented on the map as straight lines.

The inverse transformation is simply obtained from Equations 70 and 71

$$\lambda = \lambda_o + \frac{x}{RS} \tag{72}$$

$$\phi = 2.5\tan^{-1}\epsilon(0.8y/RS - 5\pi/8) \tag{73}$$

where ϵ is the base of natural logarithm.

The Miller projection has seen service in a number of atlases. This projection is quite useful for world maps.

A. Example 13

Let R = 6,378,004 m, S = 1:20,000,000, ϕ = 30°, and $\Delta\lambda$ = 30°.

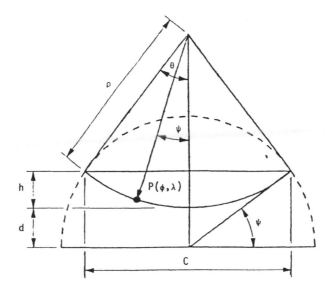

FIGURE 26. Geometry for the globular projection.

$$x = \left(\frac{6,378,004}{20,000,000}\right)\left(\frac{30}{57.2958}\right)$$

$$= 0.16700 \text{ m}$$

$$y = \left(\frac{6,378,004}{20,000,000}\right)\left[\frac{\ln\tan(45° + (0.4)(30))}{0.8}\right]$$

$$= 0.17208 \text{ m}$$

XIV. GLOBULAR[2,10]

The globular projection is a conventional means of portraying half the area of a spherical model of the Earth within a circle. In this respect, it is similar to the orthographic projection. It differs from the orthographic in that it is not defined by a specific placement of a projection point, but rather by a fairly complex mathematical scheme. While the formulas are rather complicated, the basic concept is relatively simple.

In the globular projection, the central meridian, with longitude λ_o, and half of the equator are diameters of the circle enclosing the hemisphere. The central meridian and the half of the equator are true scale.

Consider the geometry required for the derivation of the globular projection as given in Figure 26. Begin by defining a typical parallel circle at latitude ϕ. Let a typical point P have geographic coordinates ϕ and λ. Let d be the distance from the equator to the specified latitude along the central meridian.

Let c be the chord length and h be the distance from the circular arc to the chord. Let ρ be the polar radial coordinate.

The distance along the central meridian between the equator and the circle of latitude is

$$d = \frac{R \cdot \phi}{\pi/2} \tag{74}$$

From the figure,

$$c = 2R\cos\phi \tag{75}$$

$$h = R\sin\phi - \frac{R \cdot \phi}{\pi/2} \tag{76}$$

From the geometry of the circular segment,

$$c = \sqrt{4h(2\rho - h)} \tag{77}$$

Substitute Equations 75 and 76 into Equation 77 and rearrange to obtain the first coordinate ρ.

$$c^2 = 4h(2\rho - h)$$

$$2\rho - h = \frac{c^2}{4h}$$

$$2\rho = h + \frac{c^2}{4h}$$

$$\begin{aligned}
\rho &= \frac{1}{2}\left(h + \frac{c^2}{4h}\right) \\
&= \frac{1}{2}\left\{R(\sin\phi) - \frac{R\phi}{\pi/2} + \frac{4R^2\cos^2\phi}{4R\left[(\sin\phi) - \frac{R\phi}{\pi/2}\right]}\right\} \\
&= \frac{R}{2}\left[(\sin\phi) - \frac{\phi}{\pi/2} + \frac{\cos^2\phi}{(\sin\phi) - \frac{\phi}{\pi/2}}\right] \tag{78}
\end{aligned}$$

The angle, θ, also follows from the geometry of the circular segment.

$$\theta = \cos^{-1}\left(\frac{\rho - h}{\rho}\right) \tag{79}$$

Substitute Equations 76 and 78 into Equation 79.

$$\theta = \cos^{-1}\left\{\frac{\dfrac{R}{2}\left[(\sin\phi) - \dfrac{\phi}{\pi/2} + \dfrac{\cos^2\phi}{(\sin\phi) - \phi/(\pi/2)}\right] - R\left[(\sin\phi) - \dfrac{\phi}{\pi/2}\right]}{\dfrac{R}{2}\left[(\sin\phi) - \dfrac{\phi}{\pi/2}\right] + \dfrac{\cos^2\phi}{(\sin\phi) - \phi/(\pi/2)}}\right\}$$

$$= \cos^{-1}\left\{\frac{\cos^2\phi - \left[(\sin\phi) - \dfrac{\phi}{\pi/2}\right]^2}{\cos^2\phi + \left[(\sin\phi) - \dfrac{\phi}{\pi/2}\right]^2}\right\} \qquad (80)$$

The parallels are divided equally. Thus, a second polar coordinate, ψ, can be defined by the projection, using the relation $\Delta\lambda = \lambda - \lambda_o$.

$$\frac{\psi}{\Delta\lambda} = \frac{\theta}{\pi/2}$$

$$\psi = \frac{\Delta\lambda \cdot \theta}{\pi/2} \qquad (81)$$

Substitute Equation 81 into Equation 80.

$$\psi = \frac{2\Delta\lambda}{\pi}\cos^{-1}\left\{\frac{\cos^2\phi - \left[(\sin\phi) - \dfrac{\phi}{\pi/2}\right]^2}{\cos^2\phi + \left[(\sin\phi) - \dfrac{\phi}{\pi/2}\right]^2}\right\} \qquad (82)$$

The Cartesian plotting equations are

$$\left.\begin{array}{l} x = \rho\sin\psi \\[2mm] y = d + \rho(1 - \cos\psi) \end{array}\right\} \qquad (83)$$

The complicated nature of the formulas suggest that the Cartesian coordinates should be evaluated by an algorithmic scheme. The sequence of evaluation is as follows.

For a general point with $\phi \neq 0$, and $\Delta\lambda \neq 0$,

$$d = \frac{R\phi}{\pi/2}$$

$$\rho = \frac{R}{2}\left[(\sin\phi) - \frac{\phi}{2} + \frac{\cos^2\phi}{\left(\sin\phi - \dfrac{\phi}{\pi/2}\right)}\right]$$

$$\psi = \frac{2\Delta\lambda}{\pi}\cos^{-1}\left\{\frac{\cos^2\phi - \left[(\sin\phi) - \dfrac{\phi}{\pi/2}\right]^2}{\cos^2\phi + \left[(\sin\phi) - \dfrac{\phi}{\pi/2}\right]^2}\right\}$$

$$x = \rho \cdot S \cdot \sin\psi$$

$$y = S \cdot [d + \rho(1 - \cos\psi)] \qquad (84)$$

where S is the scale factor, and $\Delta\lambda$ and ϕ are in radians.

When, $\phi = 0°$, the equations are

$$\left.\begin{array}{l} x = \dfrac{R \cdot S \cdot \Delta\lambda}{\pi/2} \\[3mm] y = 0 \end{array}\right\} \qquad (85)$$

When $\Delta\lambda = 0°$, the equations are

$$\left.\begin{array}{l} x = 0 \\[3mm] y = \dfrac{RS\phi}{\pi/2} \end{array}\right\} \qquad (86)$$

Equations 84 to 86 produce the grid of Figure 27. The x axis is along the equator. The y axis is along the central meridian. The globular projection has been used extensively for atlas maps, but it is obsolete for practical use today.

XV. AERIAL PERSPECTIVE[8,9]

With the advent of aerial mapping from satellites and planes, it has become important to produce maps showing the Earth as seen from such vehicles. The aerial perspective projection is the answer to this need. This projection has the orthographic projection as its limiting case. The oblique form of the aerial perspective is derived below for the spherical model of the Earth.

Figure 28 represents the geometry for this projection. The point of perspective is taken at V, the position of the observer vehicle. Point 0, with coordinates ϕ_o and λ_o, is the nadir point of the vehicle and the point of tangency of the mapping plane. Point P on the Earth has geographic coordinates ϕ and λ. Point P' on the map is along the intersection of the ray from

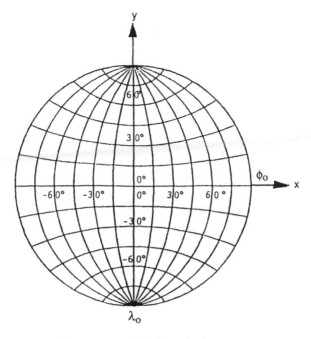

FIGURE 27. Globular projection.

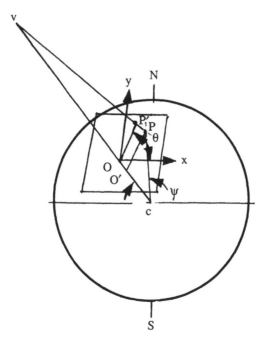

FIGURE 28. Geometry for the aerial perspective projection.

point V to point P at the mapping plane. The height of the vehicle above the Earth is given by H.

As shown in Figure 28, a set of coordinate axes is defined with an origin at 0. The x axis is in the direction of east. The y axis is taken as north.

Again, it is necessary to define two auxiliary angles, ϕ and ψ. The angle ψ is between the two radii CP and CO. From the figure, it is evident that

$$O'P = R\sin\psi \qquad (87)$$

where R is the radius of the model of the Earth, and O'P is parallel to OP' and perpendicular to CO.

A second relationship is that

$$O'C = R\cos\psi \qquad (88)$$

Also from the figure,

$$OO' = R - O'C \qquad (89)$$

Substitute Equation 88 into Equation 89 and simplify.

$$OO' = R - R\cos\psi$$
$$= R(1 - \cos\psi) \qquad (90)$$

The next step is to find a relation between the lengths of lines OP' and O'P. From the figure,

$$\tan\alpha = \frac{OP'}{H} = \frac{O'P}{H + OO'}$$

This is rearranged to obtain

$$OP' = O'P\left(\frac{H}{H + OO'}\right) \qquad (91)$$

Then, substitute Equations 87 and 90 into Equation 91.

$$OP' = R\sin\psi\,\frac{H}{H + R(1 - \cos\psi)} \qquad (92)$$

Finally, the auxiliary angle θ is introduced to orient line OP'', with respect to the coordinate axes as shown in Figure 28. This gives the Cartesian plotting coordinates.

TABLE 12
Normalized Plotting Coordinates for the Aerial Perspective Projection ($\phi_o = 45°$, $\lambda_o = 0°$, $R \cdot S = 1$, $H = 1$)

Tabulated Values: x/y

Longitude	Latitude			
	30°	40°	50°	60°
0°	0.000	0.000	0.000	0.000
	−0.250	−0.087	0.087	0.250
10°	0.144	0.131	0.110	0.084
	−0.239	−0.078	0.093	0.254
20°	0.276	0.253	0.213	0.162
	−0.207	−0.052	0.111	0.265
30°	0.388	0.356	0.302	0.231
	−0.158	−0.014	0.139	0.283
40°	0.473	0.436	0.372	0.288
	−0.098	0.035	0.174	0.306
50°	0.530	0.490	0.422	0.330
	−0.032	0.089	0.214	0.332
60°	0.560	0.520	0.452	0.258
	0.035	0.144	0.255	0.360

$$\left. \begin{array}{l} x = R\sin\psi \left[\dfrac{H}{H + R(1 - \cos\psi)} \right] \cos\theta \\[3mm] y = R\sin\psi \left[\dfrac{H}{H + R(1 - \cos\psi)} \right] \sin\theta \end{array} \right\} \quad (93)$$

The last step is to introduce the relations between the auxiliary angles θ and ψ and the geographic coordinates ϕ, λ, ϕ_o, and λ_o. This is done by means of the relations of Equations 20 to 22.

The equations required to obtain the plotting coordinates are as follows.

$\cos\psi = \sin\phi_o\sin\phi + \cos\phi_o\cos\phi\cos\Delta\lambda$

$$\left. \begin{array}{l} x = S \cdot R \cdot \left[\dfrac{H}{H + R(1 - \cos\psi)} \right] \cdot \cos\phi\sin\Delta\lambda \\[3mm] y = S \cdot R \cdot \left[\dfrac{H}{H + R(1 - \cos\psi)} \right] \cdot (\cos\phi_o\sin\phi - \sin\phi_o\cos\phi\cos\Delta\lambda) \end{array} \right\} \quad (94)$$

A sample plotting table is given in Table 12. Any particular table depends on the user's choice of H and ϕ_o. No universal table is possible.

A sample grid based on the plotting table appears as Figure 29. Note the location of the coordinate axes in the figure. This projection is limited to coverage close to the point of tangency.

Consider the multiplicative term in the second and third of Equation 94. Let $H \gg R$. Then, the term approaches a value of 1. Thus, these equations reduce to the oblique orthographic projection.

A. Example 14

Given: $\phi_o = 45°$, $\phi = 40°$, $\Delta\lambda = 40°$, $H = R = 6{,}371{,}004$ m, $S = 1{:}10{,}000{,}000$.

Find: the Cartesian coordinates for the aerial perspective projection.

$$\cos\psi = \sin\phi_o\sin\phi + \cos\phi_o\cos\phi\cos\Delta\lambda$$

$$= \sin45°\sin40° + \cos45°\cos40°\cos40°$$

$$= (0.707107)(0.642788) + (0.707107)(0.766044)(0.766044)$$

$$= 0.454520 + 0.414947 = 0.869467$$

$$x = S \cdot R \cdot \left[\frac{H}{H + R(1 - \cos\psi)}\right]\cos\phi\sin\Delta\lambda$$

$$= \left(\frac{6{,}371{,}004}{10{,}000{,}000}\right)\left[\frac{6{,}371{,}004}{6{,}3711(2 - 0.869467)}\right]\cos40°\sin40°$$

$$= \frac{(6{,}371{,}004)(0.766044)(0.642788)}{1.130533}$$

$$= 0.2775 \text{ m}$$

$$y = S \cdot R \cdot \left[\frac{H}{H + R(1 - \cos\psi)}\right](\cos\phi_o\sin\phi - \sin\phi_o\cos\phi\cos\Delta\lambda)$$

$$= \left(\frac{0.6371004}{1.130533}\right)(\cos45°\sin40° - \sin45°\cos40°\cos40°)$$

$$= \left(\frac{0.6371004}{1.130533}\right)[0.707107)(0.642788)$$

$$- (0.707107)(0.766044)(0.766044)]$$

$$= \left(\frac{0.6371004}{1.130533}\right)(0.454520 - 0.414947)$$

$$= \left(\frac{0.6371004}{1.130533}\right)(0.039573)$$

$$= 0.0223 \text{ m}$$

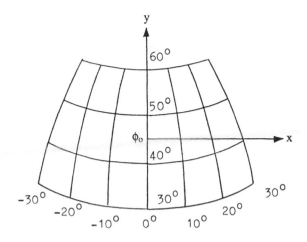

FIGURE 29. Aerial perspective projection.

XVI. VAN DER GRINTEN[8-10]

The Van der Grinten is another projection defined mathematically on a mapping plane. This projection gives the complete spherical model of the Earth within a circle. The original use of the projection entailed a graphical construction of a base map. The geometry for such a construction is given below. Then, the equations for the Cartesian coordinates are given without proof.

Consider first the graphical construction of this projection. Figure 30 contains the geometry necessary for this development. The equator is given by line VQ. Line ON is one half of the central meridian. The other meridians are circular arcs which go through the poles. The parallels are also circular arcs, the locations of which are determined by the procedure below.

Join points N and V. Locate an arbitrary latitude A' on NO, where

$$OA' = NO \frac{\phi}{\pi/2}$$

$$= 2NO \frac{\phi}{\pi} \tag{95}$$

Draw AA' parallel to VQ. The intersection of AA' with NV is B. Join points B and Q. The intersection of BQ and NO defines the point C'. Draw CC' parallel to VQ. Point C constitutes one of the necessary points of the projection. Its symmetric image about NO is a second such point, C''. The next step is to connect points A and Q. Then AQ intersects NO at D. This is the

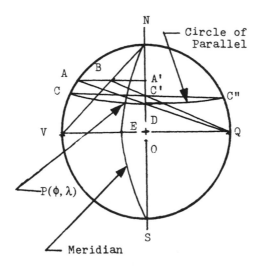

FIGURE 30.   Geometry for the Van der Grinten projection.

third point necessary to completely define a circle of parallel. A circular arc, whose radius is uniquely defined by the location of points C, D, and C″, is drawn to obtain a circle of parallel.

The next step is to find the unique circular arcs for the meridians. Since points N and S are already defined, it is only necessary to locate point E. This is simply

$$EO = VO \frac{\Delta\lambda}{\pi} \tag{96}$$

Modern computer graphics systems usually have the capability of automatically constructing circular arcs given three points. Such a system constructs one arc through points CDC″, and a second arc through points NES. The intersection of these arcs gives the location of the desired point P.

After lengthy derivation, a set of Cartesian plotting coordiantes may be obtained. First, it is necessary to define two auxiliary angles ψ and θ. These, in terms of latitude and longitude, are

$$\left. \begin{aligned} \psi &= \frac{1}{2} \left(\frac{\pi}{\Delta\lambda} - \frac{\Delta\lambda}{\pi} \right) \\ \theta &= \sin^{-1}\left(\frac{2\phi}{\pi} \right) \end{aligned} \right\} \tag{97}$$

Then, it is necessary to define five auxiliary functions.

$$f_1 = \frac{\cos\theta}{\sin\theta + \cos\theta - 1}$$

$$f_2 = \frac{f_1}{\dfrac{2}{\sin\theta} - 1}$$

$$f_3 = \psi^2 + f_1$$

$$f_4 = f_1 - f_2^2$$

$$f_5 = f_2^2 + \psi^2 \tag{98}$$

With the substitution of these auxiliary variables and functions, the Cartesian plotting equations are

$$x = \frac{\pm\,\pi \cdot R \cdot S[\psi \cdot f_4 + (\psi^2 \cdot f_4^2 - f_5 \cdot f_4)^{1/2}]}{f_5}$$

$$y = \frac{\pm\,\pi \cdot R \cdot S\{f_2 \cdot f_3 - \psi[(\psi^2 + 1)f_5 - f_3^2]^{1/2}\}}{f_5} \tag{99}$$

The sign chosen for the x coordinate depends on the sign of $\Delta\lambda$. It is required that $-180° \leqslant \Delta\lambda \leqslant 180°$. The sign chosen for the y coordinate depends on the sign of ϕ.

If $\phi = 0$ or $\Delta\lambda = 0$, Equation 99 becomes indeterminate. In these two cases, simplified equations for the coordinates are needed. If $\phi = 0$, the coordinates are given by

$$x = R \cdot S\Delta\lambda$$

$$y = 0 \tag{100}$$

If $\Delta\lambda = 0$, the coordinates are

$$x = 0$$

$$y = \pm\,\pi \cdot S \cdot R\tan(\theta/2) \tag{101}$$

A sample grid is shown in Figure 31. The equator and the central meridian are the only true-scale lines. The x axis is along the equator. The y axis is along the central meridian.

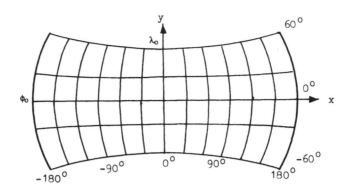

FIGURE 31. Van der Grinten projection.

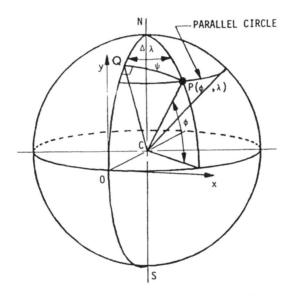

FIGURE 32. Geometry for the Cassini projection.

This projection has been used by the National Geographic Society. It gives a pleasing representation of the total surface of the Earth. The projection blends the appearance of the Mercator projection with the curves of the Mollweide.

XVII. CASSINI[8-10]

The Cassini projection is a transverse equidistant cylindrical projection. This projection is developed below for a spherical model of the Earth. The mapping coordinates follow from a consideration of distances on the sphere. In this respect, it is analogous to the azimuthal equidistant projection.

In Figure 32, let the origin of the mapping coordinate system be at the intersection of the equator and the central meridian with longitude λ_o. P is the arbitrary point on the sphere, with latitude ϕ and longitude λ, and $\Delta\lambda = \lambda - \lambda_o$. PQ is a great circle through P and perpendicular to the central meridian. Let ψ and θ be auxiliary central angles, as shown in the figure. Apply Napier's rules to the spherical triangle to obtain these angles. Recall that $z = 90° - \phi$, which is the colatitude.

$$\sin\psi = \cos(90° - \Delta\lambda)\cos\phi$$

$$= \sin\Delta\lambda\cos\phi$$

$$\psi = \sin^{-1}(\sin\Delta\lambda\cos\phi) \tag{102}$$

$$\sin(90° - \Delta\lambda) = \tan\phi\tan\theta$$

$$\tan\theta = \frac{\cos\Delta\lambda}{\tan\phi} \tag{103}$$

$$\theta = \tan^{-1}\left(\frac{\cos\Delta\lambda}{\tan\phi}\right)$$

The mapping equations follow from Equations 102 and 103 plus the radius of the sphere, using the criterion of maintaining true scale.

$$\left. \begin{array}{l} x = R \cdot S \cdot \sin^{-1}(\sin\Delta\lambda\cos\phi) \\[2mm] y = R \cdot S \cdot \left[\dfrac{\pi}{2} - \tan^{-1}\left(\dfrac{\cos\Delta\lambda}{\tan\phi}\right)\right] \end{array} \right\} \tag{104}$$

where S is the scale factor.

The second of Equation 104 can be simplified. The origin can also be transformed to an arbitrary latitude ϕ_o. The general form of the plotting equations is then

$$\left. \begin{array}{l} x = R \cdot S \cdot \sin^{-1}(\sin\Delta\lambda\cos\phi) \\[2mm] y = R \cdot S \cdot \left[\tan^{-1}\left(\dfrac{\tan\phi}{\cos\Delta\lambda}\right) - \phi_o\right] \end{array} \right\} \tag{105}$$

Figure 33 contains a sample grid for the Cassini projection. In this grid, the x axis is along the equator, and the y axis is along the central meridian. The central meridian and the equator are straight lines, as are meridians 90° from the central meridian. All other meridians and parallels are characteristic curves. Scale is true along the central meridian and along lines perpendicular to the central meridian.

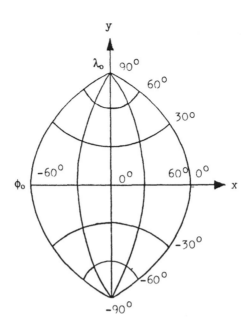

FIGURE 33.   Cassini projection.

The spherical version of the Cassini projection has had little practical usage. However, the ellipsoidal Cassini-Soldner version has been used in place of the transverse Mercator by several European countries.

A. Example 15

Given: $\lambda_o = 0°$, $\lambda = 35°$, $\phi = 35°$, $R = 6,371,007$ m, $S = 1:20,000,000$, $\phi_o = 0°$.

Find: the Cartesian coordinates for the Cassini projection.

$$\Delta\lambda = \lambda - \lambda_o$$

$$= 35 - 0 = 35°$$

$$x = R \cdot S \cdot \sin^{-1}(\sin\Delta\lambda\cos\phi)$$

$$= \left(\frac{6,371,007}{20,000,000}\right)\sin^{-1}(\sin35°\cos35°)$$

$$= (0.318550)\sin^{-1}(0.573576)(0.819152)$$

$$= (0.318550)\sin^{-1}(0.469846)$$

$$= (0.318550)\left(\frac{28.0243}{57.295}\right)$$

$$= 0.1558 \text{ m}$$

$$y = R \cdot S \cdot \left[\tan^{-1}\left(\frac{\tan\phi}{\cos\Delta\lambda}\right) - \phi_o \right]$$

$$= \left(\frac{6,371,007}{20,000,000}\right)\left[\tan^{-1}\left(\frac{\tan 35°}{\cos 35°}\right) - 0 \right]$$

$$= (0.318550)\left[\tan^{-1}\left(\frac{0.700208}{0.819152}\right) \right]$$

$$= (0.318550)[\tan^{-1}(0.854640)]$$

$$= (0.318550)\left(\frac{40.5185}{57.295}\right)$$

$$= 0.2253 \text{ m}$$

XVIII. ROBINSON[7]

The Robinson projection provides a means of showing the entire Earth in an uninterrupted form. The continents appear as units and are in relatively correct size and location. Major shearing of land units is avoided.

In order to accomplish this, a number of criteria were established. Equal area projections were ruled out in order to avoid the shearing effects. Cylindrical projections were ruled out to avoid major area-scale changes. The poles were to be represented as lines instead of points to avoid compressing the northern land masses. It was decided that straight-line, parallel parallels were essential. Finally, the individual parallels were to be evenly divided by the meridians.

Based on these criteria, this pseudocylindrical projection was developed by an iterative process. The essential geographic entities were visualized, and a mathematical description was developed. The geographic features were plotted and reviewed. This was repeated until further adjustment brought no further improvement. The final result is an equator of a length 0.8487 times the circumference of the spherical model of the Earth. The length of the central meridian is 0.5072 times the length of the equator of the projection.

The mathematical description is given as a table containing the distances of parallels from the equator and length of parallels as a function of latitude. The values given in Reference 7, as amended by the scale factor adopted by the National Geographic Society in its first world map using this projection, form the basis for the normalized plotting coordinates of Table 13. This table gives the distance of the respective parallel from the equator as ℓ_m, and the length of the parallel from the central meridian to ℓ_p, 180° E. Tabulations are in increments of 5°, a spacing adequate for practical work.

The spacing of the parallel gives the y coordinate as

$$y = R \cdot S \cdot \ell_m \tag{106}$$

TABLE 13
Parallel Spacing and Parallel
Length on the Robinson Projection
$(\lambda_o = 0°, \quad \phi_o = 0°)$

Latitude (degrees)	1_m	1_p
0	0.000	2.628
5	0.084	2.625
10	0.167	2.616
15	0.251	2.602
20	0.334	2.582
25	0.418	2.557
30	0.502	2.522
35	0.586	2.478
40	0.669	2.422
45	0.752	2.356
50	0.833	2.281
55	0.913	2.195
60	0.991	2.099
65	1.066	1.997
70	1.138	1.889
75	1.206	1.769
80	1.267	1.633
85	1.317	1.504
90	1.349	1.399

Since the meridians are evenly spaced on the parallels, the x coordinate is

$$x = R \cdot S \cdot \ell_p \frac{\lambda}{180} \tag{107}$$

where λ is the longitude in degrees. The values of ℓ_m and ℓ_p are obtained from Table 13 directly or by interpolation. For practical work, 1_m and 1_p are best interpolated by use of Stirling's interpolation formula using second differences.

The inverse relations are relatively straight forward. These are, from Equations 106 and 107,

$$\ell_m = \frac{y}{R \cdot S}$$

$$\lambda = \frac{180x}{R \cdot S \cdot \ell_p} \tag{108}$$

The procedure is to solve the first of Equations 108 for 1_m, and consult Table 13 for the value of ϕ. Then, the comparable value of 1_p is obtained. Using the second of Equations 108, the longitude is obtained.

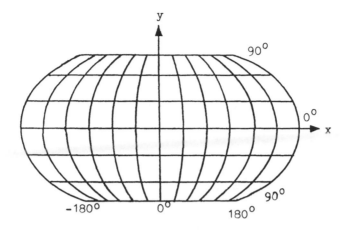

FIGURE 34. Robinson projection.

The grid for the Robinson projection is given as Figure 34. The x axis is on the equator. The y axis is along the central meridian at longitude 0°. The ±38° parallels are the standard parallels.

Three minor drawbacks of this projection may be considered. First, the projection is defined exclusively to be centered on the central meridian of 0° longitude. If one would attempt to choose another longitude for the central meridian, the balance of shapes, sizes, and relative position would be upset. Second, since at the poles, a point is distorted into a line, extreme distortion in shape and increase in size occurs. Thus, the north polar ice cap and Antarctica are not truly represented. Finally, if one wanted to portray the oceans in correct size, shape, and relative position, one would have to start over with a new position.

Despite these minor limitations, the Robinson projection is destined to replace the Van der Grinten projection as the premier projection used by the National Geographic Society. It is applicable for either a wall map or a small-scale atlas map. Because it maintains the relative size of regions, it should be a competitor to the equal area world maps for use as statistical representations.

A. Example 16

Given: $R = 6,378,000$ m, $S = 1:20,000,000$, $\phi = 42.5°$, $\lambda = 50°$. Assume linear interpolation is sufficient.

Find: x and y. From Table 13,

ϕ	ℓ_m
40°	0.669
45°	0.756

at $\phi = 42.5°$,

$$\ell_m = 0.669 + (0.756 - 0.669) \cdot \left(\frac{2.5}{5}\right)$$

$$= 0.669 + 0.087 \cdot \left(\frac{2.5}{5}\right)$$

$$= 0.669 + 0.044$$

$$= 0.713$$

$$y = R \cdot S \cdot \ell_m$$

$$= \left(\frac{6,378,000}{20,000,000}\right)(0.713)$$

$$= 0.2274 \text{ m}$$

From Table 13,

ϕ	ℓ_p
45°	2.422
45°	2.356

at $\phi = 42.5°$,

$$\ell_p = 2.422 - (2.422 - 2.356) \cdot \left(\frac{2.5}{3.0}\right)$$

$$= 2.422 - (0.066) \cdot \left(\frac{2.5}{6.0}\right)$$

$$= 2.422 - 0.033$$

$$= 2.389$$

$$X = R \cdot S \cdot \ell_p \cdot \left(\frac{\lambda}{180°}\right)$$

$$= \left(\frac{6,378,000}{20,000,000}\right) \cdot (2.389) \cdot \left(\frac{50}{180}\right)$$

$$= 0.2116 \text{ m}$$

REFERENCES

1. **Adams, O. S.,** General Theory of Polyconic Projections, Spec. Publ. 57, U.S. Coast and Geodetic Survey, U.S. Gov't. Printing Office, Washington, D.C., 1934.
2. **Deetz, C. H. and Adams, O. S.,** Elements of Map Projection, Spec. Publ. 68, U.S. Coast and Geodetic Survey, 1944.
3. **Hildebrand, F. B.,** *Introduction to Numerical Analysis,* McGraw-Hill, New York, 1956.
4. **Maling, D. H.,** *Coordinate Systems and Map Projections,* George Philip & Son, London, 1973.
5. **Nelson, H.,** Five Easy Projections, GTE-Sylvania, Mountain View, CA, 1979.
6. **Richardus, P. and Adler, R. K.,** *Map Projections for Geodesists, Cartographers, and Geographers,* North-Holland, Amsterdam, 1972.
7. **Robinson, A. H.,** *A New Map Projection: Its Development and Characteristics,* International Yearbook of Cartography, 1974.
8. **Snyder, J. P.,** Map projections used by the U.S. Geological Survey, U.S. Geol. Surv. Bull., 1532, 1983.
9. **Snyder, J. P.,** Map Projections — A Working Manual, U.S. Geol. Surv. Prof. Pap., U.S. Gov't. Printing Office, Washington, D.C., 1395, 1987.
10. **Steers, J. A.,** *An Introduction to the Study of Map Projections,* University of London, 1962.
11. Tables for a Polyconic Projection of Maps, Spec. Publ. U.S. Coast and Geodetic Survey, U.S. Gov't. Printing Office, Washington, D.C., 1946.

Chapter 7

THEORY OF DISTORTIONS

I. INTRODUCTION

Distortion is the limiting factor in map projections. Distortion prevents the realization of the ideal map, as mentioned in Chapter 1. No matter what projection technique is used, distortion occurs in length, angle, shape, area, or in a combination of these. This chapter deals with distortion in a qualitative and quantitative way. First, the different types of maps are compared in a qualitative manner. The most important cylindrical projections are placed in juxtaposition to indicate relative increases or decreases in parallel spacing. This is done for the conical, azimuthal, and world map projections also. Next, a method for comparing true-scale and distorted lengths based on simple geometrical constructions is given. Finally, a unified differential geometry approach is applied to assess the extent of distortion in a quantitative manner. Formulas are developed to quantify the distortions in length, angle, and area, as introduced in Chapter 1. Then, these formulas are applied to the most frequently used equal area, conformal, and conventional projections. Thus, we obtain a numerical means of assessing the acceptability of a map for a particular application.

Before exploring the methods for expressing distortion quantitatively, it is necessary to consider the accuracy required to estimate the effect of distortions. For the direct and inverse transformations, it was necessary to compute using a large number of significant figures. This is not the case for a useful estimation of the effect of distortion. This text suggests that two or three significant figures are all that are necessary to give a realistic estimate of distortion.

With this in mind, it is evident that a spherical model of the Earth is all that is needed to obtain this degree of accuracy. Thus, the equations for estimating distortion are greatly simplified.

II. QUALITATIVE VIEW OF DISTORTION[2]

To graphically display the effects of distortion in a qualitative manner, we will consider several grids of each type placed side by side. This is done for selected cylindrical, conical, azimuthal, and world map projections. For the first three types, the index of comparison is a grid with equal spacing of the parallels along a central meridian.

Figure 1 compares several cylindrical projections: the equatorial Mercator, the plate Carée, the cylindrical equal area, the Miller, and the perspective cylindrical. The Plate Carée is taken as the standard of comparison. The

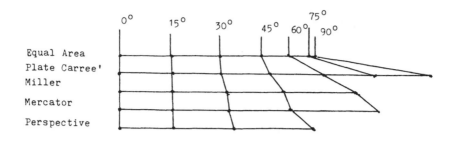

FIGURE 1.   Comparison of parallel spacing on cylindrical projections.

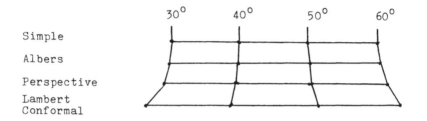

FIGURE 2.   Comparison of parallel spacing on conical projections.

spacing of the parallels along the central meridian of the Mercator, Miller, and Perspective Cylindrical increases as higher latitudes are reached. This is reversed in the cylindrical equal area projection. Here, the spacing of the parallels decreases at high latitudes. In all cases, distortion is independent of longitude.

Four conical projections are compared in Figure 2. These are the Albers, perspective, and Lambert conformal, all characterized by having one standard parallel, and the simple conic, characterized by having an infinity of standard parallels. The simple conic is taken as the reference projection since the parallels are equally spaced along the central meridian. In the Albers projection, the parallels are closer at higher latitudes and further at lower latitudes. In the Lambert conformal projection, the spacing of the parallels diverges north and south of the standard parallels, but in a gradual way. The divergence for the perspective projection is more rapid and severe north and south of the standard parallel. For the Albers, Lambert conformal, and perspective projections, the distortion is independent of longitude.

Figure 3 compares five of the azimuthal polar projections: the equal area, the equidistant, the orthographic, the stereographic, and the gnomonic. The azimuthal equidistant projection is taken as the gauge of comparison. Beginning with the orthographic, there is a steady gradation of parallel spacing, ending with the gnomonic. The orthographic projection suffers distortion as the equator is approached. The parallels are unequally spaced and are bunched

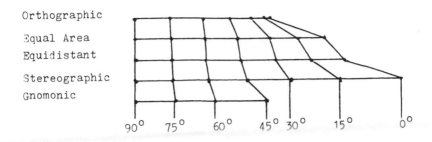

FIGURE 3. Comparison of parallel spacing on azimuthal projections.

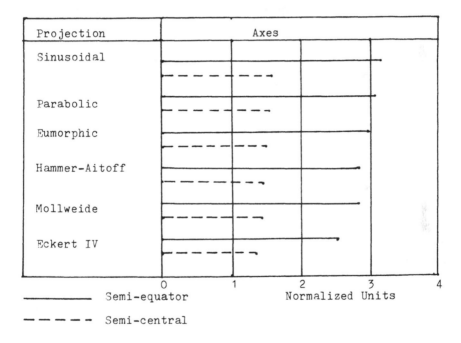

FIGURE 4. Comparison of the lengths of the axes of the equal area world maps.

together close to the equator. The compression of the parallels is not as severe for the equal area projection. The equidistant projection has equally spaced parallels. The stereographic and gnomonic projections have a divergence of the concentric parallels as the equator is approached. The distortion in the gnomonic projection is more severe, and the equator itself can never be portrayed. In all cases, distortion is independent of longitude.

Turning next to the equal area world maps, the semi-equator and semi-central meridian of the Boggs eumorphic, Hammer-Aitoff, Eckert IV, parabolic, sinusoidal, and Mollweide projections are given in Figure 4. For each projection, the area enclosed in the map is $4\pi R^2$, where R is the radius of a spherical Earth model. In order to maintain the equality of area, any increase in the length of the semicentral meridian must be offset in a decrease in the semi-equator, and vice versa. From the grids in Chapter 4, there is also a change of shape. Note also, that the spacing of the parallels must also change to conform to the equal area property. In all three of these projections, the degree of distortion is a function of longitude.

III. QUANTIZATION OF DISTORTIONS[4]

Quantitative values for distortion are needed for comparing distortions in length, area, and angle. This section presents the form in which these distortions are expressed.

In Chapter 1, local scale was introduced as an indicator of linear distortion. Local scale was defined as

$$m = \frac{\text{length of a line on the map}}{\text{length of the true length line}} \tag{1}$$

If $m = 1$, the given line is true length. If $m > 1$, the given line is larger than true scale. If $m < 1$, the given line is smaller than true scale.

Distortions in area are defined as

$$m_A = \frac{\text{area on the map}}{\text{true area on the earth}} \tag{2}$$

If $m_A = 1$, the area on the map is equal to that on the Earth, and an equal area projection is realized. If $m_A > 1$, the area on the map is greater than the true area on the reduced model of the Earth. If $m_A < 1$, the area on the map is less than the true area on the Earth model.

The equations for distortion in angle on the map yield an angular quantity α_m. This is to be compared to the true angle α_T on the Earth. This can be expressed as a percentage.

$$m_\alpha = \left(\frac{\alpha_T - \alpha_m}{\alpha_T} \right) (100) \tag{3}$$

Examinations of Equations 1 and 2 reveal that m and m_A are always positive quantities. The angular distortion m_α of Equation 3 can be either positive or negative, depending on whether α_m is smaller or larger than α_T, respectively.

IV. DISTORTIONS FROM EUCLIDEAN GEOMETRY

Distortion in length can be rather simply estimated from geometric considerations. The model of the Earth is taken as a sphere of radius R. The scale factor S shrinks the Earth so that a given line on the Earth corresponds to a true-length line on the map. The procedure to follow for cylindrical, conical, and azimuthal projections is given below. In these equations Δx and Δy are the differences in the coordinates of the ends of a straight line on the map. The great-circle distance on the Earth is defined by the symbol d.

Consider a cylindrical projection first. The distance to be compared along a parallel is Δx, which corresponds to a change in longitude of $\Delta \lambda$, in radians. The latitude of the parallel is ϕ. Then, the distortion on the map is

$$m = \frac{\Delta x}{RS\cos\phi\Delta\lambda} \tag{4}$$

Along the meridian, the distance to be compared is Δy, which coresponds to a change of latitude in radians of $\Delta\phi$. The distortion is then

$$m = \frac{\Delta y}{RS\Delta\phi} \tag{5}$$

For a general line on the map of length $\sqrt{\Delta x^2 + \Delta y^2}$, the distortion is given by

$$m = \frac{\sqrt{\Delta x^2 + \Delta y^2}}{d \cdot S} \tag{6}$$

For the conical projections, the linear polar coordinate is ρ. The distance to be compared along a parallel of latitude ϕ corresponding to a change of $\Delta\lambda$ in radians is given by $\rho\Delta\theta$, where $\Delta\theta = \Delta\lambda \cdot c$, where c is the constant of the cone. The distortion is

$$m = \frac{\rho c\Delta\lambda}{R \cdot S\cos\phi\Delta\lambda} = \frac{\rho c}{R \cdot S\cos\phi} \tag{7}$$

Along the central meridian, let the distance to be compared be Δy, corresponding to a change in latitude of $\Delta\phi$ in radians. The distortion is again given by Equation 5. Again, the distortion in a general line is given by Equation 6.

Finally, consider an azimuthal projection in which the linear polar coordinate is ρ. Since the constant of the cone is one, $\Delta\theta = \Delta\lambda$. The distance to be compared along a parallel of latitude corresponding to a change of $\Delta\lambda$ in radians is given by $\rho\Delta\lambda$. The distortion is then

$$m = \frac{\rho \Delta \lambda}{RS\cos\phi \Delta \lambda}$$

$$= \frac{\rho}{RS\cos\phi} \qquad (8)$$

Along any meridian, let the distance to be compared be $\Delta\rho$, corresponding to a change of latitude of $\Delta\phi$ in radians. In this case, the distortion is

$$m = \frac{\Delta\rho}{RS\Delta\phi} \qquad (9)$$

In the case of a general line, the distortion is again that of Equation 6.

This approach gives reasonable estimates of linear distortions, but it is not directly applicable to angle and area. The differential geometry approach presents a unified method for dealing with distortions in length, area, and angle.

V. DISTORTIONS FROM DIFFERENTIAL GEOMETRY[4]

The differential geometry approach can be conveniently applied to distortions in length, angle, and area. The concept of the fundamental quantities introduced in Chapter 2 provides the basis for the development. In order to describe distortion in length, we consider a two-dimensional plotting surface and derive terms for distortion along the parallels and meridians, as compared to true distance along a sphere. The derivation begins with the first fundamental forms of the Earth and the plotting surface. The ratio of these fundamental forms is defined to be m. From Equation 3 of Chapter 2, for an orthogonal system,

$$m^2 = \frac{E(d\phi)^2 + G(d\lambda)^2}{e(d\phi)^2 + g(d\lambda)^2} \qquad (10)$$

where the upper case letters refer to the mapping surface, and the lower case, to the sphere.

The distortion along the parametric ϕ-curve, or meridian, where $d\lambda = 0$, is from Equation 10

$$m_m = \sqrt{\frac{E}{e}} \qquad (11)$$

and the distortion along the parametric λ-curve, perpendicular to the meridian, where $d\phi = 0$, is

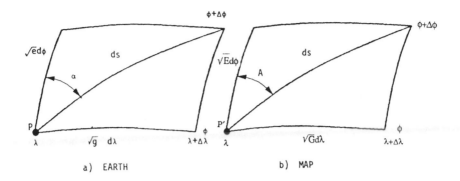

FIGURE 5. Differential parallelograms.

$$m_p = \sqrt{\frac{G}{g}} \qquad (12)$$

These distortions in distance are applied to particular projections in Sections VI to VIII.

Figure 5 is needed for the derivation of distortion in angles. The differential parallelogram in Figure 5a refers to the spherical Earth. An azimuth α is taken from point P. In Figure 5b, the representation of this azimuth on the map is given. Let A be the azimuth as portrayed on the map. The intention is to find a relation between α and A, expressed in the distortions in length m_p and m_m.

From Figure 5a, the trigonometric functions of α are

$$\left. \begin{aligned} \cos\alpha &= \sqrt{e}\,\frac{d\phi}{ds} \\[2mm] \sin\alpha &= \sqrt{g}\,\frac{d\lambda}{ds} \end{aligned} \right\} \qquad (13)$$

From Figure 5b, for the orthogonal mapping surface, the trigonometric functions of A are

$$\cos A = \sqrt{E}\,\frac{d\phi}{dS}$$

$$\sin A = \sqrt{G}\,\frac{d\lambda}{dS} \qquad (14)$$

The relations between α and A are chosen to be $\sin(A - \alpha)$ and $\sin(A + \alpha)$.

Expand

$$\sin(A - \alpha) = \sin A \cos \alpha - \cos A \sin \alpha \tag{15}$$

Substitute Equations 13 and 14 into Equation 15.

$$\sin(A - \alpha) = \sqrt{Ge}\, \frac{d\lambda}{dS} \frac{d\phi}{ds} - \sqrt{Eg}\, \frac{d\lambda}{ds} \frac{d\phi}{dS}$$

$$= (\sqrt{Ge} - \sqrt{Eg}) \frac{d\phi}{ds} \frac{d\lambda}{dS} \tag{16}$$

Substitute Equations 11 and 12 into Equation 16.

$$\sin(A - \alpha) = (m_p - m_m) \frac{d\phi}{ds} \frac{d\lambda}{dS} \sqrt{eg} \tag{17}$$

In a similar expansion,

$$\sin(A + \alpha) = (m_p + m_m) \frac{d\phi}{ds} \frac{d\lambda}{dS} \sqrt{eg} \tag{18}$$

Rearrange Equations 17 and 18 and equate.

$$\sin(A - \alpha) = \frac{m_p - m_m}{m_p + m_m} \sin(A + \alpha) \tag{19}$$

Again, we are faced in Equation 19 with a transcendental equation requiring a numerical analysis solution. For a constant m_p and m_m it can be solved by the Newton-Raphson method.[3] Write Equation 19 as

$$f(A) = 0$$

$$= \sin(A - \alpha) - \frac{m_p - m_m}{m_p + m_m} \sin(A + \alpha) \tag{20}$$

The derivative of Equation 20 is

$$\frac{df}{dA} = \cos(A - \alpha) - \frac{m_p - m_m}{m_p + m_m} \cos(A + \alpha) \tag{21}$$

The Newton-Raphson scheme for the solution of Equation 19 is then

$$A_{n+1} = A_n - \frac{f}{\frac{df}{dA}} \tag{22}$$

Substitute Equations 20 and 21 into Equation 22.

$$A_{n+1} = A_n - \frac{\sin(A_n - \alpha) - \left(\frac{m_p - m_m}{m_p + m_m}\right)\sin(A_n + \alpha)}{\cos(A_n - \alpha) - \left(\frac{m_p - m_m}{m_p + m_m}\right)\cos(A_n + \alpha)} \tag{23}$$

As an initialization, let $A_o = \alpha$. This iteration is rapidly convergent and easily computerized.

This technique is applied to the equal area and the conventional projections. As is seen in Section VII, Equation 19 has a unique solution for conformal projections.

Consider next distortion in area. The area on the map is, from Equation 13 of Chapter 2,

$$A_m = \sqrt{EG} \tag{24}$$

for an orthogonal system. The corresponding area on the model of the Earth is

$$A_e = \sqrt{eg} \tag{25}$$

The distortion in area is hereby defined for an orthogonal system as

$$m_A = \frac{A_m}{A_e}$$

$$= \sqrt{\frac{EG}{eg}} \tag{26}$$

Substitute Equations 11 and 12 into Equation 26.

$$m_A = m_p m_m \tag{27}$$

Now we are in a position to apply these formulas for distortion to selected projections. Some of the most useful of the equal area, conformal, and conventional projections are considered.

VI. DISTORTIONS IN EQUAL AREA PROJECTIONS[4]

Equal area projections, by their definition, have no distortion in area in the mapping transformation. Thus, from Equation 27,

$$m_A = m_p m_m = 1 \tag{28}$$

It is seen from Equation 28 that m_p and m_m are the reciprocals of each other. Thus, whenever it is desired to obtain linear distortions, the easier of the two evaluations is made, and the second distortion factor is simply the reciprocal of the second.

The distortions in length and angle are now considered for the Albers projections with one and two standard parallels and for the polar azimuthal and cylindrical equal area projections.

A. Distortions in Length in the Albers Projection with One Standard Parallel

Consider the first fundamental form for the map in polar coordinates.

$$(ds)_m^2 = (d\rho)^2 + \rho^2(d\theta)^2 \tag{29}$$

Next, we need the first fundamental form for the spherical Earth.

$$(ds)_e^2 = R^2(d\phi)^2 + R^2\cos^2\phi(d\lambda)^2 \tag{30}$$

Since it is necessary to have both fundamental forms in a consistent set of variables, it is necessary to relate θ and ρ to ϕ and λ for this particular projection.

Recall that $\theta = c(\lambda - \lambda_o)$. Thus,

$$d\theta = cd\lambda \tag{31}$$

Substitute Equation 31 into Equation 29.

$$(ds)_m^2 = (d\rho)^2 + c^2\rho^2(d\lambda)^2 \tag{32}$$

By observation of Equations 30 and 32, we have

$$\left.\begin{array}{l} G = c^2\rho^2 \\[2mm] g = R^2\cos^2\phi \end{array}\right\} \tag{33}$$

Substitute Equation 33 into Equation 30.

$$m_p = \sqrt{\frac{c^2\rho^2}{R^2\cos^2\phi}}$$

$$= \frac{c\rho}{R\cos\phi} \tag{34}$$

From Equation 28,

$$m_m = \frac{1}{m_p}$$

$$= \frac{R\cos\phi}{c\rho} \tag{35}$$

Recall from the derivation of the Albers projection with one standard parallel that

$$c = \sin\phi_o \tag{36}$$

and

$$\rho = \frac{R}{\sin\phi_o} \sqrt{1 + \sin^2\phi_o - 2\sin\phi\sin\phi_o} \tag{37}$$

Substitute Equations 36 and 37 into Equation 34.

$$m_p = \frac{\sqrt{1 + \sin^2\phi_o - 2\sin\phi\sin\phi_o}}{\cos\phi} \tag{38}$$

In a similar manner, substitution of Equations 36 and 37 into Equation 35 yields

$$m_m = \frac{\cos\phi}{\sqrt{1 + \sin^2\phi_o - 2\sin\phi\sin\phi_o}} \tag{39}$$

Examination of Equations 38 and 39 yields several interesting facts. First, R was divided out of the equation. Second, distortion is dependent only on latitude, and not on longitude. Third, an expansion of scale in one direction is offset by a contraction in scale in a direction orthogonal to it to maintain equal area.

B. Distortion in Length in the Albers Projection with Two Standard Parallels

Equations 34 and 35 apply in this case. The differences arise in the definition of the constant of the cone, c, and the polar coordinates, ρ. For the two standard parallel cases,

$$c = \frac{\sin\phi_1 + \sin\phi_2}{2} \tag{40}$$

From Equation 55 of Chapter 4,

$$\rho = \sqrt{\frac{\rho_2^2 + 4R^2(\sin\phi_2 - \sin\phi_1)}{\sin\phi_1 + \sin\phi_2}} \tag{41}$$

Substituting Equations 40 and 41 into Equation 34 and simplifying

$$m_p = \frac{(\sin\phi + \sin\phi_2)}{2R\cos\phi} \sqrt{\frac{4R^2\cos^2\phi + 4R^2(\sin\phi_2 - \sin\phi)}{\sin\phi_1 + \sin\phi_2}}$$

$$= \sqrt{1 + \frac{(\sin\phi_1 + \sin\phi_2)(\sin\phi_2 - \sin\phi)}{\cos^2\phi}} \tag{42}$$

Then, from Equation 28 is obtained

$$m_m = \sqrt{\frac{\cos^2\phi}{\cos^2\phi + (\sin\phi_1 + \sin\phi_2)(\sin\phi_2 - \sin\phi)}} \tag{43}$$

Note that, again, distortion in both orthogonal directions is independent of longitude.

C. Distortion in Length in the Azimuthal Projection

Equations 34 and 35 again apply. In this case, the constant of the cone is

$$c = 1 \tag{44}$$

The polar coordinate ρ is

$$\rho = R\sqrt{2(1 - \sin\phi)} \tag{45}$$

Substituting Equations 44 and 45 into Equation 34 and simplifying,

$$m_p = \frac{\sqrt{2(1 - \sin\phi)}}{\cos\phi} \tag{46}$$

From Equation 28,

$$m_m = \frac{\cos\phi}{\sqrt{2(1 - \sin\phi)}}$$

Again, distortion is independent of longitude.

D. Distortion in Length in the Cylindrical Projection

For this projection, it is necessary to return to the first fundamental forms. For the map, in Cartesian coordinates, the first fundamental form is

$$(ds)_m^2 = (dx)^2 + (dy)^2 \qquad (47)$$

The next step is to obtain a consistent set of variables. Recall that $x = R(\lambda - \lambda_o)$

Differentiating,

$$dx = Rd\lambda \qquad (48)$$

Also,

$$y = R\sin\phi$$

Differentiating,

$$dy = R\cos\phi d\phi \qquad (49)$$

Substituting Equations 48 and 49 into Equation 47,

$$(ds)_m^2 = R^2(d\lambda)^2 + R^2\cos^2\phi(d\phi)^2 \qquad (50)$$

By observation of Equations 30 and 50, we have

$$\left.\begin{array}{l} G = R^2 \\ \\ g = R^2\cos^2\phi \end{array}\right\} \qquad (51)$$

Substitute Equation 51 into Equation 30.

$$m_p = \sqrt{\frac{G}{g}}$$

$$= \sqrt{\frac{R^2}{R^2\cos^2\phi}}$$

$$= \frac{1}{\cos\phi} \qquad (52)$$

Also from observation of Equations 30 and 50, we have

$$\left.\begin{array}{l} E = R^2\cos^2\phi \\[2mm] e = R^2 \end{array}\right\} \tag{53}$$

Substitute Equation 53 into Equation 29.

$$\begin{aligned} m_m &= \sqrt{\dfrac{E}{e}} \\[2mm] &= \sqrt{\dfrac{R^2\cos^2\phi}{R^2}} \\[2mm] &= \cos\phi \end{aligned} \tag{54}$$

Comparing Equations 52 and 54, the reciprocal relationship of Equation 28 is maintained. Note also that, again, distortion is independent of longitude.

E. Distortion in Angle in the Equal Area Projections

Now that distortion in length for the selected projections has been derived, we can turn to distortion in angle.

In order to find distortion in angles, substitute Equation 28 into Equation 22.

$$\begin{aligned} \sin(A - \alpha) &= \left(\dfrac{m_p - \dfrac{1}{m_p}}{m_p + \dfrac{1}{m_p}}\right) \sin(A + \alpha) \\[3mm] &= \left(\dfrac{m_p^2 - 1}{m_p^2 + 1}\right) \sin(A + \alpha) \end{aligned} \tag{55}$$

Equation 22 can now be solved for constant m_p by the iteration method of Section V. The distortion m_p is to be considered a constant during the iteration.

A number of examples follow to indicate the method of the evaluation of the distortion equations.

F. Example 1

First, consider a polar azimuthal projection at $\phi = 60°$.

$$m_m = \frac{\cos 60°}{2(1 - \sin 60°)} = \frac{0.500}{(2)(1 - 0.866)} = 0.965$$

$$m_p = \frac{1}{m_m} = \frac{1}{0.965} = 1.036$$

G. Example 2

For the Albers projection with two standard parallels, let $\phi_1 = 35°$ N and $\phi_2 = 55°$ N. Find the linear distortion at $\phi = 45°$ N.

$$m_p = \sqrt{1 + \frac{(\sin\phi_1 + \sin\phi_2)(\sin\phi_2 - \sin\phi)}{\cos^2\phi}}$$

$$m_p = \sqrt{1 + \frac{(\sin 35° + \sin 55°)(\sin 55° - \sin 45°)}{\cos^2 45°}}$$

$$m_p = 1.145$$

$$m_m = \sqrt{\frac{\cos^2\phi}{\cos^2\phi + (\sin\phi_1 + \sin\phi_2)(\sin\phi_2 - \sin\phi)}}$$

$$m_m = \sqrt{\frac{\cos^2 45°}{\cos^2 45° + (\sin 35° + \sin 55°)(\sin 55° - \sin 45°)}}$$

$$m_m = 0.873$$

H. Example 3

Given a cylindrical equal area projection, find the linear distortion at $\phi = 35°$ N.

$$m_p = \frac{1}{\cos\phi}$$

$$m_p = \frac{1}{\cos 35°}$$

$$m_p = 1.221$$

$$m_m = \cos\phi$$

$$m_m = \cos 35°$$

$$m_m = 0.819$$

VII. DISTORTION IN CONFORMAL PROJECTIONS[5]

From Chapter 5, it is recalled that conformal projections are characterized by the fact that

$$m^2 = \frac{E}{e} = \frac{G}{g} \tag{56}$$

for an orthogonal system. Thus, at every point,

$$m = m_p = m_m \tag{57}$$

This relationship makes the work involved in any derivation of linear distortions easier, since only one ratio of the first fundamental quantities needs to be evaluated. In this section, the polar stereographic, the Lambert conformal with one and two standard parallels, and the equatorial Mercator projection are considered. These distortions are based on the sphere as the Earth model.

A. Distortion in Length for the Lambert Conformal Projection, One Standard Parallel

For the Lambert conformal projection with one standard parallel, Equations 29 and 30 are applicable. After the units are made consistent, a form of Equation 35 results:

$$m = m_p$$

$$= m_m$$

$$= \frac{\rho c}{R\cos\phi} \tag{58}$$

The next step is to apply the relations for c and ρ developed in Section IV, Chapter 5. These are

$$c = \sin\phi_o \tag{59}$$

$$\rho = R\cot\phi_o \left\{ \frac{\tan(\pi/4 - \phi/2)}{\tan(\pi/4 - \phi_o/2)} \right\}^{\sin_o} \tag{60}$$

When Equations 59 and 60 are substituted into Equation 58, the distortion in both orthogonal directions is, after simplification,

$$m = \frac{\cos\phi_o}{\cos\phi} \left\{ \frac{\tan(\pi/4 - \phi/2)}{\tan(\pi/4 - \phi_o/2)} \right\}^{\sin_o} \tag{61}$$

Again, distortion in length is independent of longitude.

B. Distortion in Length for the Lambert Conformal Projection, Two Standard Parallels

Distortion for this projection may be derived from Equation 58. This may be done sequentially from the projection equations of Section IV, Chapter 5. These are repeated below.

$$c = \sin\phi_o = \frac{\ln\left(\dfrac{\cos\phi_1}{\cos\phi_2}\right)}{\ln\left[\dfrac{\tan(\pi/4 - \phi_1/2)}{\tan(\pi/4 - \phi_2/2)}\right]} \tag{62}$$

$$\psi = \frac{R\cos\phi_1}{\sin\phi_o[\tan(\pi/4 - \phi_1/2)]^{\sin\phi_o}} \tag{63}$$

$$\rho = \psi[\tan(\pi/4 - \phi_1/2)]^{\sin\phi_o} \tag{64}$$

The sequence of evaluation is (1) Equation 62, Equation 63, (3) Equation 64, and (4) Equation 58. Note that distortion is also independent of longitude.

C. Distortion in Length for the Polar Stereographic Projections

Equation 58 is again the basis for the derivation in the polar stereographic projection. Here, $c = 1$, and it is only necessary to find a value for ρ. From Section V, Chapter 5,

$$\rho = 2R\tan(\pi/4 - \phi/2) \tag{65}$$

Substituting $c = 1$ and Equation 65 into Equation 58 simplifying, yields

$$m = 2\tan(\pi/4 - \phi/2) \cdot \frac{1}{\cos\phi} \tag{66}$$

Again, the distortion is independent of longitude.

D. Distortion in Length for the Regular Mercator Projection

Equations 30 and 47 apply for the regular Mercator projection. After the units are again made consistent, a form of Equation 52 results:

$$m = \frac{1}{\cos\phi} \tag{67}$$

Again, the distortion is independent of longitude.

E. Distortions in Area and Angle

Now that distortions in length have been derived, we can turn to distortions in area and angle. At a point on the projection, the distortion in area is given simply by

$$m_A = m^2 \tag{68}$$

For the distortion in angle at a point, substitute Equation 57 into Equation 19.

$$\sin(A - a) = \frac{(m_p - m_p)\sin(A + a)}{(m_p + m_p)}$$

$$= 0$$

$$A - a = 0$$

$$A = a \tag{69}$$

Thus, one of the properties of the conformal projection is proved. Angles are preserved in the transformation.

F. Example 4

For the regular Mercator projection, find the linear and area distortion at $\phi = 35°$.

$$m = m_p = m_m = \frac{1}{\cos\phi}$$

$$= \frac{1}{\cos 35°}$$

$$= \underline{1.22}$$

$$m_a = m^2 = (1.22)^2$$

$$= \underline{1.49}$$

G. Example 5

For the polar stereographic projection, find the linear distortion at $\phi = 80°$.

$$m = m_p = m_m = 2\tan(\pi/4 - \phi/2) \cdot \frac{1}{\cos\phi}$$

$$= 2\tan\left(45° - \frac{80°}{2}\right)\frac{1}{\cos 80°}$$

$$= \underline{1.01}$$

H. Example 6

For the Lambert conformal projection with two standard parallels at 35° N and 55° N, find the distortion at $\phi = 45°$ N. Let R = 6,378,000 m.

$$\sin\phi_o = \frac{\ln\left(\dfrac{\cos\phi_1}{\cos\phi_2}\right)}{\ln\left[\dfrac{\tan(\pi/4 - \phi_1/2)}{\tan(\pi/4 - \phi_2/2)}\right]}$$

$$\sin\phi = \frac{\ln\left(\dfrac{\cos35°}{\cos55°}\right)}{\ln\left[\dfrac{\tan(45° - 35°/2)}{\tan(45° - 55°/2)}\right]}$$

$$\sin\phi_o = 0.7108$$

$$\psi = \frac{R\cos\phi_1}{\sin\phi_o[\tan(\pi/4 - \phi_1/2)]^{\sin\phi_o}}$$

$$\psi = \frac{(6378000)\cos35°}{(0.7108) \cdot \tan\left(45° - \dfrac{35°}{2}\right)^{0.7108}}$$

$$= 11690683$$

$$\rho = \psi\left[\tan\left(\frac{\pi}{4} - \frac{\phi}{2}\right)\right]\sin\phi_o$$

$$= (11690683) \cdot \left[\tan\left(45° - \frac{45°}{2}\right)\right] 0.7108$$

$$= 6248482$$

$$= \frac{\rho\sin\phi_o}{R\cos\phi}$$

$$= \frac{(6248482) \cdot (0.7108)}{(6378000)(\cos45°)}$$

$$= 0.985$$

VIII. DISTORTION IN CONVENTIONAL PROJECTIONS

In conventional projections, $m_p \neq m_m$ and $m_p \neq 1/m_m$. Since there is no simple relation between the linear distortions along the parametric curves, it is necessary to solve for both. In addition, the conventional projections, as a general rule, defy the differential geometry approach. Thus, to quantify distortions in length it is necessary to rely on the methods of Section IV and a prior knowledge of the characteristics of the projection. This is done below for the polar gnomonic, polar azimuthal equidistant, and the polyconic projections.

A. Distortions in Length for the Polar Gnomonic Projection

Consider first the polar gnomonic projection. From the geometry of the case (Figure 1, Chapter 6), where δ is the colatitude or $\delta = 90° - \phi$,

$$m_m = \frac{R \tan\delta \, d\lambda}{R \sin\delta \, d\lambda}$$

$$= \frac{1}{\cos\delta}$$

$$= \frac{1}{\sin\phi} \tag{70}$$

For the distortion perpendicular to the meridian,

$$m_p = \frac{dS}{ds}$$

$$= \frac{1}{\sin^2\phi} \tag{71}$$

Note that linear distortion is independent of longitude.

B. Distortion in Length for the Polar Azimuthal Equidistant Projection

For the azimuthal equidistant polar projection, the distance along the meridians is true by definition. Thus,

$$m_m = 1 \tag{72}$$

The distance along the parallels on the map, as compared to those on the sphere, follows from the geometry of a circular segment.

$$m_p = \frac{2\pi R\left(\dfrac{\pi}{2} - \phi\right)}{2\pi R \cos\phi}$$

$$= \frac{\pi/2 - \phi}{\cos\phi}$$

Again, the linear distortion is independent of longitude.

C. Distortion in Length for the Polyconic Projection

The last projection to be considered is the regular polyconic. By the assumptions included in the derivation of the projection,

$$m_p = 1 \tag{74}$$

The distortion along the meridians is given by

$$m_m = 1 + \frac{\Delta\lambda^2}{2}\cos^2\Delta\lambda \tag{75}$$

In this case, distortion along the meridians is independent of latitude.

D. Distortions in Angle and Area

For any conventional projection, once the linear distortions have been found, the angular distortion can be obtained by the iterative technique of Section V. Again, the distortions m_m and m_p are considered constant during the iteration. Distortion in area again follows from Equation 27, once the values of m_m and m_p have been obtained.

Several examples follow concerning evaluation of the linear distortion equations.

E. Example 7

First, consider a polar gnomonic projection with $\phi = 60°$.

$$m_m = \frac{1}{\sin 60°} = \frac{1}{0.866} = 1.155$$

$$m_p = \frac{1}{\sin^2 60°} = \frac{1}{(0.866)^2} = 1.333$$

F. Example 8

Given: $R = 6,378,000$ m, $\phi = 45°$ N, $\lambda = 45°$ E, $\lambda_o = 10°$ E.
Find: distortions in length for a polyconic projection.

$$m_p = 1$$

$$m_m = 1 + \frac{\Delta\lambda^2}{2}\cos^2\Delta\lambda$$

$$m_m = 1 + \frac{(45° - 10°)}{2}\frac{\pi}{180}\cos^2(45° - 10°)$$

$$= 1.125$$

REFERENCES

1. **Adams, O. S.,** General Theory of Polyconic Projections, Spec. Publ. 57, U.S. Coast and Geodetic Survey, 1934.
2. **Deetz, C. H. and Adams, O. S.,** Elements of Map Projection, Spec. Publ. 68, U.S. Coast and Geodetic Survey, U.S. Gov't. Printing Office, Washington, D.C., 1944.
3. **Hildebrand, F. B.,** *Introduction to Numerical Analysis,* McGraw-Hill, New York, 1956.
4. **Richardus, P. and Adler, R. K.,** *Map Projections for Geodesists, Cartographers, and Geographers,* North-Holland, Amsterdam, 1972.
5. **Thomas, P. D.,** Conformal Projections in Geodesy and Cartography, Spec. Publ. 251, U.S. Coast and Geodetic Survey, U.S. Gov't. Printing Office, Washington, D.C., 1952.

Chapter 8

MAPPING APPLICATIONS

I. INTRODUCTION

This chapter carries us beyond the derivation of transformation equations, and beyond the relatively simple task of plugging in one type of coordinate, either geographic or Cartesian, and obtaining its analogue. The derivations are important in establishing the validity of each set of transformation equations and suggesting methods of approach if one needs to strike out for himself. A large amount of the practical work in map projections entails generating one type of coordinate from the other as part of a data generation process. Yet there are occasions when map projections are needed as part of the solution of many types of problems. This chapter is intended to extend the reader's understanding of the practical use of map projections by the examination of several applications which serve as extended examples.

The first section concerns map projections in the Southern Hemisphere. Recall that the preceding derivations produced projections directly applicable to the Northern Hemisphere. This section indicates what is needed to make the transformations valid in the Southern Hemisphere.

Next, we consider the distortion introduced in the transformation from the reference spheroid onto the authalic sphere. While this discussion is useful in itself, it also provides an example of how distances on any two surfaces can be compared.

The example of distances on the loxodrome shows how the special features of the Mercator and the gnomonic projections are used in navigation. Distances on the Earth and on the map are compared. The bearing of the vehicle is computed.

Map projection equations provide the basis for the display of data generated by tracking systems. The results of using the three most popular projections for tracking system displays, the azimuthal equidistant, the orthographic, and the gnomonic, are compared.

The final example concerns differential distances about a position. This is applicable to the equatorial Mercator projection based on a spheroidal model of the Earth. This is an approximate method for obtaining the direct and inverse transformations on a large-scale map.

II. MAP PROJECTIONS IN THE SOUTHERN HEMISPHERE

In the preceding chapters, the plotting equations were developed for the Northern Hemisphere. Only a slight revision of the formulas is needed to apply these equations to the Southern Hemisphere. The revision is necessary because of the relation of the plotting surface to the model of the Earth. The changes occur mainly in the polar azimuthal, conical, and oblique projections. The cylindrical projections do not require any changes. The world maps have formulas which automatically accommodate them to both the Northern and Southern Hemispheres.

Consider first the world maps. An example of this class is the sinusoidal projection. The plotting equations are

$$
\left.\begin{array}{l}
x = \Delta\lambda R \cdot S\cos\phi \\[12pt]
y = R \cdot S \cdot \phi
\end{array}\right\}
\tag{1}
$$

Whether the latitude is north or south, $\cos\phi$ is positive, so the x coordinate is independent of the hemisphere portrayed. In the equation for the y coordinate, if the latitude is in the north, that is, $\phi > 0$, then $y > 0$. If the latitude is in the south, that is, $\phi < 0$, then $y < 0$. Thus, the correct hemisphere is always determined.

Consider next the case for the cylindrical projections, as indicated in Figure 1. Let the point P_N and P_S have the same longitudes. Let the respective latitudes be equal in magnitude, but opposite in sign. Thus,

$$
\left.\begin{array}{l}
x_n = x_s \\[12pt]
y_n = -y_s
\end{array}\right\}
\tag{2}
$$

Generally, the cylindrical plotting equations will intrinsically accomplish this. An example is the cylindrical equal area projection with plotting equations

$$
\left.\begin{array}{l}
x = R \cdot S \cdot \Delta\lambda \\[12pt]
y = R \cdot S \cdot \sin\phi
\end{array}\right\}
\tag{3}
$$

The x coordinate is independent of latitude. For the y coordinate, if $\phi > 0$, then $y > 0$, and if $\phi < 0$, then $y < 0$.

Figure 2 compares the position of the mapping plane for polar azimuthal projections. Then, these mapping planes are shown side by side. Let P_N and P_S have the same longitudes. The latitudes of P_N and P_S are the same in magnitude, but opposite in sign. The relationship of Equations 2 again holds; but now, special logic is needed in the evaluation. Again, it is necessary to evaluate the plotting equations with the magnitude of the latitude and then change the sign of the y coordiante.

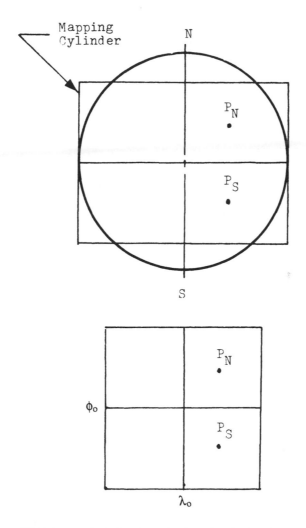

FIGURE 1. Cylindrical projections in the Southern Hemisphere.

The conical projections are treated in Figure 3. The position of the developable cones are depicted for the Northern and Southern Hemispheres. The developed cones are then compared side by side. Again, P_N and P_S have the same longitude. The latitudes have the same magnitude, but are opposite in sign. Again, the relationship of Equations 2 holds; and, again, the plotting equations must include special logic. In this situation, the latitude, ϕ_o, of the one standard parallel case, or the two latitudes, ϕ_1 and ϕ_2, of the two standard parallel case, are made positive. Then the plotting equations are evaluated with the magnitude of the southern latitude. Finally, the sign of the y coordinate is changed to obtain the correct answers.

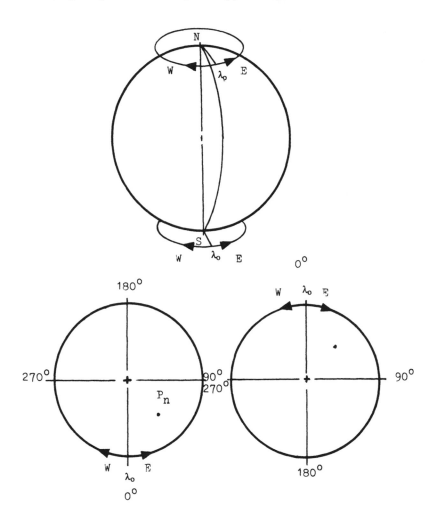

FIGURE 2.   Azimuthal polar projections in the Southern Hemisphere.

A typical oblique projection appears in Figure 4. Oblique mapping planes are tangent at T_n and T_s in the Northern and Southern hemispheres, respectively. These planes are compared side by side at the bottom of the figure. Points P_N and P_S have the same longitude. The latitudes have the same magnitude, but opposite signs. In order to make the relations of Equations 2 hold, it is necessary to evaluate the plotting equations using the magnitudes of ϕ and ϕ_o and then change the sign of the ordinate.

The procedure for dealing with the Southern Hemisphere can be summarized as follows. Most world map projections and cylindrical projections need no special treatment. For oblique, conical, and azimuthal projections, evaluate the plotting equations with the magnitude of the latitude and then change the sign of the y coordinate. This latter procedure is also necessary for the equatorial Mercator, Miller, Mollweide, and Eckert projections.

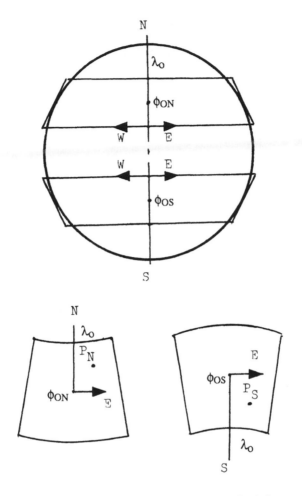

FIGURE 3.   Conical projections in the Southern Hemisphere.

The computer logic required for the implementation of these procedures is indicated in the subroutines of Chapter 9.

A. Example 1

Given: a parabolic projection, $\lambda_o = 0°$, $\phi = 30°$ S, $\lambda = 30°$ W, R = 6,378,000 m, S = 1:20,000,000.

Find: the Cartesian coordinates.

$$\Delta\lambda = \lambda - \lambda_o = -30 - 0 = -30°$$

$$y = 3.06998 \cdot S \cdot R \cdot \sin\phi/3$$

$$= (3.06998)\left(\frac{6,378,000}{20,000,000}\right)\sin\left(-\frac{30}{3}\right)$$

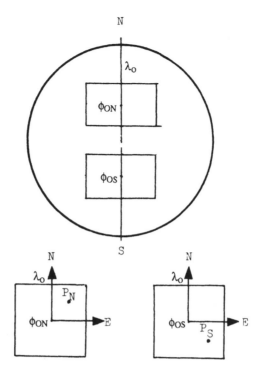

FIGURE 4.   Azimuthal oblique projections in the
Southern Hemisphere.

$$= (3.06998)(0.3189)(-0.17365)$$

$$= -0.170 \text{ m}$$

$$x = (1.53499)\frac{\Delta\lambda}{90} S \cdot R \cdot \left(2\cos\frac{2\phi}{3} - 1\right)$$

$$= (1.53499)\left(\frac{-30}{90}\right)\left(\frac{6,378,000}{20,000,000}\right)\left[2\cos\frac{(2)(-30)}{3} - 1\right]$$

$$= (1.53499)\left(-\frac{1}{3}\right)(0.3189)(0.87938)$$

$$= -0.144 \text{ m}$$

B. Example 2

Given: a polar stereographic projection, $\lambda_o = 0°$, $\phi = 75°$ S, $\lambda = 45°$
E, R = 6,378,000 m, S = 1:5,000,000.

Find: the Cartesian coordinates.

Northern Hemisphere calculations:

$$\Delta\lambda = \lambda - \lambda_o = 45 - 0 = 45°$$

$$x_n = 2RS\tan(\pi/4 - \phi/2)\sin\Delta\lambda$$

$$= (2)\left(\frac{6,378,000}{5,000,000}\right)\tan\left(45° - \frac{75}{2}\right)\sin45°$$

$$= (2)\left(\frac{6,378,000}{5,000,000}\right)(0.13165)(0.70711)$$

$$= 0.238 \text{ m}$$

$$y_n = -2RS\tan(\pi/4 - \phi/2)\cos\Delta\lambda$$

Southern Hemisphere calculations:

$$x_s = x_n = 0.238 \text{ m}$$

$$y_s = -y_n = 0.238 \text{ m}$$

C. Example 3

Given: Albers projection with 2 standard parallels, $\phi_1 = 40°$ S, $\phi_2 = 50°$ S, $\lambda_o = 0°$, $\phi = 45°$ S, $\lambda = 5°$ E, R = 6,378,000 m, S = 1:2,000,000.
Find: the Cartesian coordinates.
Northern Hemisphere calculations:

$$\Delta\lambda = \lambda - \lambda_o = 5 - 0 = 5°$$

$$c = \frac{\sin\phi_1 + \sin\phi_2}{2} = \frac{\sin40° + \sin50°}{2}$$

$$= \frac{0.64279 + 0.76604}{2} = 0.70441$$

$$\theta = \Delta\lambda c = (5)(0.70441) = 3°.5221$$

$$\rho_1 = \frac{R\cos\phi_1}{c} = \frac{(6,378,000)\cos40°}{0.70441} = 6.9361 \times 10^6 \text{ m}$$

$$\rho_2 = \frac{R\cos\phi_2}{c} = \frac{(6,378,000)\cos50°}{0.70441} = 5.8200 \times 10^6 \text{ m}$$

$$\frac{4R^2(\sin\phi_1 - \sin\phi)}{\sin\phi_1 + \sin\phi_2} = \frac{(2)(6.378 \times 10^6)^2(0.64279 - 0.70711)}{0.70441}$$

$$\frac{(2)(6.378 \times 10^6)^2(-0.06432)}{0.70441} = -7.4288 \times 10^{12} \text{ m}^2$$

$$\sqrt{\rho_1^2 + \frac{4R^2(\sin\phi_1 - \sin\phi_2)}{\sin\phi_1 + \sin\phi_2}} = \sqrt{(6.9361 \times 10^6)^2 - 7.4288 \times 10^{12}}$$

$$\sqrt{48.1094 \times 10^{12} - 7.4288 \times 10^{16}} = 6.3781 \times 10^6 \text{ m}$$

$$x = S\sqrt{\rho_1^2 + \frac{4R^2(\sin\phi_1 - \sin\phi_2)}{\sin\phi_1 + \sin\phi_2}} \sin\theta$$

$$= \left(\frac{6.3781}{2.0000}\right)\sin 3°.5221 = 0.196 \text{ m}$$

$$y = S\left\{\frac{1}{2}(\rho_1 + \rho_2) - \sqrt{\rho_1^2 + \frac{4R^2(\sin\phi_1 - \sin\phi_2)}{\sin\phi_1 + \sin\phi_2}} \cos\theta\right\}$$

$$= \frac{1/2(6.9361 + 5.8200)(10^6) - (6.3781 \times 10^6)\cos 3°.5221}{2 \times 10^6}$$

$$= \frac{6.3780 - 6.3660}{2} = \frac{0.012}{2} = 0.006 \text{ m}$$

Southern Hemisphere calculations:

$$x_s = x_n = 0.196 \text{ m}$$

$$y_s = -y_n = 0.006 \text{ m}$$

III. DISTORTION IN THE TRANSFORMATION FROM THE SPHEROID TO THE AUTHALIC SPHERE

This section investigates the amount of distortion introduced in a transformation from a reference spheroid onto the authalic sphere. Consider three positions, A_s, B_s, and C_s, on the WGS-72 reference spheroid. They are to be transformed to positions A_A, B_A, and C_A on the authalic sphere. The questions are asked, what is the difference in length between distances A_sB_s and A_AB_A, and between distances B_sC_s and B_AC_A? This is taken as an indication of the amount of distortion.

It is given that the coordinates of A_s, B_s, and C_s on the reference spheroid are as follows:

Point	Latitude	Longitude
A_s	30°	0°
B_s	60°	0°
C_s	60°	−30°

For the WGS-72 reference spheroid, from Table 1, Chapter 3, a = 6,378,135 m. For the authalic sphere corresponding to the WGS-72 spheroid, from Section III.A, Chapter 4, R_A = 6,371,004 m.

The first step is to consider the transformation of points A_s, B_s, and C_s onto the authalic sphere. The results of this transformation can be obtained from Equation 18 of Chapter 4. The latitudes of A_A, B_A, and C_A are as follows. Recall that longitude is invariant in the transformation.

Point	Latitude	Longitude
A_A	29.889°	0°
B_A	59.888°	0°
C_A	59.888°	−30°

In the next step, consider distances on the authalic sphere. For $A_A B_A$,

$$d_{1A} = R_A \phi_A$$

$$= \frac{(6,371,004)(59.888 - 29.889)}{57.29578}$$

$$= 3,335,784 \text{ m}$$

for $B_A C_A$,

$$d_{2A} = \Delta\lambda R_A \cos\phi_A$$

$$= \left(\frac{30}{57.29578}\right)(6,371,004)\cos(59.888)$$

$$= 1,673,592 \text{ m}$$

Consider next the comparable distances on the WGS-72 spheroid. From Equation 19 of Chapter 3, the radius of the parallel circle at 60° is R_o = 3,197,101 m

The distance between B_s and C_s is

$$d_{2S} = \phi R_o$$

$$= \left(\frac{30}{57.29578}\right)(3,197,101)$$

$$= 1,674,021 \text{ m}$$

From Equation 28 of Chapter 3, the average $\overline{R}_m$ at 45° N is 6,367,380 m. Approximate the distance between A_s and B_s as

$$d_{1S} = \phi\overline{R}_m$$

$$= \left(\frac{30}{57.29578}\right)(6,357,380)$$

$$= 3,333,998 \text{ m}$$

Now compare the two sets of differences between B and C.

$$\frac{d_{2A}}{d_{2S}} = \frac{1,673,592}{1,674,021} = 0.99976$$

The percentage difference is

$$(1 - 0.99976)(100) = 0.024\%$$

Between A and B,

$$\frac{d_{1A}}{d_{1S}} = \frac{3,335,784}{3,333,998} = 1.00050$$

The percentage difference is

$$(1 - 1.00050)(100) = 0.050\%$$

These differences are very small. For instance, if the scale of the map was such that a true length distance on the map between A and B was 3 in., the difference in the plotted line would be 0.0015 in. If 0 to 30° latitude had been used, rather than 30 to 60°, about twice the difference would be found. Note that if 30 to 60° is equivalent to 3 in. on a map, this is a very small scale. Thus, ellipsoid difference is negligible anyway. As a practical approach for a large-scale map, it is recommended that the ellipsoid should be used as the basis of calculations. In this latter case, the transformation to the authalic sphere should not be considered.

IV. DISTANCES ON THE LOXODROME[1]

This section gives an extended example of the practical use of the equatorial Mercator and the equatorial gnomonic projections for navigation. Each projection has unique characteristics which lend themselves to this use.

We consider two points on the spherical model of the Earth. The shortest distance between these points on the surface of the sphere is along the great circle. However, as one moves along the great circle, the angle formed by the tangent to the great circle and the tangent to the local meridian is constantly changing. This angle is the instantaneous bearing. The loxodrome is a line which has the characteristic that the angle between the tangent to the loxodrome and the tangent to the local meridian is constant. Thus, the bearing is constant. Since in navigation it is relatively easy to maintain a constant bearing, the loxodrome is the curve of choice. However, a loxodrome is of greater length than the great circle distance, unless along the equator or a meridian.

If it is desired to minimize distance traveled, the great circle distance is approximated by a series of loxodromes. Along each loxodromic segment, the bearing is held constant. The bearing changes when the next segment is reached.

On the equatorial Mercator projection, the loxodrome is a straight line. This permits the user to calculate the latitudes at which the loxodrome intersects selected meridians. On the equatorial gnomonic projection, the great circle is a straight line, but not to true scale. In a similar manner, one can calculate the latitudes at which a great circle intersects selected meridians.

In this example, a great circle distance between two given points on the spherical model of the Earth is calculated. This is compared with the distance along the loxodrome on the equatorial Mercator projection between the same two points. The great circle length on the spherical model of the Earth is also compared to the length of the great circle as portrayed on the equatorial gnomonic projection. The bearing along a single loxodrome is then calculated. The latitudes of several loxodromic crossings of selected meridians are obtained. Then, the latitudes of comparable great circle crossings are found. These define several loxodromic segments. The bearings along each of these segments is finally calculated. The geometry on the equatorial Mercator projection is given as Figure 5.

For the example, consider a spherical Earth of radius 6,378,007 m. Use $S = 1:10,000,000$. It is desired to navigate from position A ($\phi = 0°$, $\lambda = 0°$) to position D ($\phi = 30°$, $\lambda = 45°$).

First, we will find the great circle distance to map scale.

$$d_{GC} = R \cdot S \cdot \cos^{-1}(\sin\phi_A \sin\phi_D + \cos\phi_A \cos\phi_D \cos\lambda)$$

$$= \left(\frac{6,378,007}{10,000,000}\right)\cos^{-1}(\sin0°\sin30° + \cos0°\cos30°\cos45°)$$

$$= (0.6378007)[\cos^{-1}(0 + (1)(0.866025)(0.707107)] \qquad (4)$$

$$= (0.6378007)\cos^{-1}(0.612372)$$

$$= (0.6378007)\left(\frac{52.2388}{57.295}\right)$$

$$= 0.5815 \text{ m}$$

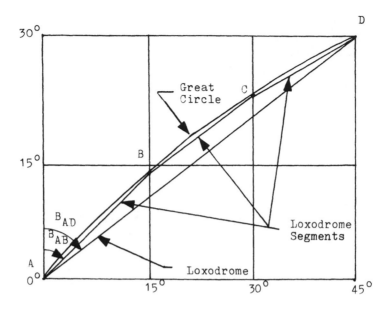

FIGURE 5.   Loxodromes and the great circle on the equatorial Mercator projection.

Next, we obtain the distance along the loxodrome from A to D on the Mercator projection. At point A, the Cartesian coordinates on the Mercator projection are $x_A = 0$, $y_A = 0$. At point D, the Cartesian coordinates are found from Equations 93 and 91 of Chapter 5.

$$x_D = R \cdot S \cdot \Delta\lambda$$

$$= \left(\frac{6,378,007}{10,000,000}\right)\left(\frac{45}{57.295}\right)$$

$$= 0.5009 \text{ m} \tag{5}$$

$$y_D = R \cdot S \cdot \ln\tan(\pi/4 + \phi_o/2)$$

$$= (0.6378007)\ln\tan\left(45° + \frac{30°}{2}\right)$$

$$= (0.6378007)(0.549306)$$

$$= 0.3503 \text{ m} \tag{6}$$

$$d_{IOX} = \sqrt{(x_D - x_A)^2 + (y_D - y_A)^2}$$

$$= \sqrt{0.5009^2 + 0.3503^2} \tag{7}$$

$$= 0.6106 \text{ m}$$

The bearing of the meridian crossing is

$$B_{AD} = N\left[\pi/2 - \tan^{-1}\left(\frac{y}{x}\right)\right]E$$

$$= N\left[90° - \tan^{-1}\left(\frac{0.3503}{0.5009}\right)\right]E$$

$$= N(90° - 34.97)E$$

$$= N\ 55°.03\ E \qquad (8)$$

Consider next the latitudes of the loxodromic meridian crossings at λ_B = 15° and λ_c = 30°. This requires use of the inverse regular Mercator formulas. By proportions

$$x_B = \frac{x_D}{3}$$

$$= \frac{0.5009}{3}$$

$$= 0.1670\ m$$

$$x_C = \frac{2}{3} \cdot x_D$$

$$= \frac{(2)(0.5009)}{3}$$

$$= 0.3339\ m$$

$$y_B = y_D\left(\frac{x_B}{x_D}\right)$$

$$= \frac{(0.3503)}{3}$$

$$= 0.11677\ m$$

$$y_C = y_D\left(\frac{x_C}{x_D}\right)$$

$$= (0.3503)\left(\frac{2}{3}\right)$$

$$= 0.23353\ m$$

$$\phi = 2[\tan^{-1}(EXP(y/R \cdot S)) - \pi/4] \qquad (9)$$

where EXP is the exponential function.

$$\phi_B = (2)(50°.21580 - 45°)$$

$$= (2)(5.21580)$$

$$= 10°.432$$

$$\phi_c = (2)(55°.26287 - 45°)$$

$$= (2)(10.26287)$$

$$= 20°.526$$

Now, find the great circle distance on the equatorial gnomonic projection with $\lambda_o = 0°$. At point A, $x_A = 0$, $y_A = 0$, and at point D,

$$x_D = a \cdot S \cdot \tan\Delta\lambda$$

$$= (0.6378007)\tan45°$$

$$= 0.6378 \text{ m} \tag{10}$$

$$y_D = \frac{a \cdot S \cdot \tan\phi}{\cos\Delta\lambda}$$

$$= \frac{(6378007)\tan30°}{\cos45°}$$

$$= \frac{(0.6378007)(0.577350)}{0.707107}$$

$$= 0.5208 \text{ m} \tag{11}$$

From Equation 7,

$$d_{GN} = \sqrt{0.6378^2 + 0.5208^2}$$

$$= 0.8234 \text{ m}$$

Note the significant differences between the lengths d_{OC}, d_{LOX}, and d_{GN}.

Now, it is necessary to use the inverse equatorial gnomonic transformation to obtain the great circle crossings at $\lambda = 15°$ and $\lambda = 30°$. First, return to Equation 9 to obtain x_B and x_C.

$$x_B = (0.6378007)\tan15°$$

$$= 0.1709 \text{ m}$$

$$x_C = (0.6378007)\tan 30°$$

$$= 0.3683 \text{ m}$$

By proportions

$$y_B = X_B\left(\frac{Y_D}{X_D}\right)$$

$$= (0.1709)\left(\frac{0.5208}{0.6378}\right)$$

$$= 0.1395 \text{ m}$$

$$y_C = X_C\left(\frac{Y_D}{X_D}\right)$$

$$= (0.3683)\left(\frac{0.5208}{0.6378}\right)$$

$$= 0.3007 \text{ m}$$

$$\phi = \tan^{-1}\left(\frac{y\cos\Delta\lambda}{R \cdot S}\right)$$

$$\phi_B = \tan^{-1}\left[\frac{y_B\cos 15°}{0.6378007}\right]$$

$$= \tan^{-1}\left[\frac{(0.1395)(0.965926)}{0.6378007}\right]$$

$$= 11°.929$$

$$\phi_c = \tan^{-1}\left[\frac{y_C\cos 30°}{0.6378007}\right]$$

$$= \tan^{-1}\left[\frac{(6.3007)(0.866025)}{0.6378007}\right]$$

$$= 22°.210 \tag{12}$$

Comparing the latitudes found in the calculation for the equatorial Mercator projection to those found for the equatorial gnomonic projection, note that the great circle is north of the loxodrome.

Next, find the y coordinates on the equatorial Mercator projection corresponding to the great circle intersections found on the equatorial gnomonic projection. Referring to Equation 6,

TABLE 1
Coordinates for
Bearing Calculation

Point	x	y
A	0	0
B	0.1670	0.1338
C	0.3339	0.2537
D	0.5009	0.3503

$$y_B = (0.6378007) \ln \tan\left(45° + \frac{11°.929}{2} \right)$$

$$= (0.6378007)(0.209721)$$

$$= 0.1338 \text{ m}$$

$$y_C = (0.6378007) \ln \tan\left(45° + \frac{22°.210}{2} \right)$$

$$= (0.6378007)(0.397727)$$

$$= 0.2537 \text{ m}$$

The coordinates to be used in the calculations for the bearing along loxodromic segments A-B, B-C, and C-D are summarized in Table 1. The bearings for each segment are finally calculated from Equation 8.

$$B_{AB} = N\left[\pi/2 - \tan^{-1}\left(\frac{0.1338 - 0}{0.1670 - 0} \right) \right] E$$

$$= N[90° - 38.71]E$$

$$= N\ 51°.29\ E$$

$$B_{BC} = N\left[\pi/2 - \tan^{-1}\left(\frac{0.2537 - 0.1338}{0.3339 - 0.1670} \right) \right] E$$

$$= N[90° - 35.68]E$$

$$= N\ 54.32\ E$$

$$B_{CD} = N\left[\pi/2 - \tan^{-1}\left(\frac{0.3503 - 0.2537}{0.5009 - 0.3339} \right) \right] E$$

$$= N[90° - 30°.05]E$$

$$= N\ 59°.95\ E$$

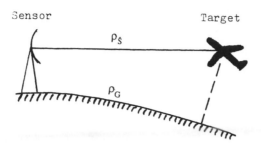

FIGURE 6.   Range between sensor and target.

The bearings become successively more easterly as the course is traversed.

V. TRACKING SYSTEM DISPLAYS

Map projection equations form the basis for displaying tracking data on electronic screens. These screens range in size from small CRTs to wall-sized large-screen displays. The most popular projections for this use are the oblique versions of the azimuthal equidistant, the gnomonic, and the orthographic. At relatively short distances between the display origin and a target, all three of these projections give comparable results. However, as the distance increases, two of these, the gnomonic and the orthographic, suffer from unacceptable amounts of distortion. This section compares the three projections at ranges where they are virtually interchangeable, that is, about 400 km.

The usual configuration for these displays is to have the tracking center at the origin of the screen. Two cases are to be considered. In the first case, a single sensor is the sole source. The second case considers a command center which acts as the compiler of data from a number of sources.

The case for a single sensor is depicted in Figure 6. The tracking device, pictured as a radar, measures slant range, ρ_s, and azimuth, α, to a target. The display shows ground range, ρ. Turning to Figure 7, a display is given which shows the tracking center and two targets. The azimuth and range from the tracking center to each target is given. Also indicated is a distance of separation, d, between the two targets. This is the basic geometry for the example below.

Since a single radar or optical system is limited in view by the horizon, there is a natural limit to the distance that can be portrayed, and the three projections to be considered are of equal value. A problem can occur in the second case, where the command center receives data from auxiliary sources, usually over the horizon from the center. Then, if the wrong projection has been chosen for the display, the display will be subject to distortions which create a false picture.

For the example, consider a large-screen display with a command center at the origin. Let this center be located at geographic coordinates $\phi_o = 45°$

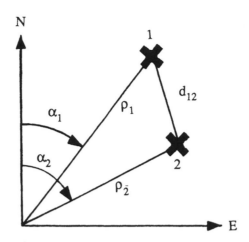

FIGURE 7.   CRT display of ranges and azimuths.

and $\lambda_o = 0°$. Consider one target located at $\phi_1 = 48°$, $\lambda_1 = 3°$, and a second target located at $\phi_2 = 48°$, $\lambda_2 = -3°$. We apply the mapping equations for the oblique azimuthal equidistant, gnomonic, and orthographic projections to determine distances and azimuths on the screen as compared to those on the spherical model of the Earth. Let this spherical model of the Earth have a radius $R = 6,378,007$ m. Let the scale of the projections be $S = 1:1,000,000$.

We begin with distances on the spherical Earth, at map scale, between the origin and targets 1 and 2. The great circle distance is defined as

$$\rho = RS\cos^{-1}(\sin\phi_1\sin\phi_2 + \cos\phi_1\cos\phi_2\cos\Delta\lambda)$$

$$\rho_1 = \rho_2 = \left(\frac{6,378,007}{1,000,000}\right)\cos^{-1}(\sin45°\sin48° + \cos45°\cos48°\cos3°)$$

$$= \left(\frac{6,378,007}{1,000,000}\right)\cos^{-1}[(0.707107)(0.743145)$$

$$+ (0.707107)(0.669131)(0.998630)]$$

$$= \left(\frac{6,378,007}{1,000,000}\right)\cos^{-1}(0.52483 + 0.472499)$$

$$= \left(\frac{6,378,007}{1,000,000}\right)\cos^{-1}(0.997983)$$

$$= \left(\frac{6,378,007}{1,000,000}\right)\frac{3.640586}{57.295}$$

$$= 0.40526 \text{ m}$$

(13)

The distance between the targets on the great circle are

$$d_{12} = \left(\frac{6,378,007}{1,000,000}\right)\cos^{-1}(\sin48°\sin48° + \cos48°\cos48°\cos6°)$$

$$= \left(\frac{6,378,007}{1,000,000}\right)\cos^{-1}[(0.743145)^2 + (0.669131)^2(0.994522)]$$

$$= \left(\frac{6,378,007}{1,000,000}\right)\cos^{-1}(0.552264 + 0.445283)$$

$$= \left(\frac{6,378,007}{1,000,000}\right)\cos^{-1}(0.99547)$$

$$= \left(\frac{6,378,007}{1,000,000}\right)\left(\frac{4.0040}{57.295}\right)$$

$$= 0.44683 \text{ m}$$

Since the range from the origin to any point on the azimuth equidistant projection is the same as on the spherical Earth, we already have that value. Since the azimuth from the origin to any other point on the azimuthal projection is the same as on the Earth, we will use the formulas for that projection to solve for the joint value.

For the azimuthal equidistant projection, ψ is angle generated in Equation 13.

$$= \cos^{-1}(0.997983)$$

$$= 3°.640580$$

$$\cos\theta = \frac{\sin\Delta\lambda\cos\phi}{\sin\psi}$$

$$= \frac{\sin3°\cos48°}{\sin3°.64058}$$

$$= \frac{(0.052336)(0.669131)}{0.6634974}$$

$$= 0.551538 \tag{14}$$

Since the quadrant is known,

$$\theta = 56°.5274$$

$$\sin\theta = 0.834150$$

$$x = R \cdot S \cdot \psi \cos\theta$$

$$= \left(\frac{6,378,007}{1,000,000}\right)\left(\frac{3.64058}{57.295}\right)(0.551538)$$

$$= 0.22352 \text{ m} \tag{15}$$

$$y = R \cdot S \cdot \psi \sin\theta$$

$$= \left(\frac{6,378,007}{1,000,000}\right)\left(\frac{3.64058}{57.295}\right)(0.34150)$$

$$= 0.33805 \text{ m} \tag{16}$$

$$\alpha_1 = \tan^{-1}\left(\frac{x}{y}\right)$$

$$= \tan^{-1}\left(\frac{0.22352}{0.33805}\right)$$

$$= 33°.473$$

$$d_{12} = (2)x$$

$$= (2)(0.22352)$$

$$= 0.44704 \text{ m} \tag{17}$$

For the orthographic projection, the mapping coordinates are as follows:

$$x = R \cdot S \cdot \cos\phi \sin\Delta\lambda$$

$$= \left(\frac{6,378,007}{1,000,000}\right)\cos 48°\sin 3°$$

$$= \left(\frac{6,378,007}{1,000,000}\right)(0.669131)(0.052336)$$

$$= 0.22336 \text{ m} \tag{18}$$

$$y = R \cdot S \cdot (\cos\phi_o\sin\phi - \sin\phi_o\cos\phi\cos\Delta\lambda)$$

$$= \left(\frac{6,378,007}{1,000,000}\right)(\cos 45°\sin 48° - \sin 48°\cos 48°\cos 3°)$$

$$= (6.378007)[(0.707107)(0.743145)$$

$$- (0.707107)(0.669131)(0.998630)]$$

$$= (6.378007)(0.52583 - 0.472499)$$

$$= (6.378007)(0.052984)$$

$$= 0.33793 \text{ m}$$

$$\rho_1 = \rho_2 = \sqrt{x^2 + y^2}$$

$$= \sqrt{0.22336^2 + 0.33793^2}$$

$$= 0.40508 \text{ m}$$

$$d_{12} = (2)x$$

$$= (2)(0.22336)$$

$$= 0.44672 \text{ m} \qquad (19)$$

From Equation 17,

$$\alpha_1 = \tan^{-1}\left(\frac{0.22336}{0.33793}\right)$$

$$= 33°.463$$

For the gnomonic projection, the azimuth is true. The Cartesian coordinates and ranges are as follows;

$$x = \frac{R \cdot S \cdot \cos\phi\sin\Delta\lambda}{\sin\phi_o\sin\phi + \cos\phi_o\cos\phi\cos\Delta\lambda}$$

$$= \left(\frac{6,378,007}{1,000,000}\right)\frac{\cos48°\sin3°}{(\sin45°\sin48° + \cos45°\cos48°\cos3°)}$$

$$= \frac{(6,378,007)(0.669131)(0.052336)}{(0.707107)(0.743145) + (0.707107)(0.669131)(0.998630)}$$

$$= \frac{(6.378007)(0.669131)(0.052336)}{0.525483 + 0.472499}$$

$$= \frac{(6.378007)(0.669131)(0.052336)}{0.997982}$$

$$= 0.22381 \text{ m} \qquad (20)$$

$$y = \frac{R \cdot S \cdot (\cos\phi_o\sin\phi - \sin\phi_o\cos\phi\cos\Delta\lambda)}{\sin\phi_o\sin\phi + \cos\phi_o\cos\phi\cos\Delta\lambda}$$

$$= \frac{(6.378007)[\cos45°\sin48° - \sin45°\cos48°\cos3°]}{0.997982}$$

TABLE 2

Comparison of Azimuths and Distances on the Earth and on Three Projections Used for Displays

Mapping basis	Azimuth (degrees)	Range to target (m)	Distance between targets (m)
Earth	33.473	0.40526	0.44683
Azimuthal equidistant	33.473	0.40526	0.44704
Orthographic	33.463	0.40508	0.44672
Gnomonic	33.473	0.40589	0.44762

$$= \left(\frac{6{,}378{,}007}{0.997982}\right)[(0.707107)(0.743145)$$

$$- (0.707107)(0.669131)(0.998630)]$$

$$= \frac{(6.378007)(0.525483 - 0.472499)}{0.997982}$$

$$= \frac{(6.378007)(0.052983)}{0.997982} = 0.33861 \text{ m}$$

$$\rho_1 = \rho_2 = \sqrt{x^2 + y^2}$$

$$= \sqrt{(0.22381)^2 + (0.33861)^2}$$

$$= 0.40589 \text{ m}$$

$$d_{12} = 2x$$

$$= (2)(0.22381)$$

$$= 0.44762 \text{ m} \qquad (21)$$

Table 2 summarizes the results from this example. At the range considered, the results are visually identical. However, even at this range, differences begin to appear. Taking the range on the model of the Earth as the basis of comparison, the portrayed ranges are shorter for the orthographic projection and longer for the gnomonic projection. As the range between origin and target on the Earth increases, these discrepancies in portrayed range become more apparent. In this example, the azimuth in the orthographic projection has a slight deviation.

The orthographic projection has enough relative computational simplicity to recommend it, even though range to target and azimuths are distorted at long range. The gnomonic projection maintains azimuth, but range becomes greatly distorted at long range. The azimuthal equidistant projection maintains true azimuth and range at all ranges. It must be kept in mind, however, that

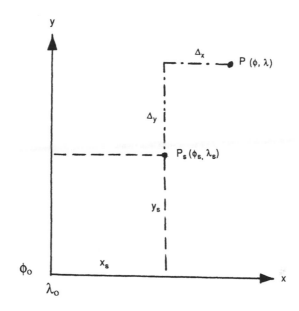

FIGURE 8. Geometry for a differential expansion.

all three suffer from distortion in distances between targets. This distortion increases with range.

VI. DIFFERENTIAL DISTANCES FROM A POSITION

One problem in mapping concerns a cluster of positions with a very small difference in latitude and longitude. Usually, this is encountered on a large-scale map. If we consider this case on an equatorial Mercator projection, with origin at the intersection of the equator and the central meridian, the difference in the plotting coordinates of two adjacent points will be the difference of two nearly equal numbers if the standard plotting equations are used. A useful alternative is to consider a differential expansion about a given geographic position as a means of obtaining the Cartesian coordinates at one point with respect to a known standard reference point.

This section considers a method for obtaining this differential expansion for the equatorial Mercator projection. By a small difference in latitude or longitude is meant a total angular difference of less than 1°. This angular difference translates into roughly 110 km on the spheroidal model of the Earth. The example in this section concludes with an estimate of the accuracy in a typical application of this method.

Consider Figure 8, which gives the geometry for the differential expansion. The origin of the projection is at ϕ_o and λ_o. The point P_s, with geographic coordinates ϕ_s and λ_s, about which the expansion is defined, is selected by the user. The Cartesian coordinates of P_s, x_s, and y_s are obtained by the evaluation of Equations 113 and 114 of Chapter 5 for the spheroidal model of the Earth.

Consider next an arbitrary point P, with geographic coordinates ϕ and λ. We will limit the differences between ϕ and ϕ_s, and λ and λ_s to about 1° each. Then,

$$\left.\begin{array}{l} \Delta\phi = \phi - \phi_s \\[2mm] \Delta\lambda = \lambda - \lambda_s \end{array}\right\} \tag{22}$$

Use the differential forms from Equations 94 and 112 of Chapter 5 to calculate the changes in position on the map.

$$\left.\begin{array}{l} \Delta x = a \cdot S\Delta\lambda \\[4mm] \Delta y = \dfrac{a \cdot S(1 - e^2)\Delta\phi}{(1 - e^2\sin^2\phi_s)\cos\phi_s} \end{array}\right\} \tag{23}$$

Note that the second of Equation 23 is evaluated with $\phi = \phi_s$. This requires only one calculation for a set of expansions about P_s and will give an acceptable approximation.

Then, the mapping coordinates of P, with respect to the origin of the coordinate system, are

$$\left.\begin{array}{l} x = x_s + \Delta x \\[2mm] y = y_s + \Delta y \end{array}\right\} \tag{24}$$

The inverse relation is also quite straight forward. Given small variations in x and with respect to x_s and y_s, the equations are

$$\left.\begin{array}{l} \Delta x = x - x_s \\[2mm] \Delta y = y - y_s \end{array}\right\} \tag{25}$$

Then, the corresponding changes in latitude and longitude are obtained by inverting Equations 23.

$$\left.\begin{array}{l} \Delta\lambda = \dfrac{\Delta x}{a \cdot S} \\[4mm] \phi = \dfrac{(1 - e^2\sin^2\phi_s)\cos\phi_s\Delta y}{a \cdot S(1 - e^2)} \end{array}\right\} \tag{26}$$

Again, the second of Equations 26 is evaluated at $\phi = \phi_s$.

The changes in latitude and longitude are applied to the latitude and longitude of the point P_s to obtain the latitude and longitude of the point under consideration.

$$\left. \begin{array}{l} \phi = \phi_s + \Delta\phi \\[2mm] \lambda = \lambda_s + \Delta\lambda \end{array} \right\} \tag{27}$$

A numerical example is now given for the direct transformation. First the Cartesian coordinates for the point about which the transformation takes place are obtained from Equations 113 and 114 of Chapter 5. Then, the differential correction is obtained from Equations 22 to 24. Then, the coordinates of P are obtained from Equations 113 and 114 of Chapter 5 to gain an estimate of the accuracy of the approximation.

For the example, let $\lambda_o = 0°$, $\lambda_s = 30°$, and $\phi_s = 30°$. Use the WGS-72 ellipsoid with a = 6,378,135 m, and e = 0.081819. Let the arbitrary point have geographic coordinates = 30°.1 and $\phi = 30°.1$. Use S = 1:100,000.

First, find the Cartesian coordinates of the point P_s, about which the expansion is to take place.

$$x_s = aS\Delta\lambda$$

$$= \left(\frac{6,378,135}{100,000}\right)\left(\frac{30}{57.295}\right)$$

$$= 33.3963 \text{ m}$$

$$y_s = aS\ln\left[\tan\left(\frac{\pi}{4} + \frac{\phi}{2}\right)\left(\frac{1 - e\sin\phi}{1 + e\sin\phi}\right)^{e/2}\right]$$

$$= \frac{6,378,135}{100,000}\ln\left\{\tan\left(45° + \frac{30°}{2}\right) \times \left[\frac{1 - (0.081819)\sin30°}{1 + (0.081819)\sin30°}\right]^{0.081819/2}\right\}$$

$$= (63.78135)\ln\left\{(1.73205)\left[\frac{1 - (0.081819)(0.500000)}{1 + (0.081819)(0.500000)}\right]^{0.04091}\right\}$$

$$= (63.78135)\ln\left\{(1.73205)\left(\frac{1 - 0.04091}{1 + 0.04091}\right)^{0.04091}\right\}$$

$$= (63.78135)\ln\left\{(1.73205)\left(\frac{0.95909}{1.04091}\right)^{0.04091}\right\}$$

$$= (63.78135)\ln\{(1.73205)(0.996656)\}$$

$$= (63.78135)(0.545957) = 34.82184 \text{ m}$$

Next, consider the differential latitudes and longitudes for point P. From Equations 22,

$$\Delta\phi = \phi - \phi_s$$

$$= 30.1 - 30 = 0.1°$$

$$\Delta\lambda = \lambda - \lambda_s$$

$$= 30.1 - 30 = 0.1°$$

Apply Equations 23 to obtain the differential corrections.

$$\Delta x = aS\Delta\lambda$$

$$= \left(\frac{6,378,135}{100,000}\right)\left(\frac{0.1}{57.295}\right)$$

$$= 0.11132 \text{ m}$$

$$\Delta y = \frac{aS(1 - e^2)\Delta\phi}{(1 - e^2\sin^2\phi_s)\cos\phi_s}$$

$$= \left(\frac{6,378,135}{100,000}\right)\left(\frac{0.1}{57.295}\right)\frac{[1 - (0.081819)^2]}{[1 - (0.081819)^2\sin^230°]\cos30°}$$

$$= \frac{(0.11132)(1 - 0.006694)}{[1 - (0.006694)(0.250000)](0.866025)}$$

$$= \frac{(0.11132)(0.993306)}{(0.998326)(0.866025)}$$

$$= 0.12789 \text{ m}$$

Thus, the total coordinates with respect to the map origin are, from Equations 24,

$$x = x_s + \Delta x$$

$$= 33.3963 + 0.1113$$

$$= 33.5076 \text{ m}$$

$$y = y_s + \Delta y$$

$$= 34.8218 + 0.1279$$

$$= 34.9497 \text{ m}$$

Finally, let us compare these results to the values obtained by an evaluation of the regular Mercator mapping equations at the desired point.

$$x = aS\Delta\lambda$$

$$= \left(\frac{6,378,135}{100,000}\right)\left(\frac{30.1}{57.295}\right)$$

$$= 33.5076 \text{ m}$$

$$y = \left(\frac{6,378,135}{100,000}\right)\ln\left\{\tan\left(45° + \frac{30°.1}{2}\right)\right.$$

$$\times \left[\frac{1 - (0.081819)\sin30°.1}{1 + (0.081819)\sin30°.1}\right]^{0.04091}\right\}$$

$$y = (63.78135)\ln\{(1.73555)$$

$$\times \left[\frac{1 - (0.081819)(0.501511)}{1 + (0.081819)(0.501511)}\right]^{0.04091}\right\}$$

$$= 63.78135\ln\left\{(1.73555)\right.$$

$$\times \left[\frac{1 - 0.04103}{1 + 0.04103}\right]^{0.04091}\right\}$$

$$= 63.78135\ln\{(1.73555)(0.996646)\}$$

$$= (63.78135)(0.547965)$$

$$= 34.9499 \text{ m}$$

As would be expected from a linear function, the results for x are exact in either approach. Since the calculation in the second of Equations 23 is a nonlinear function, and an approximation was made by evaluating ϕ at ϕ_s, it is not expected that the result would be exact. Still, the result is within acceptable accuracy for mapping.

This procedure can be equally applied to polar coordinates. The only requirement is that the required differential relations be available.

REFERENCES

1. **Bowditch, N.,** *American Practical Navigator,* Defense Mapping Agency Hydrographic Center, Washington, D.C., 1977.
2. **Hershey, A. V.,** The Plotting of Maps on a CRT Printer, Naval Surface Weapon Center Rep. No. 1844, Dahlgren, VA, 1963.

Chapter 9

COMPUTER APPLICATIONS

I. INTRODUCTION

At this point in the study of map projections, it should be obvious that the practical use of the plotting equations is feasible only if a computer is available. A desk calculator may be useful for a single evaluation of a set of equations, but a computer is essential for any large-scale computational scheme.

To this end, this chapter illustrates some of the computer programs and subroutines necessary to various aspects of mapping. First subroutines are given which accomplish the direct transformation for 18 of the most useful projection schemes. This is followed by subroutines for a selection of inverse transformations.

A program is introduced which evaluates single points or generates an entire grid for the direct transformation. This program calls up the mapping subroutines selected by the user. The program also carries out the inverse transformation for selected projections.

Then, a program is presented which applies to state plane coordinates. Direct and inverse transformations are included for both the transverse Mercator and Lambert conformal projections. Following this is a brief program which generates coordinates in the universal transverse Mercator (UTM) grid system.

Finally, since a map is a visual experience, rather than simply a set of numbers, some consideration is given to the computer display of mapping grids and positions.

II. DIRECT TRANSFORMATION SUBROUTINES[5]

The majority of the practical applications of map projection transformations calls for the evaluation of the direct plotting equations developed in Chapters 4, 5, and 6. In this case, geographic position is known, and the desired model of the Earth and the desired scale are chosen. For a selected projection, the Cartesian plotting coordinates are then calculated.

This section presents a series of 12 subroutines which calculate the Cartesian coordinates. The most useful and practical projections have been programmed as procedures in TURBO PASCAL.[8] The listings of the procedures are given in Figure 1. Each subroutine depends on the availability of the radius of the spherical Earth model and the scale selected for the transformation. In the discussion of the subroutine, it is pointed out if the transformation is automatically applicable to Southern Hemisphere application, or if special attention is needed to make it applicable to that hemisphere, as noted in Section II, Chapter 8.

```
(********************************************************************)
PROCEDURE Dir_Azimuthal_Equidistant (a,s,Phi,Lamda,PhiO,LamO:Real;
                                     VAR X,Y :Real);

VAR   Psi,CosPsi,SinPsi,CosTheta,Dlamda,SinTheta : Real;

  BEGIN
    Dlamda := Lamda - LamO;
    CosPsi :=  Sin(PhiO)*Sin(Phi) + Cos(PhiO)*Cos(Phi)*Cos(DLamda);
    PSI := ArcCos(CosPSI);
    CosTheta := Sin(DLamda)*Cos(Phi)/Sin(PSI);
    SinTheta := (Cos(PhiO)*Sin(Phi) - Sin(PhiO)*Cos(Phi)*Cos(DLamda))/Sin(PSI);
    X := A*S*PSI*CosTheta;
    Y := A*S*Psi*SinTheta;
    IF (X = -0.0) Then   X := 0.0;
    If (Y = -0.0) then   Y := 0.0;
  END; (of procedure)

(********************************************************************)
PROCEDURE Dir_EqArea_Azimuthal(a,s,Phi,Lamda,PhiO,LamO:Real; VAR X,Y:Real);

VAR   Dlamda,TempPhi : Real;

  BEGIN
    Dlamda := Lamda - LamO;
    TempPhi := ABS(Phi);
    X := a*s*SQRT(2*(1.0 - Sin(TempPhi)))*Cos(Dlamda);
    Y := a*s*SQRT(2*(1.0 - Sin(TempPhi)))*Sin(Dlamda);
    IF (X = -0.0) THEN X := ABS(X);
    IF (Y = -0.0) THEN Y := ABS(Y);
    IF (Phi < 0) THEN
       Y := -Y;
  END; (of procedure)

(********************************************************************)
PROCEDURE Dir_EqArea_Conical (a,s,Phi,Lamda,PhiO,LamO,Phi1,Phi2:Real;Var X,Y:Real
);

VAR  Dlamda,TempPhi,TempPhi1,TempPhi2,Rho1,Rho2,Term1,Term2,Term3 : Real;
     Term4,Term5 :Real;

  BEGIN
    Dlamda := Lamda-LamO;
    TempPhi := ABS(Phi);
    TempPhi1 := ABS(Phi1);
    TempPhi2 := ABS(Phi2);
    Term1 := Sin(TempPhi1) + Sin(TempPhi2);
    Rho1 := 2*a*Cos(TempPhi1) / Term1;
    Rho2 := 2*a*Cos(TempPhi2) / Term1;
    Term2 := 0.5*(Rho1+Rho2);
    Term3 := 4.0*Sqr(a)*(Sin(TempPhi1)-Sin(TempPhi))/Term1;
    Term4 := Dlamda*0.5*Term1;
    Term5 := Sqrt(Sqr(Rho1)+Term3);
    X := s*Term5*Sin(Term4);
    Y := s*(Term2 - Term5*Cos(Term4));
    IF (Phi < 0) THEN
       Y := -Y;
  END; (of procedure)

PROCEDURE Dir_EqArea_Cylindrical (a,s,Phi,Lamda,LamO:Real; VAR X,Y:Real);

VAR   Dlamda : Real;

  BEGIN
    Dlamda := Lamda - LamO;
    X := a*s*Dlamda;
    Y := a*s*Sin(Phi);
  END;
```

FIGURE 1.   Listings of direct transformation procedures. (Courtesy of B. McNamara, U.S. Coast and Geodetic Survey, Sacramento, CA.)

```
{***********************************************************************}
PROCEDURE Dir_Equatorial_Mercator(a,s,Phi,Lamda,Lam0:Real;VAR X,Y:Real);

VAR    Dlamda,TempPhi,Term1 : Real;

  BEGIN
    Dlamda := Lamda - Lam0;
    TempPhi := ABS(Phi);
    Term1 := Tan(Pi/4.0 + TempPhi/2.0);
    IF ABS(Term1) > 1000000.0 THEN
      Term1 := 1000000.0;
    X := a*s*Dlamda;
    Y := a*s*LN(Term1);
    IF (Phi < 0) then
      Y := -Y;
  END; {of procedure}


{***********************************************************************}
PROCEDURE Dir_Gnomic(a,s;Phi,Lamda,Phi0,Lam0:Real; VAR X,Y:Real);

VAR    Dlamda,Term1,Term2 : Real;

  BEGIN
    Dlamda := Lamda - Lam0;
    Term1 := Sin(Phi0)*Sin(Phi) + Cos(Phi0)*Cos(Phi)*Cos(Dlamda);
    Term2 := Cos(Phi0)*Sin(Phi) - Sin(Phi0)*Cos(Phi)*Cos(Dlamda);
    X := a*s*Cos(Phi)*Sin(Dlamda) / Term1;
    Y := (a*s*Term2) / Term1;
  END; {of procedure}


{***********************************************************************}
PROCEDURE Dir_LambertConformal(a,s,Phi,Lamda,Lam0,Phi1,Phi2:Real;VAR Phi0,X,Y:Re
al);

VAR    Dlamda,TempPhi,TempPhi1,TempPhi2,Term1,SinPhi0,Rho1,Theta,Psi,Rho:Real;

  BEGIN
    Dlamda := Lamda - Lam0;
    TempPhi := ABS(Phi);
    TempPhi1 := ABS(Phi1);
    TempPhi2 := ABS(Phi2);
    Term1 := Ln(Tan(Pi/4-TempPhi1/2) / Tan(Pi/4-TempPhi2/2));
    SinPhi0 := Ln(Cos(TempPhi1)/Cos(TempPhi2)) / Term1;
    Phi0 := ArcSin(SinPhi0);
    Rho1 := a*Cos(TempPhi1)/SinPhi0;
    Theta := Dlamda*SinPhi0;
    Psi := a*Cos(TempPhi1)/ (SinPhi0*Tan(Pi/4-TempPhi1/2));
    Rho := Psi*Tan(Pi/4-TempPhi/2);
    X := s*Rho*Sin(Theta);
    Y := s*(Rho1-Rho*Cos(Theta));
    IF (Phi < 0) THEN
      Y := -Y;
  END; {of procedure}


{***********************************************************************}
PROCEDURE Dir_Mollweide(a,s,Phi,Lamda,Lam0:Real; VAR X,Y:Real);

VAR    Dlamda,Theta,DTheta,Limit : Real;  I : Integer;

  BEGIN
    Dlamda := Lamda - Lam0;
    Theta := ABS(Phi);

    IF (Theta <= 89.9*PI/180.0) THEN
     BEGIN
      FOR I := 1 TO 50  DO
        BEGIN
         DTheta := (Pi*Sin(Theta)-2*Theta-Sin(2*Theta))/(1+Cos(2*Theta))/2;
         Theta := Theta + DTheta;
        END; {for-loop}
     END; {if-then}

    X := a*s*Dlamda*Cos(Theta)/1.11072;
    Y := a*s*Sin(Theta)*1.41421;
    IF (Phi <= 0) THEN
      Y := -Y;
  END; {of procedure}
```

FIGURE 1 continued.

```
(************************************************************************)
PROCEDURE Dir_Parabolic(a,s,Phi,Lamda,Lam0:Real; VAR X,Y:Real);
VAR    Dlamda : Real;

  BEGIN
    Dlamda := Lamda - Lam0;
    X := Dlamda*a*s*(2*Cos(0.666667*Phi)-1.0) / 1.02332;
    Y := a*s*Sin(Phi/3.0)*3.06998;
  END; {of procedure}

(************************************************************************)
PROCEDURE Dir_Polyconic(a,s,Phi,Lamda,Phi0,Lam0:Real; VAR X,Y:Real);

VAR    Dlamda,TempPhi,Term1,Term2 : Real;

  BEGIN
    Dlamda := Lamda - Lam0;
    TempPhi := ABS(Phi);
    IF (Phi = 0.0) THEN
      BEGIN
        X := a*s*Dlamda;
        Y := 0.0
      END
    ELSE
      BEGIN
        Term1 := Sin(Dlamda*Sin(TempPhi));
        Term2 := Cos(Dlamda*Sin(TempPhi));
        X := a*s*Cos(TempPhi)*Term1/Sin(TempPhi);
        Y := a*s*(TempPhi-Phi0+Cos(TempPhi)*(1-Term2)/Sin(TempPhi));
      END;

    IF (Phi < 0) THEN
      Y := -Y;
  END; {of procedure}

(************************************************************************)
PROCEDURE Dir_Sinusoidal(a,s,Phi,Lamda,Lam0:Real; VAR X,Y:Real);

VAR    Dlamda : Real;

  BEGIN
    Dlamda := Lamda - Lam0;
    X := a*s*Dlamda*Cos(Phi);
    Y := a*s*Phi;
  END; {of procedure}

(************************************************************************)
PROCEDURE Dir_Stereographic(a,s,Phi,Lamda,Lam0:Real; VAR X,Y:Real);

VAR    Dlamda,TempPhi,Term1 : Real;

  BEGIN
    Dlamda := Lamda - Lam0;
    TempPhi := ABS(Phi);
    Term1 := Tan(Pi/4 - TempPhi/2);
    X := 2.0 * a*s*Term1*Sin(Dlamda);
    Y := -2.0 * a*s*Term1*Cos(Dlamda);
  END; {of procedure}
```

FIGURE 1 continued.

The origin of the coordinate system, defined by the central meridian, λ_o, and some standard parallel, ϕ_0, is indicated for each projection. In every case, λ_o is to be selected by the user. For some projections, ϕ_0 is at the discretion of the user. In others, ϕ_0 is implicitly defined by the derivation of the transformation.

Four equal area world projections are included in the set: equal area cylindrical, parabolic, sinusoidal, and Mollweide. In all four of these, the equator, $\phi_o = 0^o$, is the standard parallel. The coordinate axes are the central meridian, chosen by the user, and the equator. For the cylindrical, parabolic, and sinusoidal projections, the plotting equations are straightforward, and a southern latitude, denoted by a negative number, leads to a negative y coordinate. In the Mollweide projection, the iteration converges for positive or negative ϕ. Thus, no additional logic is required.

Two conical projections are included. These are the two standard parallel cases for the Albers (equal area conical) and the Lambert conformal. The two standard parallel cases are preferred in order to better distribute distortion. The latitudes of the two standard parallels are chosen by the user. The central meridian is the y axis, and x axis intersects the central meridian at the standard parallel closer to the equator. The techniques of Section II, Chapter 8, are needed to make the plotting coordinates applicable to the Southern Hemisphere. This has been automatically included in the algorithms.

Two polar projections are given in Figure 1. These are the azimuthal equal area and the stereographic. For both, $\phi_o = 90°$ and is the origin of the coordinate system. The positive x axis is at longitude 90°. The positive y axis is at longitude 180°. The central meridian is at longitude 0°. Again, special consideration, built into the procedures, permit their application to the Southern Hemisphere.

The azimuthal equidistant and the gnomonic projections are both oblique azimuthal. It is the user's choice for both the central meridian and the latitude of tangency. The x and y axes emanate from this point of tangency. Special consideration is required to make these projections applicable to the Southern Hemisphere.

The equatorial Mercator projection has the central meridian as the y axis, and the equator, $\phi_o = 0°$, as the x axis. The equation for the y coordinate can be evaluated with positive or negative values of ϕ. Thus, no special logic is needed to make the equations applicable to the Southern Hemisphere.

In the polyconic projection, the central meridian, λ_o and ϕ_o are up to the user. The central meridian again defines the y axis, and the x axis intersects the central meridian at ϕ_o.

With the exception of the Mollweide projection, all of these procedures entail a direct evaluation of the plotting equations without any iteration or approximation. In general, precise values of ϕ and λ are available. This leads to precise values of x and y for any plotting applications and for use in subsequent calculations.

III. INVERSE TRANSFORMATION SUBROUTINES[5]

Of importance in many aspects of practical work is the availability of inverse transformations algorithms. Nine of these are displayed in Figure 2.

For the inverse transformations, the x and y coordinates available are of limited accuracy, such as those scaled from a map or obtained by digitizing a map. Thus, the resulting ϕ and λ are of limited accuracy, even though the

```
(***********************************************************************)
PROCEDURE Inv_EqArea_Azimuthal(X,Y,a,s,Lam0:Real; VAR Phi,Lamda:Real);

VAR   Dlamda,SinPhi : Real;

  BEGIN
    Dlamda := ArcTan(Y/X);
    Which_Quadrant(X,Y,Dlamda);
    Lamda := DLamda + LamO;
    SinPhi := 1.0 - (Sqr(X) / (2*SQR(a)*SQR(s)*SQR(Cos(Dlamda))));
    Phi := ArcSin(SinPhi);
  END; (of procedure)

(***********************************************************************)
PROCEDURE Inv_EqArea_Conical(X,Y,a,s,Lam0:Real; VAR Phi,Lamda:Real);

VAR   Term1,Term2,Term3,Rho1,Rho2,Dlamda,SinPhi : Real:

  BEGIN
    Term1 := (Sin(Phi1)+Sin(Phi2)) / 2.0;
    Rho1 := a*Cos(Phi1)/Term1;
    Rho2 := a*Cos(Phi2)/Term1;
    Term2 := (Rho1+Rho2)/2.0;
    Dlamda := ArcTan(X/s/(Term2-Y/s))/ Term1;
    Lamda := Dlamda + LamO;
    Term3 := Sin(Dlamda*Term1);
    SinPhi := Sin(Phi1)-Term1/2/a/a*(Sqr(X)/Sqr(s)/Term3/Term3-Sqr(Rho1));
    Phi := ArcSin(Phi);
  END; (of procedure)

(***********************************************************************)
PROCEDURE Inv_EqArea_Cylindrical(X,Y,a,s,Lam0:Real; VAR Phi,Lamda:Real);

VAR   Dlamda,SinPhi : Real;

BEGIN
  Dlamda := X/(a*s);
  Lamda := Dlamda + LamO;
  SinPhi := Y/(a*s);
  Phi := ArcSin(SinPhi);
END; (of procedure)

(***********************************************************************)
PROCEDURE Inv_Equatorial_Mercator(X,Y,a,s,Lam0:Real; VAR Phi,Lamda:Real);

VAR   Dlamda : Real;

BEGIN
  Dlamda := X/(a*s);
  Lamda := Dlamda + LamO;
  Phi := 2.0 * (ArcTan(Exp(Y/(a*s))) - PI/4.0);
END; (of procedure)

(***********************************************************************)
PROCEDURE Inv_LambertConformal(X,Y,a,s,Lam0,Phi1,Phi2:Real; VAR Phi,Lamda:Real);

VAR   Term1,SinPhi0,Rho1,Theta,Dlamda,Psi : Real;

BEGIN
  Term1 := Tan(PI/4 - Phi1/2);
  SinPhi0 := Ln(Cos(Phi1)/Cos(Phi2));
  SinPhi0 := SinPhi0/Ln(Term1/Tan(PI/2 - Phi2/2));
  Rho1 := a*Cos(Phi1)/SinPhi0;
  Theta := ArcTan(X/s/(Rho1-Y/s));
  Dlamda := Theta/SinPhi0;
  Lamda := Dlamda + LamO;
  Psi := a*Cos(Phi1)/(SinPhi0*Term1);
  Phi := PI/2 - 2*ArcTan(X/(s*Psi*Sin(Theta)));
END; (of procedure)
```

FIGURE 2.   Listing of inverse transformation procedures. (Courtesy of B. McNamara, U.S. Coast and Geodetic Survey, Sacramento, CA.)

```
(******************************************************************)
PROCEDURE Inv_Mollweide(X,Y,a,s,Lam0:Real; VAR Phi,Lamda:Real);

VAR    SinTheta,SinPhi,Theta,Dlamda : Real;

BEGIN
   SinTheta := Y/a/s/1.41421;
   Theta := ArcSin(SinTheta);
   SinPhi := (2*Theta + Sin(2*Theta))/Pi;
   Phi := ArcSin(SinPhi);
   Dlamda := X * 1.11072 /a/s/Cos(Theta);
   Lamda := Dlamda + Lam0;
END; (of procedure)

(******************************************************************)
PROCEDURE Inv_Parabolic(X,Y,a,s,Lam0:Real; VAR Phi,Lamda:Real);

VAR    SinPhi,Dlamda: Real;

BEGIN
   SinPhi := Y/a/s/3.06998;
   Phi := 3.0 * ArcSin(SinPhi);
   Dlamda := X*1.02332/a/s;
   Dlamda := Dlamda/(2*Cos(Phi*0.666667)-1);
   Lamda := Dlamda + Lam0;
END; (of procedure)

(******************************************************************)
PROCEDURE Inv_Sinusoidal(X,Y,a,s,Lam0:Real; VAR Phi,Lamda:Real);

VAR    Dlamda : Real;

BEGIN
   Phi := Y/a/s;
   Dlamda := X/a/s/Cos(Phi);
   Lamda := Dlamda + Lam0;
END; (of procedure)

(******************************************************************)
PROCEDURE Inv_StereoGraphic(X,Y,a,s,Lam0:Real; VAR Phi,Lamda:Real);

VAR    Dlamda,Term1,Term2 : Real;

BEGIN
   Dlamda := ArcTan(Y/X);
   Which_Quadrant(X,Y,Dlamda);
   Lamda := Dlamda + Lam0;
   Term1 := SQRT((SQR(X) + SQR(Y))/4/SQR(a)/SQR(s));
   Term2 := ArcTan(Term1);
   Phi := PI/2 - 2*Term2;
END; (of procedure)
```

FIGURE 2 continued.

algorithms themselves have not detracted from the accuracy. This is a caveat that the user must keep in mind.

There are four types of direct plotting equations. In the first, the coordinates are uncoupled. That is, x is a function of λ alone, and y is a function of ϕ alone. The inverses to these are found rather simply. Examples of these are the equal area cylindrical and the equatorial Mercator. In the second, there is only one degree of coupling. For instance, y is a function of ϕ alone, and x is a function of both ϕ and λ. In this case, the relation for a single dependence is solved first, and then the relation for a double dependence is solved. Examples of this are the parabolic and sinusoidal projections. The third type has both coordinates dependent on both ϕ and λ. However, the coupling may be removed by completing the square of a sine and cosine or by division to produce a tangent function. Examples of these are the two-parallel cases of the Albers and Lambert conformal, the equal area azimuthal and stereographic polar cases.

The procedures given in Figure 2 are based on the above three types. In all cases there is a closed-form transformation to obtain ϕ and λ from x and y.

The last type has both coordinates dependent on ϕ and λ in a manner that defines trigonometric simplification. Then, two approaches are possible. The first is a series expansion. An example of this is given in Section VII, Chapter 5, for the State Plane coordinate system for the transverse Mercator and the Lambert conformal projections. The second is a numerical technique such as relaxation. Both of these are costly in terms of computer storage space and computing time.

IV. CALLING PROGRAM FOR SUBROUTINES[6]

The TURBO PASCAL[8] procedures of Sections II and III must be incorporated into a main program in order to produce useful results. In many applications, the procedures may be only a small segment of a very large program, but they may be the main feature of a program dedicated solely to mapping. In this section, the discussion of the architecture of such a program is based on the flow chart of Figure 3.

At the beginning of such a program, the constants for the Earth model are included. The logical values to exercise the options and direct the logical flow are also required. The initial decision is whether a direct transformation or an inverse transformation is required. If the inverse transformation is chosen, a series of x and y coordinates are converted to geographic. Logic must be provided to utilize one of the procedures of Figure 2 to accomplish this transformation. The resulting geographic coordinates are then stored either on hard copy or in a computer data base. At this point, this particular application is concluded.

If the direct transformation option is chosen, the next decision relates to whether the user wants tables or graphics. If the choice is tables, then the selected latitudes and longitudes are converted to x and y coordinates through the vehicle of the selected projection of Figure 1. Usually it is only of academic interest to calculate an entire grid while exercising this option. The useful method is to transform selected geographic positions of interest. The resulting Cartesian coordinates are to be saved as either a printout or as information in a computer data base. Once this is done, this particular application is ended.

If the graphics option is chosen, the next decision concerns the need for a grid. In graphics, the grid may be of considerable use in the utilization of the finished product. If the answer to the preceding question is yes, a grid is needed, the grid is then generated using the selected direct transformations of Figure 1. Information must be supplied to define the maximum extent and

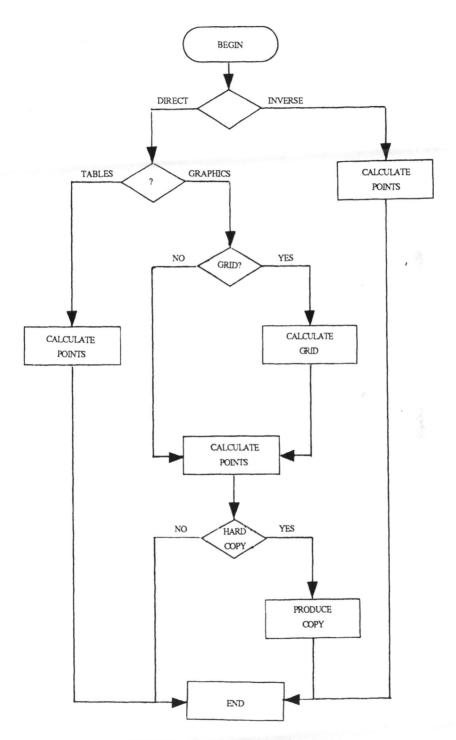

FIGURE 3. Flow chart for a mapping program.

the fineness of the grid. Whether a grid is chosen or not, the next step is to transform the selected points of interest using the procedures of Figure 1. Besides individual points, this may include such things as boundaries and topographic features retrieved from a data base.

The final question is whether the completed map is to be produced on paper or simply stored in computer memory. If the answer to this question is yes, a hard copy is desired, then this copy is produced. In the other option, the data are simply stored for the next usage. Thus, this final logic path is completed.

While it is conceivable that such a main program could include procedures based on direct and inverse transformations of all the major map projections, practical applications generally narrow down the list of projections considerably. Section VI, Chapter 10, gives recommended projections for a variety of uses. Usually, a user will select only the projection or limited number of projections applicable to his work. In each case, the user must consider the requirements for having both the direct and inverse transformation available.

V. STATE PLANE COORDINATES[4]

The program, STAPLN, is based on the four transformation algorithms given in Section VIII, Chapter 5. The variables are defined. The input sequence is given. A listing of the program is shown in Figure 4. The output format is also included. The program is written in FORTRAN IV.

The definition of the input variables is as follows.

Variable	Format	Definition
NTRAN	I5	Projection option 0 = transverse Mercator 1 = Lambert conformal
TC (6)	E14.8	Transverse Mercator constants (NTRAN = 0)
LC (11)	E14.8	Lambert conformal constants
ALFA	E14.8	Latitude (NOPT = 0) (degrees) X coordinate (NOPT = 1) (ft)
BETA	E14.8	Longitude (NOPT = 0) (degrees) Y coordinate (NOPT = 1) (ft)
SCALE	E14.8	Feet to inches (NOPT = 0) Inches to feet (NOPT = 1)
NOPT	I5	0 = direct transformation 1 = inverse transformation
NCASE	I5	0 = case to follow 1 = last case

The input sequence is

1. NTRAN
2. TC or LC
3. ALFA, BETA, SCALE, NOPT, NCASE
•
•
•
n. ALFA, BETA, SCALE, NOPT, NCASE

Each data point entered will produce a line of output consisting of latitude, longitude, x coordinate, and y coordinate. Once the type of projection is chosen, the data entry may be a mix of direct and inverse transformations. The calculations will continue until the last case flag is sensed.

VI. UTM GRIDS

This section introduces a subroutine written in BASIC A which accomplishes the direct transformation in the UTM grid system, as discussed in Section IX, Chapter 5. Several important features of the implementation of the algorithm are highlighted. The subroutine itself appears as Figure 5.

First, it is assured that the ellipsoidal constants needed by the subroutine have been defined in the main program. Also, it has been assumed that the main program has located the correct zone for the calculation, that is, the provided $\Delta\lambda$ is appropriate for the calculations which follow. The subroutine is entered with the coordinates ϕ and $\Delta\lambda$ for the point to be transformed.

The false easting is set automatically. The false northing depends on the sign of ϕ. The absolute value of ϕ is required to evaluate the equations leading to the Cartesian coordinates.

The radius of curvature in the prime vertical (Rp) is obtained from Equation 21 of Chapter 3. The distance along the meridian, from the equator to latitude ϕ, is given by Equation 37 of Chapter 3. After the auxiliary variable η^2 is defined, Equations 118 of Chapter 5 are evaluated for x and y. These values are then applied to Equations 117 of Chapter 5 to obtain the northing and easting of the transformed point. In the evaluation of the northing equation, it is again important to test ϕ to obtain the correct hemisphere for the calculation.

This subroutine can be incorporated into a main program which generates a grid and then locates individual positions within the grid.

VII. COMPUTER GRAPHICS[6,7]

The modern approach to mapping is through the use of computer graphics. This section deals with computerized treatments of the grid and geographic

```
      SUBROUTINE TRNMER(ALFA,BETA,NOPT,TC,X,Y,PHI,LAM,SCALE)
      REAL LAM,LAMS
      DIMENSION TC(6)
      DTR=57.29578
      IF(NOPT.EQ.1) GO TO 1
C     ANGLE TO CARTESIAN
      PHI=ALFA
      LAM=BETA
      PHIS=PHI*3600.
      LAMS=LAM*3600.
      SNP=SIN(PHI/DTR)
      CSP=COS(PHI/DTR)
      FAC=(TC(2)-LAMS)/10000.
      S1=30.922417   *CSP/(1.-.0067666658*SNP*SNP)**0.5
      S1=S1*(TC(2)-LAMS-3.9174*FAC*FAC*FAC)
      FAC=S1/100000.
      SM=S1+4.0931*FAC*FAC*FAC
      FAC=3.2808333 *SM*TC(5)/100000.
      X=TC(1)+3.2808333*SM*TC(5)+CAC*FAC*FAC*TC(6)
      FAC=SM/100000.
      FAC=25.52381*FAC*FAC
      PHI1=PHIS+FAC*((1.-.006768658*SNP*SNP)**2.)*SNP/CSP
      PHI1D=PHI1/3600.
      SNP1=SIN(PHI1D/DTR)
      PHI2=PHIS+FAC*((1.-.006768658*SNP1*SNP1)**2.)*SNP1/COS(PHI1D/DTR)
      PHI2D=PHI2/3600.
      CSP2=COS(PHI2D/DTR)
      FAC=PHI2-60.*TC(3)-TC(4)
      DUM=4.463344-.02352*CSP2*CSP2
      Y=101.27941*TC(5)*(FAC-(1052.8934-DUM*CSP2*CSP2)*CSP
     1*SIN(PHI2D/DTR))
      X=X*SCALE
      Y=Y*SCALE
      RETURN
C     CARTESIAN TO ANGLE
    1 X=ALFA
      Y=BETA
      FAC=(X-TC(1))/100000.
      SG1=X-TC(1)-TC(6)*FAC*FAC*FAC
      FAC=SG1/100000.
      SM=(X-TC(1)-TC(6)*FAC*FAC*FAC)*.3048006061/TC(5)
      OMS=TC(4)+.0098736756*Y/TC(5)
      OMD=TC(3)/60.+OMS/3600.
      CSO=COS(OMD/DTR)
      PHIPM=TC(3)
      PHIPS=OMS+(1047.54671+(6.19276+.050912*CSO*CSO)
     1*CSO*CSO)*CSO*SIN(OMD/DTR)
      PHIPD=PHIPM/60.+PHIPS/3600.
      SNPP=SIN(PHIPD/DTR)
      FAC=SM/100000.
      PHIS=PHIPS-25.5231*((1.-.006768658*SNPP*SNPP)**2.)*FAC*FAC*SNPP
     1/COS(PHIPD/DTR)
      PHI=PHIPM/60.+PHIS/3600.
      SNP=SIN(PHI/DTR)
      SA=SM-4.0831*FAC*FAC*FAC
      FAC=SA/100000.
      S1=SM-4.0331*FAC*FAC*FAC

      DELAM1=S1*((1.-.006768658*SNP*SNP)**.5)/30.922417/COS(PHI/DTR)
      FAC=DELAM1/10000.
      DELAMA=DELAM1+3.9174*FAC*FAC*FAC
      FAC=DELAMA/100000.
      LAM=(TC(2)-DELAM1-3.9174*FAC*FAC*FAC)/3600.
      RETURN
      END
```

FIGURE 4.   Listing of state plane coordinate system program.

```
      SUBROUTINE LAMCON(ALFA,BETA,NOPT,LC,X,Y,PHI,LAM,SCALE)
      REAL LAM,LC,LAMS
      DIMENSION LC(11)
      DTR=57.29570
      IF(NOPT.EQ.1) GO TO 1
C     ANGLE TO CARTESIAN
      PHI=ALFA
      LAM=BETA
      LAMS=LAM*3600.
      PHIM=PHI*60.
      PHIMH=FLOAT(IFIX(PHIM))
      PHIS=(PHIM-PHIMH)*60.
      CSP=COS(PHI/DTR)
      S=101.27941*(60.*(LC(7)-PHIMH)+LC(8)-PHIS+(1052.8939
     1-(4.483344-.02352*CSP*CSP)*CSP*CSP)*CSP*SIN(PHI/DTR))
      FAC=S/100000000.
      R=LC(3)+S*LC(5)*(1.+FAC*FAC*(LC(9)-FAC*LC(10)+FAC*FAC*LC(11)))
      THT=LC(6)*(LC(2)-LAMS)/3600.
      X=LC(1)+R*SIN(THT/DTR)
      SNTH=SIN(THT/2./DTR)
      Y=LC(4)-R+2.*R*SNTH*SNTH
      X=X*SCALE
      Y=Y*SCALE
      RETURN
C     CARTESIAN TO ANGLE
    1 X=ALFA*SCALE
      Y=BETA*SCALE
      FAC=X-LC(1)
      DUM=LC(4)-Y
      FAC=FAC/DUM
      THT=DTR*ATAN(FAC)
      LAM=LC(2)/3600.-THT/LC(6)
      R=(LC(4)-Y)/COS(THT/DTR)
      SNTH=SIN(THT/2./DTR)
      S1=(LC(4)-LC(3)-Y+2.*R*SNTH*SNTH)/LC(5)
      FAC=S1/100000000.
      S2=S1/(1.+FAC*FAC*LC(9)-FAC*FAC*FAC*LC(10)
     1+FAC*FAC*FAC*FAC*LC(11))
      FAC=S2/100000000.
      S3=S1/(1.+FAC*FAC*LC(9)-FAC*FAC*FAC*LC(10)
     1+FAC*FAC*FAC*FAC*LC(11))
      FAC=S3/100000000.
      S=S1/(1.+FAC*FAC*LC(9)-FAC*FAC*FAC*LC(10)
     1+FAC*FAC*FAC*FAC*LC(11))
      OMM=LC(7)-600.
      OMS=36000.+LC(8)-.0098736756*S
      OMD=OMM/60.+OMS/3600.
      CSO=COS(OMD/DTR)
      PHIM=LC(7)-600.
      PHIS=OMS+(1047.5467+(6.19276+.050912*CSO*CSO)*CSO*CSO)
     1*CSO*SIN(OMD/DTR)
      PHI=PHIM/60.+PHIS/3600.
      RETURN
      END
```

FIGURE 4 continued.

```
      PROGRAM STAPLN(INPUT,OUTPUT,TAPE5=INPUT,TAPE6=OUTPUT)
      REAL LAM,LC
      DIMENSION LC(11),TC(6)
      READ(5,11) NTRAN
   11 FORMAT(I5)
C     READ CONSTANTS
      IF(NTRAN.EQ.1) GO TO 12
      DO 13 I=1,6
   13 READ(5,6) TC(I)
    6 FORMAT(E14.F)
      GO TO 14
   12 DO 15 I=1,11
   15 READ(5,6) LC(I)
C     WRITE HEADINGS
   14 WRITE (6,3)
    3 FORMAT(1H1)
      WRITE (6,4)
    4 FORMAT(17X,1HY,17X,1HY,10X,8HLATITUDE,7X,9HLONGITUDE)
    1 X=Y=LAM=PHI=0.
      READ(5,2) ALFA,BETA,SCALE,NOPT,NCASE
    2 FORMAT(3E14.8,2I5)
C     CALL TRANSFORMATIONS
      IF(NTRAN.EQ.1) CALL LAMCON(ALFA,BETA,NOPT,LC,X,Y,PHI,LAM,SCALE)
      IF(NTRAN.EQ.1) GO TO 8
      CALL TRNMER (ALFA,BETA,NOPT,TC,X,Y,PHI,LAM,SCALE)
C     WRITE OUTPUT
    8 WRITE(6,9) X,Y,PHI,LAM
    9 FORMAT(4(4X,E14.8))
      IF(NCASE.EQ.0) GO TO 1
      STOP
      END
```

FIGURE 4 continued.

```
400 'sub utm
410 PHIC=ABS(PHIR)
420 SNP=SIN(PHIC)
430 CSP=COS(PHIC)
440 TNP=TAN(PHIC)
450 ESQ=E*E
460 FLE=500000!
470 FLN=0!
480 IF PHI<0! THEN FLN=1E+07
490 RP=A/SQR(1!-ESQ*SNP*SNP)
500 FAC1=(1!-.25*ESQ-.046875*ESQ*ESQ)*PHIC
510 FAC2=(.375*ESQ+.093758*ESQ*ESQ)*SIN(2!*PHIC)
520 FAC3=(.058594*ESQ*ESQ)*SIN(4!*PHIC)
530 DM=A*(FAC1-FAC2+FAC3)
535 ETA2=ESQ*CSP*CSP/(1!-ESQ)
540 FACX=1!-TNP*TNP+ETA2
550 FACY=5!-TNP*TNP+9!*ETA2
560 DLAM2=DLAMR*DLAMR
570 DLAM3=DLAM2*DLAMR
580 DLAM4=DLAM2*DLAM2
590 CSP3=CSP*CSP*CSP
600 X=RP*(DLAMR*CSP +DLAM3*CSP3*FACX/6!)
610 Y=DM+RP*(DLAM2*SNP*CSP/2!+DLAM4*SNP*CSP3*FACY/24!)
620 IF PHI<0! THEN Y=-Y
630 N=.9996*Y+FLN
640 E=.9996*X+FLE
650 RETURN
```

FIGURE 5.   Listing of UTM program.

features. The scale for any graphic display can be determined automatically. A means of accomplishing this is given. Finally, an example is included which ties together all of these concepts.

As stated in Section V, many applications require the generation of a grid. As can be seen from the figure of the grids in Chapters 4, 5, and 6, these are simply collections of points, horizontal and vertical lines, and inclined or curved lines. All commercially available graphics systems contain the ability to locate points and draw lines, as well as certain special figures such as circles, circular arcs, squares, triangles, and rectangles. The use of the plotting equations, however, makes it useful to depend only on the point and line options.

It is possible to rely only on the line option. In this case, the length of a meridian or parallel is approximated by a large number of segments in all cases. However, this is inefficient. It is not desirable to locate a point by hundreds of calculations. As an example, consider the origin in any of the polar azimuthal projections. Logic must be inserted to recognize that the origin is a point and can be located by a single calculation. In projections, such as the equatorial Mercator, with simple horizontal and vertical lines, it is more efficient to locate the extremities of the line and draw the entire parallel or meridian as a single computation. Only the inclined and curved lines need to be approximated by a series of linear segments. Then, the quality of the line generated will depend on the length of segments chosen and ultimately on the resolution of the screen.

Of course, the values of the meridians and parallels chosen for the grid need to be denoted. It should be the user's option to decide whether each is to be numbered or which alternate ones need to be. If a user is working in the same geographic area over an extended period of time, it makes sense to save a particular grid. Then, it can be recalled as needed.

Once the grid has been generated, we come to the geographic features to be displayed. Specific positions can be represented, obviously, by a point or a symbol, often accompanied by a title or a coded number. Geographic outlines, such as coastlines or rivers, depend on the density of the data base in relation to the scale of the map. When dealing with boundaries, such as political or property, if they are east-west or north-south, the extremities can define a line. In other cases, a series of short segments is needed.

For the maximum utilization of the screen, the largest x extent or y extent of the map should govern. It is necessary to provide the logic to accomplish this as part of the calculations. Then, a linear graphic scale must accompany the computer-generated map to give a fair idea of distance.

On many occasions it is necessary to make a hard copy of a screen image of a map. In some instances, this may only entail a screen dump into a dot matrix printer. Major computer graphics systems have software to produce the desired copy, very often of a quality permitting further reproduction.

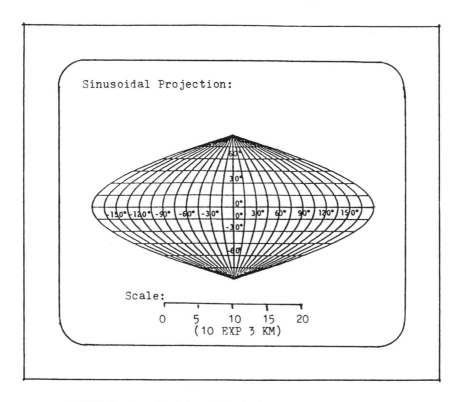

FIGURE 6.   Example of sinusoidal projection grid displayed on a monitor.

Figure 6 is an example of a screen dump of the sinusoidal projection. The North and South Poles are points, best reproduced in a single computation for each. The parallels are straight lines each, produced by a reference to a line computation. The meridians, except for the central meridian, are sine curves, produced by a sequence of small segments. Note the slight raggedness since the segments are not quite short enough to guarantee a smooth curve. For this particular screen, the east-west extent determines the scale of the figure. A linear graphic scale gives an estimate of distance along the equator.

REFERENCES

1. **Enders, B. and Peterson, B.,** *BASIC Primer for the IBM PC & XT*, Plume-Waite, New York, 1984.
2. **Fuori, W. M., Ganghran, S., Giora, L., and Fuori, M.,** *FORTRAN 77, Elements and Programming Style, Hasbrouck Heights, NJ,* Hayden, 1986.
3. **Hershey, A. V.,** Fortran IV Programming for Cartography, Tech. Rep. TR-2339, Naval Surface Warfare Center, Dahlgren, VA, 1969.

4. **Pearson, F. F.,** STAPLN: A Program for Direct and Inverse Transformation of State Plane Coordinates, Tech. Rep. TN82-439, Naval Surface Warfare Center, Dahlgren, VA, 1982.
5. **Pearson, F. F.,** *Map Projection Software,* Sigma Scientific, Blacksburg, VA, 1984.
6. **Pearson, F. F.,** *User's Guide to Program Map,* PSC, Warsaw, VA, 1984.
7. **Ryan, D. L.,** *Principles of Automated Drafting,* Marcel Dekker, New York, 1984.
8. **Swan, T.,** *Mastering TURBO PASCAL,* Hayden, Hasbrouck Heights, NJ, 1986.

Chapter 10

USES OF MAP PROJECTIONS

I. INTRODUCTION

This concluding chapter leads to recommended map projections for particular applications. To this end, two important considerations are advanced. The first is the ability of the chosen projection to closely represent geometric shapes on the Earth by corresponding shapes on the map. A second consideration is the degree of curvature of the parallels and meridians tolerated for a given use. After these topics are investigated, various types of maps are defined. Then, some specific considerations in the choice of a projection are advanced. The limits of coverage for specific projections lead to a recommended set of map projections for most every need. Then, in the final section of this chapter, a brief summary is given of the most important points covered on the way to understanding the theory of map projection and applying this theory as an art.

II. FIDELITY TO FEATURES ON THE EARTH[6]

A basic and obvious requirement is that geometric features on the model of the Earth be faithfully represented on the map. As noted in Chapters 1 and 7, this is impossible to do for all features on a single map. This section investigates five specific features which require comment. The first is the representation of the North and South Poles. The second is the representation of equal areas as a function of longitude. The third is the representation of corresponding areas as a function of latitude. The fourth is the matter of orthogonality of the system of parallels and meridians. The final consideration is the treatment of general angles. For the discussion, it is necessary to consult Figure 1, Chapter 1, for the basic cartographic shapes.

First of all, the North and South Poles are points, located at a finite distance from the equator. Any projection that represents them as lines or locates them at infinity has regions of extreme distortion that may be too great for some applications. Projections with a faithful representation of a single pole are any of the polar azimuthal: equal area, stereographic, polar gnomonic, polar azimuthal equidistant, and polar orthographic. An acceptable representation of two poles on a single map is found on the sinusoidal, parabolic, Mollweide, and transverse Mercator. Projections in which the poles are represented as lines of finite length are the equal area cylindrical, plate Carrée, and Eckert IV. Projections in which the poles are at infinity are the equatorial Mercator and the equatorial gnomonic. Clearly, this last class will not faithfully represent extremes of latitude.

Consider next $10° \times 10°$ quadrangles on the Earth at the same latitude, and placed side by side in longitude. On the model of the Earth, each will enclose the same area and have the same shape. Consider their display on various projections. On the cylindrical projections: equal area, equatorial Mercator, and plate Carrée, the relative area and shape is maintained. This is also true for the conical projections: Albers, Lambert conformal, and simple, and the azimuthal polar projections: equal area, stereographic, orthographic, gnomonic, and azimuthal equidistant. On the Bonne projection and the equal area world maps: sinusoidal, parabolic, Mollweide, Eckert IV, Boggs, and Hammer-Aitoff, the area remains the same by definition, but the shape of adjacent quadrangles changes. The situation is different for the equatorial gnomonic and stereographic projections. For these, shape and area change for adjacent quadrangles.

If the $10° \times 10°$ quadrangles are at the same longitude, but are placed adjacent at varying latitudes, those closer to the equator enclose more area than those closer to the poles. On the map, this relation is true for all of the equal area projections, even if distortion of shape occurs. In a projection such as the plate Carrée, all of these quadrangles have the same area on the map. In the equatorial, Mercator, gnomonic, and stereographic projections, the quadrangles closer to the poles are represented as having greater area than those closer to the equator.

On the model of the Earth, the system of meridians and parallels is orthogonal. By definition, this is true for the conformal projections, and is a feature of all polar azimuthal projections. This is also true for the cylindrical projections such as the equal area, Miller, and plate Carrée. This is not true for the equal area world maps and the polyconic, except at the central meridian.

No projection is perfect for handling the general angle of greatest practical importance, the azimuth, in all cases. For large-scale maps, the transverse Mercator, and the Lambert conformal, particularly in the state plane coordinate system, are generally adequate, but corrections may be necessary as the scale diminishes. All of the azimuthal projections give correct azimuths to and from the pole, but this is of limited practicality. The oblique azimuthal equidistant projection gives the correct azimuth from the origin of the projection, but not from any other points. For large-scale maps, the oblique gnomonic and orthographic projections also give acceptable azimuths from the origin.

III. CHARACTERISTICS OF PARALLELS AND MERIDIANS[2]

For certain uses, it is important to the cartographer to have parallels and meridians of a specific geometric form. Whether parallels or meridians are straight or curved or have a particular type of curvature depends on the projection chosen for use. Tables 1 and 2 are a convenient reference to the characteristics of parallels and meridians as a function of the particular projection to be used.

TABLE 1
Map Projections with Straight-Line Meridians

Meridians	Parallels	Projection
Parallel	Straight lines	Cylindrical equal area
	Widest apart at equator	Simple cylindrical
	Equidistant	Perspective cylindrical
	Widest apart at poles	Mercator
		Miller
Radial	Concentric circles	Polar gnomonic
	Widest apart away from poles	Polar stereographic
	Equidistant	Polar azimuthal equidistant
	Widest apart at poles	Polar azimuthal equal area
		Polar orthographic
Converging	Concentric arcs	Simple conics
		Albers
		Lambert conformal

TABLE 2
Map Projections with Curved Meridians

Meridians	Parallels	Projection
Equally spaced along a given parallel	Arcs	Bonne
	Concentric	Polyconic
	Not concentric	Hammer-Aitoff
	Straight lines	Sinusoidal
	Equally spaced	Mollweide
	Not equally spaced	Parabolic
		Eumorphic
Not equally spaced along a given parallel	Straight lines	Equatorial orthographic
		Oblique & equatorial azimuthal equal area
	Arcs	Transverse Mercator
		Oblique & equatorial stereographic
		Oblique & equatorial gnomonic
		Oblique azimuthal equidistant
		Globular

Table 1 considers projections with straight-line meridians arranged in parallel, radial, or converging patterns. The parallels are either straight lines, concentric arcs, or concentric circles. The relative spacing of the parallels is the key characteristic of the projections. For the cylindrical equal area projection, the parallels are farthest apart at the equator. In the simple cylindrical projections, the parallels are equally spaced throughout the grid. In the perspective cylindrical, the equatorial Mercator and the Miller parallels are farthest apart at the poles.

In the polar forms of the gnomonic and stereographic projections, the concentric circles, which form the parallels, are farthest apart toward the equator. The parallels are equally spaced in the polar azimuthal equidistant projection. For the polar azimuthal equal area and polar orthographic projection, the parallels are farthest apart at the poles.

For the conical projections with parallels represented as concentric arcs, the relative spacing depends on whether a secant or tangent configuration is chosen.

In Table 2, projections with curved meridians are given. The first criterion is whether or not the curved meridians are equally spaced along a given parallel. In the case of meridians equally spaced along a given parallel, the parallels may be either arcs or straight lines. In the Bonne and Werner projections, the arcs are concentric. In the polyconic and Hammer-Aitoff projections, the arcs are not concentric. For the sinusoidal projection, the straight-line parallels are equally spaced. The straight-line parallels are not equally spaced in the Mollweide, parabolic, and eumorphic projections.

For the cases where the meridians are not equally spaced along a given parallel, the parallels may be straight lines or arcs. The equatorial orthographic, and oblique and equatorial azimuthal equal area projections have straight-line parallels. The transverse Mercator, the oblique and equatorial cases of the stereographic and gnomonic, oblique azimuthal equidistant, and globular projections have curved parallels.

IV. CONSIDERATIONS IN THE CHOICE OF A PROJECTION

Associated with the theory and practice of map projection methods is the need to use map projections correctly. It must be kept in mind that the use of map projections is an art. The rule, of course, is to choose a system that minimizes distortion to an acceptable level in the region of interest.

Several considerations must be factored into the choice of the proper projection to suit a user's need. These include type of map, any special features that must be preserved, and the types of data to be portrayed. The scale of the map and the accuracy required are also major considerations. Finally, the map as an art form may need to be considered. Types of maps may be associated with their intended uses. Some different types of maps are defined below.

A thematic map[3] is a special purpose map which portrays a single topic. It gives the geographic distribution of some physical or cultural entities. These entities may range from peanuts to people or to anything else. Included in this category are statistical and enumerated data. Topographic maps give the general topographic features of a locality. Included under topographic maps are hydrographic maps of specific bodies of water. Political maps outline political jurisdictions and may range in scale from county size to nation size, to the whole world. Navigational mapping provides a means of indicating routes of travel including distance and direction. Also of importance are political and topographic considerations. The military uses of maps range from ground, air, and ocean support of armed forces to intelligence work aimed at observing the actions of an enemy or a potential enemy. Astrographic applications range from local snapshots of the sky to complete atlases of the heavens. In all cases, the intended use of the map must be considered in choosing the correct projection.

Consider next, two of the major categories of projections, the equal area and the conformal projections. The equal area projections, as a class, are excellent for the accurate display of statistical data. The equal area property faithfully maintains proper proportions on the map. However, this is often purchased by extreme distortions in shape. Thematic maps are almost universally equal area projections. On the other hand, the conformal projections maintain shape relatively well, but at the expense of size. The use of conformal projections to display statistical data brings about a misleading impression.

The property of conformality is of great interest to those working in the field of photogrammetry and remote sensing. The Lambert conformal and the transverse and oblique Mercator projections are most often used by imagery analysts. Also, surveying systems make use of the Lambert conformal and transverse Mercator projections on large-scale maps. In fact, for large-scale maps and relatively short distances, any one of the conformal maps, and also the Bonne projection, can be used for relatively good approximations of distances and azimuths.

Air and sea navigation have required the use of the equatorial Mercator projection with the loxodrome and the great circle and the gnomonic projection with the great circle. In both of these projections, the navigation is between two arbitrary points and over great distances. In Figure 6, Chapter 5, the great circle route and the accompanying loxodrome are displayed. Recall that on the spherical model of the Earth, the great circle is the shortest line between two points. Distortion on the Mercator projection interchanges the relative magnitude of the great circle vs. the loxodrome. The gnomonic projection, such as the equatorial case of Figure 3, Chapter 6, is used in conjunction with the Mercator. On this projection, the great circle is a straight line, although not in true scale, while the loxodrome is a curved line of greater length.

If, however, one needs true distances and azimuths from a single arbitrary point to other points anywhere in the world, then the oblique azimuthal equidistant projection can accommodate. Such a projection is given in Figure 7, Chapter 6. However, it must be cautioned that the distances between any two of the terminal points is not true scale, and the azimuths between terminal points are not true.

Another consideration is the display of enumerative data. In contrast to the display of statistical data,[8] where relative area is important, one can wish to display enumerative data where only the number of points is important. Then any one of the projections in this text can be arbitrarily chosen. However, the plate Carrée can be a particularly convenient projection for this use. Figure 23, Chapter 6, can contain a base map of the world on the plate Carrée projection. If there is no concern for effects of distortion, this projection recommends itself due to the simplicity of the direct and inverse transformations.

Scale is a factor in the choice of a projection for a specific application. As shown in the extended example on projections for a CRT display in Chapter 8, when the scale is large, a number of projections will give nearly equivalent results. Recall in that example the azimuthal equidistant, the orthographic, and the gnomonic projections centered at a tracking site give essentially imperceptible differences in their representation of the tracking data. If greater distances were displayed on the same screen area, that is, the scale became smaller, the differences would become more and more obvious.

Another consideration relates to the accuracy required. Some applications require considerable accuracy. An example of this is navigation or state plane coordinate calculations. Other applications do not require as much accuracy to convey the necessary information. As an example of the latter, consider an interactive computer graphics system in which a map is digitized and selected points are transformed from Cartesian to geographic coordinates. There is a limit on the accuracy obtained in this transformation. The resolution of the digitizing and imperfections in the map due to such things as humidity or methods of reproduction will limit the accuracy of the final result. A spherical model of the Earth with an appropriate radius is probably all that is required. Only when extreme accuracy is required is it necessary to use the appropriate spheroidal model. The user must keep the required level of accuracy in mind.

Maps can be considered as art forms in two ways. First, a well-designed map is expected to be aesthetically pleasing as well as utilitarian. Finally, keep in mind that maps are often used as a decorative motif. More often in this case, the more exotic looking projections, such as the interrupted projections or the equal area world maps, become elements in interior decorating or a company logo. In all cases, the specific application to be addressed by the projection may be the overriding factor in the choice of projection.

V. RECOMMENDED AREAS OF COVERAGE

The precept for choosing the correct map for the required job is helped by the apparent natural adaptability of particular types of projections to certain areas of the Earth. By consulting the grids in this text, it is seen that the azimuthal polar projections can easily and with minimum distortion handle areas immediately adjacent to the poles. Likewise, the cylindrical projections are natural for the regions above and below the equator. The areas at mid-latitudes are conveniently spanned by conical projections of one or two standard parallels. Recall that in these projections, distortion is not a function of longitude.

Maps may be classified into those of world coverage and those of local coverage. For local coverage, a set to choose from would include the polar azimuthal, a conical projection, and a cylindrical projection. In the choice of a conical projection, the two standard parallel cases are always more desirable in order to limit the effects of distortion and to adequately distribute it in a map.

Specifically, for local conformal coverage anywhere in the world, a natural set is the Mercator between $\pm 30°$ latitude, a Lambert conformal from $\pm 30°$ to $\pm 60°$, and the polar stereographic from $\pm 60°$ to $\pm 90°$. For an equal approach, the same regions can be covered by a cylindrical, an Albers, and the Lambert azimuthal, respectively.

Figure 1a indicates a set of either conformal or equal area maps applicable to local coverage in the Northern and Southern Hemispheres for a restricted difference in longitude. If a greater difference in longitude is required, it can be incorporated on the same map, or adjacent sheets can be used together. Recall that in these projections distortion is independent of longitude.

A different situation occurs if it is necessary to use maps side by side at changing latitude. Latitude does have an effect on distortion. One way to overcome this is to build in some overlap between the various projections to obtain some semblance of continuity. An example of this is given in Figure 1b, where the basic set is modified for 10° of overlap between projections of the three basic types. Consider point P_1 at a latitude of 30°, as it appears on both the cylindrical and the conical projection. Also shown is a point P_2 of latitude 60° on both the conical and the azimuthal projection. In this way some continuity may be maintained on going from sheet to sheet.

If there is a desire to include an entire sphere on a map, the world statistical maps are the best approach. However, one must learn to live with the excessive distortion or use interrupted variations. Note that only equal area and conventional projections are generally used for a map of the entire Earth. Extreme distortions at the periphery of the conformal projections treated in this text will not permit this.

Tables 3, 4, and 5 are included to summarize regions of coverage for equal area, conformal, and conventional projections, respectively. The tables are marked with an "X" for those regions that are best portrayed on a particular projection. The regions that are considered are the polar, the mid-latitudes, the equatorial, a transverse zone, and the entire world.

VI. RECOMMENDED SET OF MAP PROJECTIONS

The question arises as to which projection to use. This must always be conditioned by the use for the projection. With the use firmly in mind, the author recommends a basic set of projections that will see the user through the vast majority of applications. These are given in Table 6.

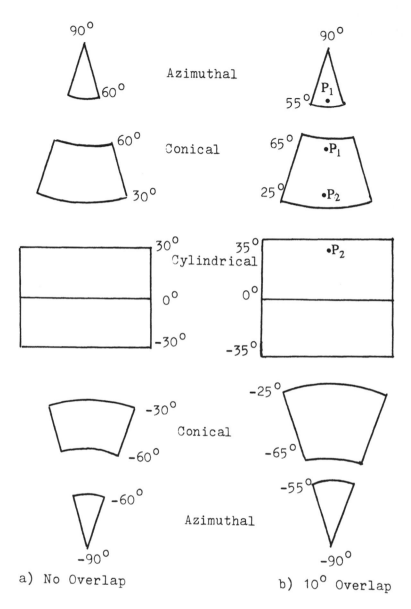

FIGURE 1.   Basic set of maps for local coverage.

For a local area of coverage, the Earth should be zoned according to whether an azimuthal, conical, or cylindrical projection is appropriate. If the conical projection applies, the two standard parallel cases should be used for a better distribution of distortion. Recall that in all of these recommended projections, the distortion is independent of longitude. Entries 1a, 1b, and 1c in the table are the recommended equal area and conformal projections for each local area.

TABLE 3
Regions of Coverage of Equal Area Projections

Projection	Polar	Mid-latitude	Equatorial	Transverse	World
Conical, 1 standard parallel		×			
Conical, 2 standard parallels		×			
Polar azimuthal	×				
Oblique azimuthal		×			
Equatorial azimuthal			×		
Bonne		×			
Cylindrical			×	×	×
Mollweide			×		×
Parabolic			×		×
Hammer-Aitoff			×		×
Eumorphic			×		×

TABLE 4
Regions of Coverage of Conformal Projections

Projection	Polar	Mid-latitude	Equatorial	Transverse
Equatorial Mercator			×	
Oblique Mercator		×	×	
Transverse Mercator				×
Lambert conformal, 1 standard parallel		×		
Lambert conformal, 2 standard parallels		×		
Polar stereographic	×			
Oblique stereographic		×		
Equatorial stereographic			×	

If the local area is of major north-to-south extent, a transverse Mercator projection is recommended. For this projection, distortion is independent of latitude.

For worldwide coverage, the choice is between equal area and conventional maps for practical use. Out of the equal area choices, the Mollweide and the Boggs Eumorphic give the most acceptable form. Distortion is not as pronounced in these as it is in the other potential choices. Of the conventional choices, the Robinson is the best.

All the conventional projections of great utility have a special feature which recommends their use for special projects. The polyconic is well adapted for areas of a significant north-to-south extent. In many applications it may replace the transverse Mercator. The gnomonic is characterized by all great circles appearing as straight lines. However, the length of the great circles is not distortion free. The azimuthal equidistant maintains true range and azimuth from the origin of the projection. In addition, this range is portrayed by a

TABLE 5
Regions of Coverage of Conventional Projections

Projection	Polar	Mid-latitude	Equatorial	Transverse	World
Polar gnomonic	×				
Oblique gnomonic		×			
Equatorial gnomonic			×		
Oblique azimuthal equidistant		×	×		
Polar azimuthal equidistant	×				
Polar orthographic	×				
Oblique orthographic		×	×		
Simple conic		×			
Polyconic		×	×	×	
Miller			×		
Plate Carrée			×		×
Van der Grinten		×	×		
Globular	×	×	×		
Aerial perspective	×	×	×		
Robinson					×

TABLE 6
Recommended Set of Projections

Latitude range	Equal area	Conformal
Local area coverage		
60 to 90°	Polar azimuthal	Polar stereographic
−60 to −90°		
30 to 60°	Albers	Lambert
−30 to −60°	2 standard parallels	2 standard parallels
−30 to 30°	Cylindrical	Equatorial Mercator
−30 to 30° transverse band	—	Transverse Mercator
World coverage		
	Mollweide	—
	Boggs eumorphic	—
Conventional		
Polyconic	Transverse strip	
Gnomonic	Great circles	
Azimuthal equidistant	True range & azimuth	
Plate Carrée	Simplicity	
Orthographic	Photogrammetry & remote sensing	
	Aerial perspective	
Robinson	World coverage	

straight line. The plate Carrée has the characteristic of extreme simplicity in the direct and inverse transformation. When no greater sophistication is required, it can be useful for local area and world-wide coverage. Finally, the orthographic and aerial perspective projections have been useful in representing the images obtained by photogrammetry and remote sensing.

VII. CONCLUSION

This text has brought together in a single volume a fairly complete set of map projection equations of use in modern applications. The three major classes of map projections, equal area, conformal, and conventional have been considered one by one. For each class, the subclasses of azimuthal, conical, and cylindrical projections have been investigated. In the case of the equal area and conventional projections, world maps were also considered.

Of major importance to users of maps, the derivations for the direct transformation from geographic coordinates resulted in a collection of planar Cartesian mapping equations. These can be programmed for any user requirements. The derivations for the inverse transformations also resulted in equations which are easy to program for the computer. The inverse transformation was given for a selection of important projections.

Of nearly equal importance were the methods for obtaining quantitative estimates of distortion. The intelligent user of maps needs to know how much distortion is subverting the concept of an ideal map. It is not expected that many users will start with a blank sheet of paper and a geographic data base and create his or her own map. The usual procedure is to start with a base map of known projection and scale, and use the projection equations to fill in the desired detail.

The intent of this volume is to give the user the understanding to choose the base map with a projection that will adequately suit his or her needs. Since a map is a form of communications, the projection chosen must communicate what the user intends it to.

The reason that there are so many map projections is that there have been so many attempts to reach an acceptable representation of the spherical or spheroidal surface on a flat surface. Thus, map projections provide a variety of methods of transforming from an undulating Earth to a flat piece of paper, and obtaining in the end, a fairly reliable representation. It is the user's duty to choose the map projection scheme that best satisfies his or her particular requirements.

REFERENCES

1. **Bowditch, N.**, *American Practical Navigator,* Defense Mapping Agency Hydrographic Center, U.S. Gov't Printing Office, Washington, D.C., 1977.
2. **Deetz, C. H.**, Cartography, A Review and Guide, Spec. Publ. 205, U.S. Coast and Geodetic Survey, U.S. Gov't Printing Office, Washington, D.C., 1962.
3. **Dent, B. D.**, *Principles of Thematic Map Design,* Addison-Wesley, Reading, MA, 1985.
4. **Lillesand, T. M. and Kiefer, R. W.**, *Remote Sensing and Image Interpretation,* John Wiley & Sons, New York, 1987.
5. **Low, J. W.**, *Plane Table Mapping,* Harper & Row, New York, 1952.
6. **Snyder, J. P.**, Map projections — a working manual, *U.S. Geol. Surv. Prof. Pap.,* 1395, U.S. Gov't Printing Office, Washington, D.C., 1987.
7. **Wolf, P. R.**, *Elements of Photogrammetry,* McGraw-Hill, New York, 1974.
8. *SAS/GRAPH User's Guide,* Statistical Analysis Systems, Raleigh, NC, 1981.

INDEX

N

Printed and bound by CPI Group (UK) Ltd, Croydon, CR0 4YY

23/10/2024

01778237-0013